LOGIC FOR CONCURRENCY AND SYNCHRONISATION

TRENDS IN LOGIC
Studia Logica Library

VOLUME 18

SCOPE OF THE SERIES

Trends in Logic is a bookseries covering essentially the same area as the journal *Studia Logica* – that is, contemporary formal logic and its applications and relations to other disciplines. These include artificial intelligence, informatics, cognitive science, philosophy of science, and the philosophy of language. However, this list is not exhaustive, moreover, the range of applications, comparisons and sources of inspiration is open and evolves over time.

Volume Editor
Heinrich Wansing

The titles published in this series are listed at the end of this volume

LOGIC FOR CONCURRENCY AND SYNCHRONISATION

Edited by

RUY J.G.B. DE QUEIROZ
Universidade Federal de Pernambuco, Recife, Brazil

SPRINGER-SCIENCE+BUSINESS MEDIA, B.V.

A C.I.P. Catalogue record for this book is available from the Library of Congress.

ISBN 978-94-017-4000-5 ISBN 978-0-306-48088-1 (eBook)
DOI 10.1007/978-0-306-48088-1

Printed on acid-free paper

Contents

List of Figures xi

List of Tables xv

Foreword xvii
Johan van Benthem

Preface xix

Contributing Authors xxi

Part I From a Structural Perspective

1
Geometry of Deduction via Graphs of Proofs 3
Anjolina Grisi de Oliveira and *Ruy J. G. B. de Queiroz*

1 Motivation 4
1.1 Relevance to proof theory: towards the geometry of natural deduction 6
1.2 Relevance to theoretical computer science: complexity and concurrency 8
2 The idea of studying proofs as geometric objects 12
2.1 Representing proofs in a graph 12
2.2 Some results 16
2.2.1 Elementary results on isomorphisms 16
2.2.2 On the genus of a derivation 16
2.2.3 On the relationship between the notions of direct proof, normal proof and proof, and the subformula property 17
3 Proof-nets 17
3.1 Criticisms of natural deduction 18
3.1.1 Classical symmetry 18
3.1.2 Global rules 18
3.1.3 The commutative conversions 19
3.2 Analyzing the sequent calculus 19
3.2.1 The non-determinism of Hauptsatz 19
3.2.2 Redundancies in proofs 20
3.2.3 Non-unicity of proofs 20
3.3 Definition of links 21

	3.4	Proof-structures	22
	3.4.1	Examples	23
	3.5	The soundness criterion	24
	3.5.1	No short trip condition	24
	3.5.2	Danos–Regnier criterion	28
	3.5.3	Asperti criterion	30
	3.6	The soundness of the criteria	30
	3.7	Cut-elimination	34
4	Logical flow graphs		34
	4.1	Motivation	35
	4.2	Defining logical flow graphs	36
	4.2.1	The sequent calculus	36
	4.2.2	Building a logical flow graph	36
	4.2.3	A more precise definition	38
	4.2.4	Examples	40
	4.3	Analyzing the cut-elimination process	40
	4.3.1	An example of expansion after cut-elimination	42
	4.4	Cycles in proofs	44
	4.5	A combinatorial model	45
	4.5.1	Optical graphs	45
	4.5.2	The duplication operation	46
	4.5.3	Connections with cut-elimination	48
5	Multiple-conclusion classical calculi		50
	5.1	Lack of symmetry in natural deduction	51
	5.2	Kneale's tables of development	53
	5.2.1	Classical and intuitionistic logic	55
	5.2.2	Some examples	55
	5.2.3	The inadequacy	56
	5.3	Shoesmith and Smiley refinements	57
	5.3.1	Developments as graph arguments	57
	5.3.2	Solving the inadequacy	59
	5.3.3	Second solution: multiple junction	62
	5.3.4	Third solution: identification of premises and conclusions	65
	5.4	Ungar's system	67
	5.4.1	The problem of substitution	68
	5.4.2	Solving the problem of substitution	69
6	Finale		70
	6.1	Towards a new perspective: N-graphs	71
	6.1.1	Proof-graphs	72
	6.1.2	Unsound proof-graphs	74
	6.1.3	Global soundness criterion	78
	6.1.4	The soundness of the criterion	79
	6.1.5	Some examples	80

2

Chu's Construction: A Proof-Theoretic Approach 89

Gianluigi Bellin

1	Preface		89
2	The trip translation		93
	2.1	Languages, intuitionistic and classical MALL	93
	2.2	The functorial trip translation for MALL	95
3	Chu's construction		98
4	Proof-nets, trips and translations		100

4.1 Proof-nets: basic definitions 100
4.2 Trips and linear λ terms 103
4.3 Reduced translations from **CMLL**$^-$ to **IMLL**$^-$ 103
4.4 Chu's construction in **MLL** with units 106

3
Two Paradigms of Logical Computation in Affine Logic? 111
Gianluigi Bellin
1 Introduction 112
2 Sequent calculus of MAL + Mix 119
2.1 Mix and weakening permutations 120
2.2 Weakening reductions and expansions 121
3 Additive mix 123
3.1 Properties of additive mix 123
3.2 Additive mix: reductions and permutations 124
4 Proof-nets for MAL + Mix 126
4.1 Proof-structures and proof-nets 126
4.2 Computation and elimination of irrelevance 128
4.3 Examples and properties of irrelevance elimination 131
4.4 Mix-elimination 134
5 Cut-elimination modulo irrelevance 138
6 Symmetric reductions require Mix 141

4
Proof Systems for π-Calculus Logics 145
Mads Dam
1 Introduction 146
2 Preliminaries on the π-calculus 148
3 A π-μ-calculus 153
4 Example specifications 159
5 Proof system, modal fragment 162
5.1 Basic judgments 164
5.2 A proof system for the modal fragment 166
5.3 Structural rules 166
5.4 Logical rules 168
5.5 Rules for equality and inequality 168
5.6 Rules for transition modalities 170
5.7 Rules for input/output modalities 174
5.8 Examples 177
6 Soundness and completeness for the modal fragment 179
7 Proof rules for recursive formulas 180
7.1 Rules for fixed point unfolding and approximation 181
7.2 Rule of discharge 183
8 Finite control completeness 188
9 Natural numbers 192
10 Buffers 194
11 Conclusion 196

Part II From a Descriptive Perspective

5
A Tutorial Introduction to Symbolic Model Checking 215
David Déharbe

1 Introduction 215
2 Kripke structures 216
2.1 Definitions 216
2.2 Computation tree 217
3 Temporal logic model checking 218
3.1 Computation tree logic 218
3.1.1 Syntax 219
3.1.2 Semantics 219
3.2 Algorithm 221
3.3 Experimental results and conclusion 225
4 Symbolic model checking 225
4.1 Kripke structures and propositional logic 226
4.1.1 Representing states and transitions 226
4.1.2 State space traversal 226
4.2 Binary decision diagrams 227
4.3 Algorithms 229
4.3.1 Fundamentals of lattices and fixpoints 229
4.3.2 Fixpoint characterization of CTL operators 230
4.3.3 Results and extensions to symbolic model checking 232
5 Conclusion 233

6
Modal Logics for Finite Graphs 239
Mario R. F. Benevides

1 Introduction 239
2 Finite directed graphs 241
2.1 Language and models 241
2.2 Proof theory 242
2.3 Soundness and completeness results 243
3 Finite acyclic directed graphs 244
3.1 Proof theory 244
3.2 Canonical models 245
3.3 Completeness for finite acyclic directed graphs 248
4 Undirected graphs 257
5 Loopless undirected graphs 257
6 Modal definability 258
6.1 Colouring 259
6.2 Hamiltonian graphs 260
6.3 Eulerian graphs 260
6.4 Connectivity 261
6.5 Planarity 261
7 k-Colourable graphs 261
7.1 Completeness for k-colourable graphs 263
8 Conclusions 266

7
Bisimulation and Language Equivalence 269
Colin Stirling
1 Introduction 269
2 Background 270
3 Caucal's hierarchy 274
4 Richer logics 276
5 Finite model theory 278

List of Figures

1.1	Graph representing a proof	14
1.2	Identity links	21
1.3	Modus ponens	22
1.4	Proof-structure	23
1.5	Proof-structure	23
1.6	Proof-structure	24
1.7	Formula as a box	24
1.8	Axiom link	25
1.9	Terminal formula	26
1.10	Times link switched on "L"	26
1.11	Times link switched on "R"	27
1.12	Par link switched on "L"	27
1.13	Par link switched on "R"	28
1.14	D-R graphs	29
1.15	D-R graph	30
1.16	Proof-net of $\vdash C^{\perp}, C \otimes (A \wp B), A^{\perp} \otimes B^{\perp}$	32
1.17	Proof-nets P_1 and P_2, respectively	33
1.18	Connecting P_1 and P_2 through a times link	33
1.19	Connecting P_1 and P_2 through a par link	33
1.20	The orientation of the edges in a LFG	37
1.21	Logical flow graph to the formulas p and q	38
1.22	Logical flow graph to the formula A	40
1.23	Logical flow graph to the formulas A and B	41
1.24	An example of expansion	42
1.25	The building block B_{j+1}	43
1.26	Building blocks linked	43
1.27	General scheme of a cycle in a proof	45
1.28	Split of a cycle in a proof	45
1.29	Optical graphs	46

1.30 Duplication operation 48
1.31 An exponential growth 48
1.32 Focussing point followed by a defocussing one in the *LFG* of a proof 49
1.33 Focussing point followed by a defocussing one 49
1.34 An operation of duplication breaking a cycle 50
1.35 Development of $\neg P/P \to Q$ 54
1.36 Development of $\neg P/P \to Q$ 55
1.37 A fallacy 55
1.38 Development versus Graph argument 58
1.39 Development with a corner 59
1.40 Corner at B 59
1.41 Junction 59
1.42 Non-standard proof from R to Q 60
1.43 Graph arguments Π_1 and Π_2 61
1.44 Graph arguments Π'_1 and Π'_2 61
1.45 Junction of Π'_1 and Π'_2 62
1.46 Invalid arguments 63
1.47 Circuit in an invalid argument 63
1.48 Circuit in a valid argument 64
1.49 Multiple-junction 64
1.50 "key example" with multiple-junction 65
1.51 Multiple-junction 65
1.52 Identification of premises and conclusions 66
1.53 Identification of premises and conclusions: "key example" 66
1.54 The problem of substitution 68
1.55 An attempt to derive $A \vee A \vdash A \wedge A$ 69
1.56 Combination on graphs 70
1.57 Links in a proof-graph 72
1.58 Logical links 75
1.59 Structural links 76
1.60 Proof-graphs 77
1.61 Expansion link in proof-graphs 78
1.62 Contraction link in proof-graphs 78
1.63 N-graph of $(A \wedge B) \to (A \wedge B \wedge C) \vee \neg C$ 81
1.64 N-graph of $\vdash ((A \to B) \to A) \to A$ 82
2.1 Links 101
2.2 Girard's trips 101
2.3 Linear λ-terms 103

2.4 Admissible orientations. 104
3.1 Links 126
3.2 Prune 132
3.3 Big prune 133
3.4 Cover 138
3.5 Cut 140
3.6 Example 33 141
3.7 Mix 142
4.1 Buffer with reclaimable cells 151
4.2 Fixed point interference 183
5.1 State transition diagram for Kripke structure $ABPsender$ 217
5.2 Partial state transition diagram of $ct(ABPsender)$ 219
5.3 Illustration of universal CTL operators 220
5.4 Labeling algorithm for CTL formulas 222
5.5 Labeling algorithm for $\mathbf{EX}f$ formulas 222
5.6 Labeling algorithm for $\mathbf{E}[f\mathbf{U}g]$ formulas 223
5.7 Labeling algorithm for $\mathbf{EG}f$ formulas 224
5.8 Binary tree for $g \wedge \neg s \wedge \neg w$. 228
5.9 BDD for $g \wedge \neg s \wedge \neg w$. 228
5.10 Algorithm for symbolic model checking CTL formulas 230
5.11 Algorithm for $\mathbf{EX}f$ formulas 231
5.12 Algorithm for $\mathbf{E}[f\mathbf{U}g]$ formulas 231
5.13 Algorithm for $\mathbf{EG}f$ formulas 232
5.14 Trace of function call $SmcEG(ABPsender, \mathbf{EG}\neg s \wedge \neg w)$ 232
6.1 vR_tw and wR_fv. 240
6.2 Graph G, where a formula A is true at vertex v. 241
6.3 f-unsatisfied point 249
6.4 Abbreviation for undirected graphs 259
6.5 Square 1,2,3,4 is 4-colourable and triangle a,b,c is not. 259
6.6 Graph 1,2,3,4,5 is Hamiltonian and graph a,b,c,d is not. 260
6.7 Graph 1,2,3,4,5 is Eulerian and graph a,b,c,d is not. 260
6.8 Planar Graph. 262
6.9 Formation rule for the graph presented in Figure 6.8 . 262
6.10 Graph **K5**. 262
7.1 Simple transition graphs 270
7.2 Quotiented transition graph 272
7.3 A pushdown automaton 274
7.4 Ingredients of the graphs G and G' 281

List of Tables

1.1	Basic development schemata for propositional fragment	54
1.2	Inadequacy of Kneale's developments	57
1.3	Ungar's system	67
1.4	Circuits in a multiple-conclusion calculus	69
4.1	Transition rules	150
4.2	Formula semantics	155
4.3	Proof rules for parallel composition vs. transition modalities	172
4.4	Monotonicity rules for io-modalities	175
4.5	Rules for parallel composition vs. io-modalities	176

Foreword

The study of information-based actions and processes has been a vibrant interface between logic and computer science for several decades now. Indeed, several natural perspectives come together here. On the one hand, logical systems may be used to describe the dynamics of arbitrary computational processes – as in the many sophisticated process logics available today. But also, key logical notions such as model checking or proof search are themselves informational processes involving agents with goals. The interplay between these descriptive and dynamic aspects shows even in our ordinary language. A word like "proof" denotes both a static 'certificate' of truth, and an activity which humans or machines engage in. Increasing our understanding of logics of this sort tells us something about computer science, and about cognitive actions in general.

The individual chapters of this book show the state of the art in current investigations of process calculi such as linear logic, π-calculus, and μ-calculus – with mainly two major paradigms at work, namely, linear logic and modal logic. These techniques are applied to the title themes of concurrency and synchronisation, but there are also many repercussions for topics such as the geometry of proofs, categorial semantics, and logics of graphs. Viewed together, the chapters also offer exciting glimpses of future integration, as the reader moves back and forth through the book. Obvious links include modal logics for proof graphs, labeled deduction merging modal and linear logic, Chu spaces linking proof theory and model theory, and bisimulation-style equivalences as a tool for analyzing proof processes.

The combination of approaches and the pointers for further integration in this book also suggests a grander vision for the field. In classical computation theory, Church's Thesis provided a unification and driving force. Likewise, modern process theory would benefit immensely from a synthesis bringing together paradigms like modal logic, process algebra, and linear logic – with their currently still separate worlds of bisimulations, proofs, and normalisation. If this Grand Synthesis is ever going to happen, books like this are needed!

JOHAN VAN BENTHEM, ILLC Amsterdam & CSLI Stanford

Preface

The contributions published in this volume arose in the context of the project *Logic for Concurrency and Synchonisation* (*LOCUS*) which was concerned with the relationship between proof theory (*à la* Curry–Howard-like calculi) and concurrency theory (*à la* π-calculus, μ-calculus), as well as the application of those formalisms to the verification of group-based protocols.

The project also sought to investigate the possibility of defining a unifying methodology (algebraic methods vs. logical methods) for the formalisation of distributed systems, concurrency and synchronisation, using the most recent techniques coming from mathematical logic (in particular, labelled deduction, type theory, and modal logic), proof theory and semantics of concurrent processes.

Four institutions participated in the project: Universidade Federal de Pernambuco (UFPE), Universidade Federal de Alagoas (UFAL), Universidade Federal da Bahia (UFBA), and Universidade Federal do Rio de Janeiro (UFRJ).

Outline

Chapter 1 reviews a collection of recent and less recent work around graph-theoretical tools used in proof theory, leading to some ideas for bringing together the old (e.g., Kneale's symmetric proof system) and the new (Girard's graph-theoretic criterion to check soundness of graphs of proof) in order to further enhance the tools for the understanding of natural deduction (ND): the geometry of interaction of ND-proofs, their lack of symmetry and their proof complexity.

In Chapter 2, Bellin argues that the essential interaction between classical and intuitionistic features in the system of linear logic is best described in the language of category theory. The main result is to show that the intuitionistic translations induced by Girard's trips determine the functor from the free $*$-autonomous category $\mathcal{A}$ on a set of atoms $\{P, P', \ldots\}$ to $\mathcal{C} \times \mathcal{C}^{op}$, where $\mathcal{C}$ is the free monoidal closed category with products and coproducts on the set of atoms $\{P_O, P_I, P'_O, P'_I, \ldots\}$ (a pair P_O, P_I in $\mathcal{C}$ for each atom P of $\mathcal{A}$).

In Chapter 3, Bellin proposes a notion of *symmetric reduction* for a system of proof-nets for *Multiplicative Affine Logic with Mix* (**MAL** + Mix) (namely, multiplicative linear logic with the mix-rule the unrestricted weakening-rule), and proves that such a reduction has the strong normalisation and Church–Rosser properties.

In Chapter 4, Dam studies the problem of verifying general temporal and functional properties of mobile and dynamic process networks, cast in terms of the π-calculus.

In Chapter 5, Déharbe gives a tutorial introduction to CTL model checking and its symbolic BDD-based version implementation.

In Chapter 6, Benevides presents modal logics for four classes of finite graphs: finite directed graphs, finite acyclic directed graphs, finite undirected graphs and finite loopless und irected graphs.

In Chapter 7, Stirling looks at the relationships between bisimulation equivalence and language equivalence.

Acknowledgements. The project *Lógica para Concorrência e Sincronização* (*LOCUS*) received a 2-year (December 1996 to December 1998) funding from the Brazilian national council for scientific and technological development *CNPq* (*Conselho Nacional de Desenvolvimento Científico e Tecnológico*), under the *Programa Temático Multi-institucional em Ciência da Computação – Fase III* (*ProTeM-CC*), grant *680067/95-7*.

I should also like to thank the following people who helped to keep the standards high: A. Asperti, M. Benevides, P. Blackburn, A. Carbone, S. Dal-Zilio, R. Goldblatt, O. Grumberg, D. Hughes, M. Kracht, M. Otto, D. Vakarelov.

Last, but not least, I am most grateful to thank Heinrich Wansing (member of the editorial board of the *Trends in Logic* series) for his great care in dealing with the present volume. Let me also thank the Kluwer staff involved, in particular, Jolanda Voogd (Publishing Assistant) and Charles Erkelens (Acquisitions Editor), Humanities Unit – History and Philosophy of Science.

RUY J.G.B. DE QUEIROZ
Recife, February 2003

Contributing Authors

Gianluigi Bellin is a *Ricercatore* at the Facoltà di Scienze of the Università degli Studi di Verona, and a Lecturer at the School of Mathematics of Queen Mary and Westfield College, University of London.

Mario R. F. Benevides is a *Professor Adjunto* at the Instituto de Matemática and Coordenação de Programas de Pós-Graduação em Engenharia de Sistemas (COPPE) of the Universidade Federal do Rio de Janeiro (UFRJ).

Mads Dam is a Researcher at the Swedish Institute of Computer Science (SICS), and *Docent* at Royal Technical University, Stockholm, Sweden.

David Déharbe is a *Professor Adjunto* in the Departamento de Informática e Matemática Aplicada of the Universidade Federal do Rio Grande do Norte (UFRN) in Natal, Brazil.

Anjolina Grisi de Oliveira is a *Professor Adjunto* at the Centro de Informática (CIn) of the Universidade Federal de Pernambuco (UFPE) in Recife, Brazil.

Ruy J.G.B. de Queiroz is a *Professor Adjunto* at the Centro de Informática (CIn) of the Universidade Federal de Pernambuco (UFPE) in Recife, and a *Pesquisador 1C* of the Conselho Nacional de Desenvolvimento Científico e Tecnológico (CNPq), Brazil.

Colin Stirling is a *Professor* at the School of Informatics, University of Edinburgh, Scotland.

I

FROM A STRUCTURAL PERSPECTIVE

Chapter 1

GEOMETRY OF DEDUCTION VIA GRAPHS OF PROOFS

Anjolina Grisi de Oliveira*
Centro de Informática
Universidade Federal de Pernambuco
ago@cin.ufpe.br

Ruy J. G. B. de Queiroz†
Centro de Informática
Universidade Federal de Pernambuco
ruy@cin.ufpe.br

Abstract We are here concerned with the study of proofs from a geometric perspective. By first recalling the pioneering work of Statman in his doctoral thesis *Structural Complexity of Proofs* (1974), we review two recent research programmes which approach the study of structural properties of formal proofs from a geometric perspective: (i) the notion of *proof-net*, given by Girard in 1987 in the context of linear logic; and (ii) the notion of *logical flow graph* given by Buss in 1991 and used as a tool for studying the exponential blow up of proof sizes caused by the cut-elimination process, a recent programme (1996–2000) proposed by Carbone in collaboration with Semmes.

Statman's geometric perspective does not seem to have developed much further than his doctoral thesis, but the fact is that it looks as if the main idea, *i.e.* extracting structural properties of proofs in natural deduction (ND) using appropriate geometric intuitions, offers itself as a very promising one. With this in mind, and having at our disposal some interesting and rather novel techniques developed for *proof-nets* and *logical flow graphs*, we have tried to focus our investigation on a research for an alternative proposal for looking at the geometry of ND systems. The lack of symmetry in ND presents a challenge for such a kind of study. Of course, the obvious alternative is to look at multiple-conclusion

*Research partially funded by a grant from PROPESQ/UFPE under the *Projeto Enxoval*.
†Research partially funded by a CNPq *Bolsa de Produtividade em Pesquisa* ("Pesquisador 1-C"), grant 301492/88-3.

R.J.G.B. de Queiroz (ed.), Logic for Concurrency and Synchronisation, 3–88.

calculi. We already have in the literature different approaches involving such calculi. For example, Kneale's (1958) *tables of development* (studied in depth by Shoesmith & Smiley (1978)) and Ungar's (1992) multiple-conclusion ND.

After surveying the main research programmes, we sketch a proposal which is similar to both Kneale's and Ungar's in various aspects, mainly in the presentation of a multiple conclusion calculus in ND style. Rather than just presenting yet another ND proof system, we emphasise the use of 'modern' graph-theoretic techniques in tackling the 'old' problem of adequacy of multiple-conclusion ND. Some of the techniques have been developed for *proof-nets* (e.g. splitting theorem, soundness criteria, sequentialisation), and have proved themselves rather elegant and useful indeed.

Keywords: proofs as graphs, natural deduction, multiple-conclusion, geometry of deduction

1. Motivation

In 1980's various studies in "Logic and Computation" were pursued with the intention of giving a logical treatment of computer programming issues. Some of these studies have brought in a number of interesting proof-theoretic developments, such as for example:

- the functional interpretation of logical connectives[1] via deductive systems which use some sort of labelling mechanism:

  (i) Martin-Löf's *Intuitionistic Type theory* [53], which contributed to a better understanding of the foundations of computer science from a type-theoretic perspective, drawing on the connections between constructive mathematics and computer programming;

  and

  (ii) the *Labelled Deductive Systems*, introduced by Gabbay [34], which, arising from the need of computer science applications to handle "meta-level" aspects of logical system in harmony with object-level, helped providing a more general alternative to the "formulae-as-types" paradigm;

- *Linear Logic*, introduced by Girard in [38]. Since then it has become very popular in the theoretical computer science research community. The novelty here is that the logic comes with new connectives forming a new logical system with various interesting features for computer science, such as the possibility of interpreting a sequent as the state of a system and the treatment of a formula as a resource.

In recent years, linear logic has been established as one of the most widely used formalisms for the study of the interface between logic and computation. One of its key aspects represents a rather interesting novelty for studying the geometry of deductions: the concept of *proof-nets*. The theory of proof-nets developed out of a comparison between the sequent calculus and natural deduction (ND) Gentzen systems [36] as well as from an analysis of the importance of studying the structural properties of proofs through a *geometrical* perspective.

Another recent work which also presents a *geometrical* analysis in the study of structural properties of proofs has been developed by Carbone in collaboration with Semmes [14, 15, 16, 17, 18, 22]. Again in the context of "Logic and Computation", the analysis of Carbone and Semmes is motivated by questions which involve the middle ground between mathematical logic and computational complexity. In the beginning of the 1970's, Cook used the notion of satisfiability (a concept from logic) to study one of the most fundamental dichotomies in theoretical computer science: **P** versus **NP**. By the end of the decade Cook and Reckhow had established an important observation which puts emphasis on a relevant direction in complexity theory: **NP** is closed under complementation iff there is a propositional proof system in which all tautologies have a polynomial size proof [27]. This represents an important result linking mathematical logic and computational complexity since it relates classes of computational problems with proof systems. Motivated by questions such as the length of proofs in certain classical proof systems (in the style of Gentzen sequent calculus), Carbone set out to study the phenomenon of expansion of proofs, and for this purpose she found in concept of *logical flow graphs*, introduced by S. Buss [13], a rather convenient mathematical tool. Using the notion of *logical flow graph*, Carbone was able to obtain results such as, for example, providing an explanation for the exponential blow up of proof sizes caused by the cut-elimination process. With appropriate geometrical intuitions associated with the concept of *logical flow graph*, Carbone and Semmes developed a combinatorial model to study the evolution of graphs underlying proofs during the process of cut-elimination.

Now, if on the one hand we have

- Girard's proposal of studying the geometry of deductions through the concept of *proof-nets*, (in [40] he presents various arguments in defense of his programme, emphasizing the importance of "finding out the geometrical meaning of the *Hauptsatz*, *i.e.* what is hidden behind the somewhat boring syntactical manipulations it involves"),

on the other hand, there is

- Carbone's systematic use of *logical flow graph* in a geometrical study of the cut-elimination process, yielding a combinatorial model which uncovers the exponential expansion of proofs after cut-elimination.[2]

Although with different ends and means these two works concern the study of structural features of proofs by a geometric perspective. Back in the 1970's we find the work of Statman [65] proposing the idea of studying proofs as geometric objects for the analysis of structural properties. Even though some of the ideas of invertibility of rules in ND go back to Kneale's tables of development [49], it seems fair to say that Statman pioneered this kind of study in his doctoral thesis [65], where he systematically analyses the global structural complexity of proofs in ND systems.

One wonders why Statman's work did not develop much further than his doctoral thesis, but the fact is that it looks as if the main idea, *i.e.* extracting structural properties of proofs in ND using appropriate geometric intuitions, offers itself as a very promising one.

1.1 Relevance to proof theory: towards the geometry of natural deduction

Before laying out the whole theory of proof-nets and with the intention of justifying the development of such a theory, Girard makes criticisms both to sequent calculus and to ND systems. It seems that Girard's intention was to construct a framework joining good features of natural deduction and sequent calculus. The idea was to start defining *proof structures*, which he called "the natural deduction of linear sequent calculus", via the notion of *links*, *i.e.* a relation between formula occurrences. Those *proof structures* which represented a logically correct proof would then be called *proof nets*. In order to recognise a proof-net Girard defined a soundness criterion based on a certain geometrical interpretation of the logical connectives and this could be done due to the symmetry enjoyed by the proof rules (*i.e.* links). Alternatives to the so-called *no shorttrip condition* have been proposed over the years, and as a result, there arose more interesting connections between logic and computation within the context of linear logic: the paradigm of 'proofs as (distributed) processes'.

The lack of symmetry in ND systems is one of the aspects pointed out by Girard. In fact, ND does not have a well defined environment in terms of symmetries, like sequent calculus where the left and right side of the 'turnstile' ($\vdash$) have the following dualities:

negative	$\vdash$	positive
conjunction	$\vdash$	disjunction

And indeed, it seems to be the case that the lack of symmetry in ND systems turns out to be its "Achilles' heel" since its formulation by Gentzen [36]. In order to prove his fundamental theorem (cut-elimination theorem, the so-called Gentzen's *Hauptsatz*) about the structure of proofs, Gentzen abandoned the ND system and formulated the sequent calculus. In [58, 59] Prawitz gives an important result for ND systems by presenting the *normal form* theorems, which correspond to the *Hauptsatz*. Again, because of the lack of symmetry, Prawitz' formulation of ND classical system does not include the constants $\vee$ and $\exists$ as primitive. This has motivated various studies for a presentation of a proof of normal form theorems for full classical logic, using a formulation including the connectives $\vee$ and $\exists$ as primitive. Statman's work [65] was the first presentation of the normalisation theorem for full first order classical ND systems. Although with different and simpler methods to prove the normal form theorems for full first order classical logic, Stalmarck [64] as well as L.C. Pereira and C. Massi [56] follow Statman in some technical details and present an alternative solution to the normal form theorems for full classical logic. These works, together with the proposal of Andou [4], approach ND classical first order system as formalised by Prawitz, *i.e.* through the inclusion of the classical absurdity rule (Λ_C). On the other hand, the works of J. Seldin [62] and J. von Plato [57] give a proof of the normal form theorems for full ND classical systems formulated by the use of the *law of excluded middle*.

We believe that the problem of the lack of symmetry in ND systems does not have a satisfactory solution yet. As a consequence, ND does not have a uniform treatment for both intuitionistic and classical logic. For a start, the derivations which use the Λ_C inference rule are not exactly what one might call "natural", such as for example the derivation of $A \vee \neg A$:

$$\dfrac{\dfrac{\dfrac{\dfrac{\dfrac{[A]^1}{A \vee \neg A} \quad [\neg(A \vee \neg A)]^3}{\Lambda}}{\neg A}\,1 \qquad \dfrac{\dfrac{\dfrac{[\neg A]^2}{A \vee \neg A} \quad [\neg(A \vee \neg A)]^3}{\Lambda}}{A}\,2}{\Lambda}}{A \vee \neg A}\,3$$

There are plenty of other examples, such as the ND derivation of Peirce's law, which in addition to not being so natural have given rise to Curry's formulation of an additional rule to the system to cope with positive implicational classical logic [24]. Moreover, with the *classical absurdity rule* the normalisation procedure for full classical logic turns out to be not unproblematic. For all these reasons, we firmly believe that a "natural" solution for the classical case should go back to Gentzen in the sense that the *law of the excluded middle* is introduced as a primitive instead of the *classical absurdity rule*, as proposed by Prawitz, the latter becoming a derived rule. Together with this, one should seek to formulate a symmetric proof system to accommodate the dualities of classical logic.

We have already mentioned a number of works which approach the classical case by the use of the *law of excluded middle*. In our case, we wish to incorporate the *law of excluded middle* in the context of a framework based on a geometrical perspective. Let us recall some of those approaches to normalisation for full classical logic which use the *law of excluded middle*:

- The rule for the *law of excluded middle* proposed by Tennant [66] and von Plato [57] has its advantages for the classical case.

- The rule proposed by Seldin is very similar to the *reductio ad absurdum* (*i.e.* Λ_C) and the normalisation procedure is not as simple as Prawitz', although it is technically sound.

We claim that the attempt to look at the problem of symmetry in ND systems by bringing in some geometrical intuitions offers itself as a richer perspective.

Another alternative to deal with the problem of lack of symmetry in ND systems is to look at multiple-conclusion calculi. We already have in the literature different approaches to deal with it. We mention for example the books by Shoesmith & Smiley [63] and the one by Ungar [67]. In a footnote in [58, p. 44] Prawitz talks about the importance of such kind of calculi:

> "... one may therefore consider modifications of the Gentzen-type system for classical logic. One rather natural modification is to make the system more symmetrical with respect to $\wedge$ and $\vee$. In the present system, the deduction forks only upwards, and the forking is so to say conjunctively; one could now allow the deductions to fork also downwards, disjunctively, so that the deduction is allowed to fork from a disjunction $A \vee B$ into two branches, starting with A and B respectively. Such a system was presented by the author in a colloquium in Los Angeles in 1964. (A suggestion to such a system may be found in Kneale[3] but the rules are there stated without sufficient restrictions)..."

In fact, the suggestion of Kneale & Kneale (also in [50]) as originally presented needs appropriate restrictions and proper metamathematical foundations. In [63] the authors refine Kneale's proposal by presenting alternative definitions of formal proof for use in multiple conclusion calculi. However, neither the soundness criteria for proof graphs, nor the normalisation procedure, are so well developed as Girard's proof-nets via the soundness criteria and the sequentialisation theorem. As a follow-up to the present survey, we wish to draw up a proposal similar in spirit to Kneale's, though using the techniques developed in the context of linear logic (soundness criteria, sequentialisation, etc.). Moreover, we want to deal with classical symmetries without necessarily restricting the calculus to a multiple-conclusion. The idea is that to a proof graph which is multiple-conclusion there should be a proof graph whose conclusions are turned into a single formula, even if for this one has to group the formula occurrences into 'conclusion classes' (a notion which is dual to the notion of 'assumption classes'). A future development should also include the study of the complexity of ND derivations in the spirit of the work of Alessandra Carbone for the sequent calculus.

1.2 Relevance to theoretical computer science: complexity and concurrency

NP-completeness and proof systems. In a recent retrospective of the development and the influence of the concept of NP-completeness, C. Papadimitriou [55] asks himself what is the nature and the extent of the impact of NP-completeness on theoretical computer science and computer science in general, and why did it become such a pervasive and influential concept. He then adds that

> one reason of the immense impact of NP-completeness has to be the appeal and elegance of the class P, that is, of the thesis that "polynomial worst-case time" is a plausible and productive mathematical surrogate of the empirical concept of "practically solvable computational problem." But, obviously, NP-completeness also draws on the importance of NP, as it rests on the widely conjectured contradistinction between these two classes. In this regard, it is crucial that NP captures vast domains of computational, scientific, and mathematical endeavor, and seems to roughly delimit what mathematicians and scientists had been aspiring to compute feasibly. True, there are domains, such as strategic analysis and counting, which have been within our computational ambitions, and still seem to lie outside NP; but they are exceptions rather than the rule. NP-completeness has thus become a valuable intermediary between the abstraction of computational models and the reality of computational problems, grounding complexity theory to computational practice.

The landmark for all this was S. Cook's [25] theorem stating that "P=NP iff there is a deterministic algorithm to recognise the tautologies of propositional logic." In elementary logic one learns that the method of truth tables is the most direct way of testing whether a formula is a tautology, even though the test may cost (in the worst-case) up to 2^n where n is the number of propositional variables in the formula. No "shortcut" method is known that does any better than this exponential growth rate. On the other hand, as proof theory, the branch of symbolic logic which looks at *proofs* as their objects of study, developed to the extent of turning its focus of attention from the "validity of proofs" to the "structural complexity of proofs", it made sense to ask (as did Cook and Reckhow in the 1970s) whether tautologies have proofs of polynomial-size lengths in any proof system. This motivated the development of the concept of *polynomial time simulation* which formed the basis for the study of *proof systems* and their complexity.

Since then, a lot of work has been done with the intention of classifying the relative complexity of various proof systems by proving lower bounds for the well known proof systems. For instance, in 1974 Cook and Reckhow already define a hierarchy of various propositional proof systems. They placed *resolution* into the class of "known not be super" (a proof system is said to be "super" if it admits polynomial length proofs of all tautologies). Many years later, A. Haken [46] showed that *resolution* requires exponential time on the pigeonhole formula. A recent classification of various proof systems can be found in [61] and [68]. (See also [51]).

Proof theory and the deviation via non-classical logics. If, on the one hand, the work on classifying the relative complexity of proof systems usually took for granted that the notion of *proof system* (and the logic it was supposed to be a proof system for) was well settled and very much a unanimously accepted platform, on the other hand, much work on proof theory since 1960s (following the work of Prawitz, Martin-Löf, Girard, Howard, Tait, and others) and its *structural* aspects (mainly emphasised by Statman), together with its due impact on theoretical computer science via λ-calculus and combinatory logic, have given rise to refinements of various classical systems such as *natural deduction* and the *sequent calculus*. One such refinement has almost turned into a school of thinking about the interactions between logic (*qua* proof theory) and computation: J.-Y. Girard's linear logic. Indeed, a novel perspective on proof theory and the information flow arising out of cut-elimination is given by Girard in his research programme entitled *geometry of interaction*. Drawing on an analysis of the geometrical properties of Gentzen's *Cut-Elimination Theorem*, the programme offers a highly innovative account of the connections between *computation* and *deduction*.

One can distinguish two main aspects of the programme, namely: (1) a geometrical view of deductions and polarities, thus of invertibility of proof rules by duality;[4] and (2) the characterisation of cut-elimination in terms of the iteration of a single operator, together with a kind of 'normal form theorem' (in the style of Kleene's own).[5] While the first aspect relies heavily on the sequent calculus formulation of linear logic (together with the device of *proof nets*), the second one draws on strong normalisation results for the system F of second-order λ-calculus.

Back to classical logic and to *natural deduction*, but bringing in some geometrical intuitions. The study of structural properties of proofs is part of the "theory of proofs" as pointed out by Statman in his doctoral thesis *Structural Complexity of Proofs*" [65]. Unlike "proof theory" which is concerned with the study of the validity of proofs, the study of structural properties of proofs refers to "meta-level" considerations, having the proof itself as the object of study. Thus, it involves questions like: (i) how long is a proof of a given formula; (ii) how to identify two proofs of the same theorem; (iii) what happens if we invert a given derivation (*i.e.* swap hypotheses and conclusions); (iv) what is the meaning of this inversion operation; etc. A classical general result about the structure of proofs is the cut-elimination theorem (the *Hauptsatz*) for *sequent calculus* given by Gentzen in [36] and quoted here:

> "Every LJ- LK-derivation can be transformed into an LJ or LK-derivation with the same endsequent and in which the inference figure called a 'cut' does not occur."

LJ refers to the intuitionistic logic while LK to the classical logic.

Prawitz [58, 59] extended this result for *natural deduction* systems defining the *normal form* theorems. These theorems are very important because they assure that if there is a derivation of a given formula, this proof, called *normal form* or *proof without cuts*, has a specific form, besides having certain properties, for instance the *subformula property*.

The methods used in proving such theorems rely very heavily on syntactic manipulations. On the other hand, the work of Girard has pointed to a rather interesting direction which is to turn to the geometry of deductions, and use geometric intuitions to look at classical results about normal forms and normalisation.

Proofs as processes. In an attempt to turn the hints and speculations in Girard's work in linear logic about its applicability to concurrent computation into a fully realised connection, S. Abramsky set out to extend the "proposition-as-types" paradigm (encompassing the Curry–Howard interpretation of intuitionistic logic) to concurrency. The idea was to try and lay out the foundations of typed concurrent programming in such a way that concurrent processes, rather than functional programs, would become the counterpart of proofs. The key element in the whole approach had already appeared in an earlier paper by Abramsky himself on "Computational Interpretations of Linear Logic" [1], and it had to deal with the question of how to give a computational interpretation of the duality in classical linear logic. The main observation was that the cut rule of classical linear logic

$$\frac{\vdash \Gamma, A \qquad \vdash \Delta, A^{\perp}}{\vdash \Gamma, \Delta}$$

is completely symmetric, and the formulas A and $A^{\perp}$ could be seen as communicating processes (input and output) such as in Milner's π-calculus.

In an attempt to carry the idea further G. Bellin and P. Scott [10] presented some fundamental facts about Milner's process calculus, the π-calculus, and about Girard's representation of proofs in linear logic as proof-nets, detailing Abramsky's "proofs-as-processes" paradigm [2] for interpreting classical linear logic (CLL) into a "synchronous" version of Milner's π-calculus, the translation being given at the abstract level of proof-structures. In that paper they also give a detailed treatment of the information flow in proof-nets and show how to mirror various evaluation strategies for proof-normalisation. Bellin and Scott give soundness and completeness results for the process-calculus translations of various fragments of CLL.

Of course, for the extension of the propositions-as-types paradigm to concurrency to be fully achieved, it was necessary to look at the other half of the paradigm, *i.e.* "processes-as-proofs". As already pointed out by Abramsky, this would require showing how a process calculus, sufficiently expressive to allow a reasonable range of concurrent programming examples to be handled, could be exhibited as the computational correlate of a proof system. Abramsky claimed that this had been achieved with his subsequent work on interaction categories [3]. More recently,[6] Bellin has raised important questions which seem to remain unanswered such as: To which extent Abramsky's slogan "processes-as-proofs" can be realised? Under which conditions a formal computation in a process calculus can be called a proof? How close logical computation can be to the dynamics of concurrent processes?

2. The idea of studying proofs as geometric objects

We can say that the idea of studying proofs as geometric objects is motivated by an analysis of its structural properties. In his doctoral thesis [65], Statman begins this kind of study, where he analyzes the global structural complexity of proofs in natural deduction systems. We quote from his thesis:

> "One way to proceed from the study of the validity of proofs, which occupies most of current proof theory, to the study of their structure is to view them as geometric objects. Differences between proofs "previously judged only by aesthetic criteria of elegance or convenience" become now principal objects of study."

He then proposes a representation of proofs through graphs, in which the relation between formula-occurrences is represented by incidence relation on the vertices of the graphs. His aim was to study the global structural complexity of proofs in natural deduction systems. From his geometric model, Statman analyzes the effect of the elimination of derived rules on the complexity of proofs; studies the complexity of equational proofs and also presents a structural analysis of a normalisation procedure for classical logic.

Our intention here is to bring out some of the pioneering ideas in Statman's work. It is rather unfortunate that the work did not develop much further than his doctoral thesis.

2.1 Representing proofs in a graph

The relations between formula-occurrences in proofs of natural deduction systems are defined in the following two ways:

Relation 1 The relation between premise and conclusion of the inference; and

Relation 2 An assumption occurrence with the conclusion which cancels it.

Statman notes that Relation 2 results in non planar figures when proofs are represented as graphs. From graph theory, we know that planar figures G are easily embeddable in a surface S in such a way that lines representing edges of G intersect only at points of S representing vertices of G. However, non planar figures need more complicated surfaces to be embedded in, for example, the surface known in graph theory as "torus". A classical example is the Kuratowski graph ($K_{3,3}$), which is embeddable in the torus. Consequently, to measure the non planar feature caused by Relation 2, Statman uses the notion of *genus* $\gamma(G)$ of a graph G. Firstly, we give the notion of what an embeddable graph G is [65]:

Definition 1 (embeddable G) A graph G is embeddable in a closed orientable surface M if G can be drawn on M in such a way that lines representing edges of G intersect only at points of M representing vertices of G.

Definition 2 (genus of a graph) The genus $\gamma(G)$ of a graph G is the minimum number of "handles" which must be added to a sphere so that the graph G is embeddable in the resulting surface.

The complexity of Relation 2 in a derivation D is measured by the genus of the graph which represents D.

Example 3 The following derivation is represented as a graph shown in Figure 1.1:

$$\dfrac{\dfrac{\dfrac{\dfrac{(4, q \to r) \qquad \dfrac{(2, p \to q) \quad (1, p)}{(3, q)} \to E}{(5, r)} \to E}{(6, p \to r)} 1 \to I}{(7, (p \to q) \to (p \to r))} 2 \to I}{(8, (q \to r) \to ((p \to q) \to (p \to r)))} 4 \to I$$

In Figure 1.1 the labels of the edges can assume two forms: (a, b, c) or (a, c). As we will see in its formal definition, the first element, *i.e.* a, refers to the number of the premise; c indicates the inference rule; and b refers to the number of the assumption cancelled by the rule.

A propositional language (PL) is defined. It consists of:

- propositional variables: $Var = \{p, q, r, \ldots\}$;
- propositional constants: $Cte = \{C_1, \ldots, C_m\}$;
- connectives constants: $Oper = \{O_1, \ldots, O_n\}$;
- formulas: built up by operations $F_1, \ldots, F_{k_i} \to O_i F_1, \ldots, F_{k_i}$ (the set of formulas is denoted by Fm). Formula occurrences are members of $\mathbb{N} \times Fm$.

Definition 4 (substitution) A substitution θ is a function $\theta : Var \to Fm$ such that $dom\theta = \{p \in Var : \theta p \neq p\}$ is finite.

The notation of *rules* and *inferences* is given as follows:

Rules

$$R = \frac{\begin{pmatrix} & < [A^i_j] : 1 \leq j \leq a_i > & \\ & P_i & : 1 \leq i \leq p \end{pmatrix}}{C}$$

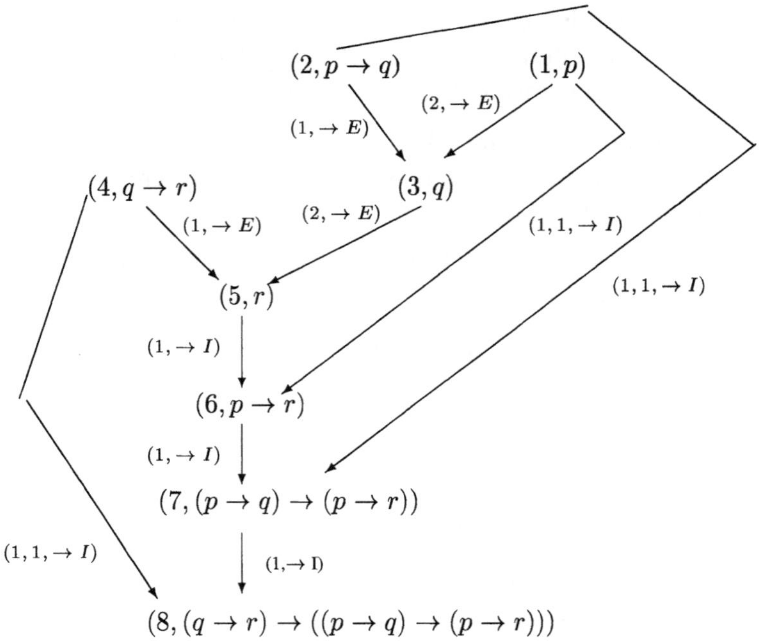

Figure 1.1. Graph representing a proof

Inferences

$$\theta R = \frac{(\qquad \overset{< [\theta A^i_j] : 1 \leq j \leq a_i >}{\theta P_i} \qquad : 1 \leq i \leq p)}{\theta C} R$$

Where A^i_j, C and P_i are formulas, θ is a substitution, $1 \leq p$ and $0 \leq a_i$ (p, a are given parameters).

The edges of graphs representing proofs as proposed by Statman are labelled from a set defined as follows.

Definition 5 (set of labels) A set of labels L is defined as follows:

$$L = \{(i, R) : 1 \leq i \leq p^R, \ R \in N\} \bigcup \{(i, j, R) : 1 \leq i \leq p^R, \ 1 \leq j \leq a^R_i, \ R \in N\}$$

where:

- N is a finite set of rules R with parameters p^R, a^R;
- p^R is the number of premises of the rule R;
- a^R is the number of assumptions of the rule R;

- the index i refers to the premises while j to the assumptions of the rules.

In Example 3 we can see the edges labelled in this way.
Now, we give a precise definition of graphs of proofs.

Definition 6 (directed graph) Let L be a finite set. A directed graph D with labels from L is a pair (V, E), where :

- V is a finite set of vertices of D and $V \subseteq \mathbb{N} \times Fm$;
- E is a subset of $V \times V \times L$ and represents the edges of D.
- The notation $V(D)$ is used to denote the set of vertices of D. Similarly, $E(D)$ denotes the set of edges of D.

Let $\mathbb{D}$ be the set of all directed graphs D as defined here. Then, if $D \in \mathbb{D}$ the following notation is given:

1. $D = \genfrac{}{}{0pt}{}{D}{x}$ if $x \in V(D)$ and x is not the initial vertex of any member of $E(D)$. In terms of natural deduction systems, x represents the conclusion of the derivation. Once the natural deduction system has only one conclusion and the *outdegree* of x is zero, *i.e.* it is not a premise of any rule.
2. $D = \genfrac{}{}{0pt}{}{[H]}{D}$ if $[H] \subseteq V(D)$ and $y \in [H]$ implies that y is not the final vertex of any member of $E(D)$. Clearly, $[H]$ represents the set of hypotheses of a given derivation;
3. $D = \genfrac{}{}{0pt}{}{< [H_i] : 1 \leq i \leq t >}{D}$ of $D = \genfrac{}{}{0pt}{}{[H_i]}{D}$ for $1 \leq i \leq t$ and $i \neq j \rightarrow [H_i] \cap [H_j] = \emptyset$.

Definition 7 (the set of derivations) The set of N derivations, denoted by Der_N, is the smallest set $X \subseteq \mathbb{D}$ satisfying:

1. if $F \in F_m$ and $n \in \mathbb{N}$ then $(\{(n, F)\}, \emptyset) \in X$ (in this case, D contains only one formula; the set of edges is empty);
2. if for $1 \leq i \leq p^R$ and $R \in N$

$$D_i = \begin{array}{c} < [\theta A^i_j] : 1 \leq j \leq a^R_i > \\ D_i \\ (n_i.\theta P_i) \end{array} \in X$$

   and $i \neq j \Rightarrow D_i \cap D_j = \emptyset$ and whenever $x \in [\theta A^i_j]$, x is not the initial vertex of an edge with label of the form (k, l, R'), for $R' \in N$, then if $1 \leq i \leq p^R \Rightarrow (n, \theta C) \notin V(D_i)$ we have

$$\frac{\left(\begin{array}{c} < [\theta \not{A}^i_j] : 1 \leq j \leq a^R_i > \\ D_i \\ (n_i, \theta P_i) \end{array} : 1 \leq i \leq p^R \right)}{(n, \theta C)} R \in X$$

2.2 Some results

2.2.1 Elementary results on isomorphisms.

Definition 8 (isomorphism between derivations) If D_0, $D_1 \in \mathbb{D}$ we say that D_0 is isomorphic to D_1, denoted by $D_0 \approx D_1$, if there is a bijection $b : V(D_0) \to V(D_1)$ and $b : E(D_0) \to E(D_1)$ satisfying $b((x, y, l)) = (b(x), b(y), l)$. If b also satisfies, $b(x)$ is an occurrence of the same formula as x, we write $D_0 \equiv D_1$.

Proposition 9 *Isomorphisms between derivations are unique.*

Proposition 10 *For each $D_0 \in Der_N$ there is a D_0^* such that whenever $D_0 \approx D_1$ there is a $\theta \in Sub$ such that $D_1 \equiv \theta D_0^*$.*

2.2.2 On the genus of a derivation. Statman proves that for each number n there is derivation D such that $\gamma(D) = n$. On the other hand, such measures of "graph theoretic" complexity as chromatic number, connectivity, arboricity, etc. which are not topological invariants tend to be absolutely bounded on derivations.

First he considers explicit definitions, and although his main results on complexity of proofs concerning explicit definitions are purely geometrical (topological), a logical analysis is still needed: for a theory of proofs such an analysis asks for (i) derived rules associated with the defined notion; (ii) a procedure for the elimination of such derived rules. If elimination procedures ϵ commute with substitutions, then for usual measures μ there will correspond functions f_μ such that $\mu(\epsilon D) \geq f_\mu(\mu(D))$ for all derivations D. Moreover, for any reasonable μ, $\mu(D) \leq \mu(\epsilon D)$, and from this it easily follows that $\gamma(D) \leq \gamma(\epsilon D)$. Finally, an interesting dichotomy is established: for several structural properties $\mathcal{P}$, $\gamma(\epsilon D) = \gamma(D)$ implies that $\mathcal{P}(\epsilon D)$ and, if, for some D, $\neg\mathcal{P}(\epsilon D)$ then there is no f such that for D, $\gamma(\epsilon D) \leq f(\gamma(D))$.

Using graph-theoretic results of Tutte and Nordhaus, Statman is able to establish the exact relation between the genus of a derivation and the number of its cancelled assumption occurrences:

Proposition 11 *Suppose G has a Hamiltonian path. Then there is a $D \in Der_{\mathrm{I}}$ such that $\beta_1(G) = \beta_1(|D|)$ (where $\beta_1(G)$ is the first Betti number of G) and G is isomorphic to a contraction of $|D|$.*

Corollary 12 *Suppose $N \supseteq$ I and $\#D \stackrel{\text{def}}{=}$ the number of cancelled assumption occurrences in D, then for each $D \in Der_N$, $\#(D) \leq \frac{\#D}{2}$ and for each positive rational r there is a $D \in Der_N$, such that $\frac{\#D}{2+r} < \#(D)$.*

("I" is Prawitz' natural deduction system for intuitionistic logic.)

2.2.3 On the relationship between the notions of direct proof, normal proof and proof, and the subformula property.

Definition 13 If F occurs in D, let $G(D)\langle< F >$ be the subgraph of G induced by those members of $V(D)$ containing F as a subformula. We say that a proof of D, *i.e.* a derivation without uncancelled assumptions, is direct if whenever F occurs in D it is connected to the endformula of D in $G(D) < F >$.

Proposition 14 *For any derivation $D \in Der_{\mathrm{M}}$, D is normal if and only if whenever F occurrs in D the connected component C of $G(D) < F >$ containing this occurrence contains also an uncancelled assumption occurrence, or the end formula, of D.*

("M" is Prawitz' natural deduction system for minimal logic.)

Corollary 15 *If D is an M proof then D is normal iff D is direct.*

3. Proof-nets

Proof-nets were introduced by Girard in his seminal paper [38] on the presentation of Linear logic. Since then the notation and results have evolved and changed. The idea was to provide a graph-theoretic representation of deductions in linear logic. The latter is a *substructural* logic because its sequent calculus does not include the *structural rules* as originally given by Gentzen [36] for classical logic sequent calculus. The *contraction* and *weakening* structural rules are dropped. Other features of the system are: the distinction between multiplicative and additive connectives; the introduction of *exponentials* (allowing contraction and weakening in a controlled form); and the introduction of linear negation. Depending on what of those features are required, one may have various fragments of linear logic such as, multiplicative linear logic (MLL), additive linear logic (ALL), multiplicative additive linear logic (MALL), etc.

Proof-nets represent the main tool for Girard's intention of studying the geometry of deductions. As we will see in Section 3.5 the *soundness criterion* for proof-nets is based on a purely geometrical analysis of the structure of the graphs representing proofs. The theory is simple and works well for multiplicative linear logic (MLL), although it becomes more complicated for the full calculus. Our presentation here will refer only to the multiplicative linear logic without constants (MLL^- fragment).

Girard defends his programme of geometry of interaction by making criticisms to ND and sequent calculus [38, 44, 40, 45], here shown in Section 3.1. The claim is that his framework joins good features of those logical systems. Girard defines *proof-structures*, *i.e.* graphs of proofs, which he calls "the natural deduction of linear sequent calculus" [38] through the notion of *links*, *i.e.* a relation between formula-occurrences. Those proof-structures which represent a proof logically correct are called *proof-nets* , or *nets* for short.

Here we shall try to explain the main issues involved in the criticisms to natural deduction and sequent calculus made by Girard. After that, we present his considerations for studying the geometry of deductions which result in the definition of the *links* that form the proof-nets. Next we give the definition of proof-structures as well as how to recognise which proof-structures are proof-nets by showing some criteria of soundness. We conclude this section by introducing the cut-elimination in the context of proof-nets. (For further details, see the two chapters by G. Bellin in this volume: 'Chu's Construction: A Proof-Theoretic Approach' [8], and 'Two paradigms of logical computation in affine logic?' [9].)

3.1 Criticisms of natural deduction

3.1.1 Classical symmetry. In [38, 40, 45], Girard has pointed out that natural deduction is unsuitable to deal with classical symmetries. In fact, natural deduction does not have a well defined environment in terms of symmetries, like sequent calculus where the left and right side of the 'turnstile' ($\vdash$) have the following dualities:

negative	$\vdash$	positive
conjunction	$\vdash$	disjunction

The lack of symmetry in ND systems has motivated a formulation of multiple-conclusion calculi as we will see in the section after the next one.

Proof-nets can be considered as a natural deduction for linear sequent calculus which accept several conclusions, and thus can deal with the classical symmetries.

3.1.2 Global rules. Natural deduction has rules which do not apply to formulas, but to the whole deduction. The introduction rule for $\rightarrow$ is an example of such a rule:

$$\frac{\begin{array}{c}[A]\\ \vdots\\ B\end{array}}{A \rightarrow B} \rightarrow\text{-}I$$

This kind of inference rule is called *improper rule* by Prawitz [59] because it discharges hypotheses. Besides the $\rightarrow$-*introduction*, there are other improper rules: $\vee$ and $\exists$-*elimination*, and $\forall$-*introduction*.

Proof-nets as a sort of "natural deduction for linear sequent calculus" do not include rules with such a feature.

3.1.3 The commutative conversions. In ND, the rules of $\vee-$*elimination* and $\exists$-*elimination* are framed as follows:

$$\frac{A \vee B \quad \begin{array}{c}[A]\\ C\end{array} \quad \begin{array}{c}[B]\\ C\end{array}}{C} \vee\text{-}E$$

$$\frac{\exists x P(x) \quad \begin{array}{c}[P(t)]\\ C\end{array}}{C} \exists\text{-}E$$

Girard calls "C" as "an extraneous formula" and says that because of this the theory of normalisation "becomes extremely complex and algorithmically awkward" [44], once we have the "commutative conversions".[7]

The presence of commutative conversions is another feature of natural deduction system which proof-nets do not include.

In summary, Girard claims that he has constructed a system like natural deduction, but which only keeps its good features.

3.2 Analyzing the sequent calculus

Through examples, this subsection illustrates some features of sequent calculus which are the result of Girard's investigations when he developed linear logic and the concept of proof-nets.

3.2.1 The non-determinism of Hauptsatz. The algorithm of cut-elimination is non-deterministic, once we face situations like the one shown in the following example [44]:

Example 16 The usual Gentzen's cut-elimination procedure consists of pushing the cut upwards. Hence, when the following fragment of proof is found:

$$\frac{\dfrac{\vdash \Gamma, A}{\vdash \Gamma', A}(\mathbf{r}) \qquad \dfrac{\vdash A^{\perp}, \Delta}{\vdash A^{\perp}, \Delta'}(\mathbf{s})}{\vdash \Gamma', \Delta'}$$

two choices arise:

$$\cfrac{\cfrac{\cfrac{\vdash \Gamma, A \qquad \vdash A^{\perp}, \Delta}{\vdash \Gamma, \Delta}\,cut}{\vdash \Gamma', \Delta}\,(\mathbf{r})}{\vdash \Gamma', \Delta'}\,(\mathbf{s})$$

$$\cfrac{\cfrac{\cfrac{\vdash \Gamma, A \qquad \vdash A^{\perp}, \Delta}{\vdash \Gamma, \Delta}\,cut}{\vdash \Gamma, \Delta'}\,(\mathbf{s})}{\vdash \Gamma', \Delta'}\,(\mathbf{r})$$

3.2.2 Redundancies in proofs. The stages of the derivations in the sequent calculus carry along many formulas (*i.e.* the context) which do not participate of the rule which has being applied. It causes unnecessary redundancies in proofs as illustrated in the following example given in [45]:

Example 17

$$\cfrac{\vdash C, D, \Delta}{\vdash C \wp D, \Delta}\,\wp$$

The context Δ is rewritten without any change.

As we will see here, the context is dropped in a proof-net.

3.2.3 Non-unicity of proofs. The sequent calculus (also the linear sequent calculus) may allow different proofs of the same sequent when the inference rules are applied in a different order [35, 45].

Example 18 The sequent

$$\vdash (A \otimes B) \otimes C, A^{\perp} \wp B^{\perp}, C^{\perp}$$

has the following two proofs

$$\cfrac{\cfrac{\cfrac{\vdash A, A^{\perp} \qquad \vdash B, B^{\perp}}{\vdash A \otimes B, A^{\perp}, B^{\perp}}\,\otimes}{\vdash A \otimes B, A^{\perp} \wp B^{\perp}}\,\wp \qquad \vdash C, C^{\perp}}{\vdash (A \otimes B) \otimes C, A^{\perp} \wp B^{\perp}, C^{\perp}}\,\otimes$$

and

$$\cfrac{\cfrac{\cfrac{\vdash A, A^{\perp} \qquad \vdash B, B^{\perp}}{\vdash A \otimes B, A^{\perp}, B^{\perp}}\,\otimes \qquad \vdash C, C^{\perp}}{\vdash (A \otimes B) \otimes C, A^{\perp}, B^{\perp}, C^{\perp}}\,\otimes}{\vdash (A \otimes B) \otimes C, A^{\perp} \wp B^{\perp}, C^{\perp}}\,\wp$$

Using proof-nets these two proofs are identified. An important feature of proof-nets: the proof of a theorem is dealt with as a unique object.

3.3 Definition of links

The main intention of Girard when he provides the concept of proof-nets is to study the geometry of deductions. The idea is to replace a natural deduction proof

$$\begin{array}{c}\Gamma\\ \vdots\\ A\end{array}$$

by a proof-net with several conclusions $\Gamma^\perp, A$, using the negation symbol $(\perp)$ every time a formula is turned upside down (*i.e.* a premise is turned into a conclusion and vice-versa) [40]. He then defines two *identity links* - *axiom link* and *cut link* (see Figure 1.2) - which allow one to replace a hypothesis with a conclusion and vice-versa.

Figure 1.2. Identity links

The symbol *CUT* in a cut link is not a formula, but a place-holder. The *axiom link* has no premise. Two more links, *par* and *times*, are defined as follows.

Par link The following inference in natural deduction:

$$\cfrac{\begin{array}{c}[A]\\ \vdots\\ B\end{array}}{A \to B}$$

is replaced by

$$\frac{A^\perp \qquad B}{A^\perp \wp B}$$

Consequently, the following binary rule (*par link*) is obtained:

$$\frac{C \qquad D}{C \wp D}$$

The formulae C and D are premises of the link, while $C \wp D$ is the conclusion.

Times link The *times* link is defined as:

$$\frac{C \qquad D}{C \otimes D}$$

The formulae C and D are premises of the link while $C \otimes D$ is the conclusion.

The traditional rule for elimination of $\rightarrow$ (*modus ponens*) in natural deduction:

$$\frac{A \qquad A \rightarrow B}{B}$$

is replaced by the inference shown in Figure 1.3.

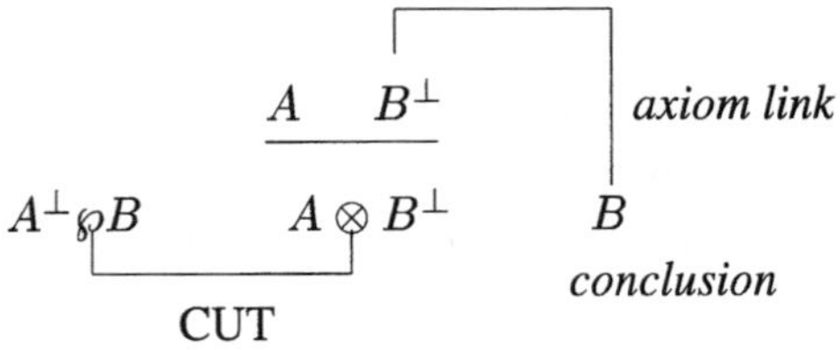

Figure 1.3. Modus ponens

Note that B and $A \rightarrow B$ are turned upside down in Figure 1.3.

3.4 Proof-structures

Using the links defined in the previous subsection, proof-structures for the MLL^- fragment are constructed with several conclusions, like in sequent calculus, but with a notation similar to natural deduction. The class of those proof-structures which represent logically correct proofs is called proof-nets. In the next subsection we will study how to define this class by applying a global soundness criterion on the proof-structures.

Definition 19 (Proof-structures [11]) A proof-structure for MLL^- consists of (i) a nonempty set of formula-occurrences together with (ii) a set of links between these formula-occurrences. These links are as defined in the previous subsection: *axiom link*, *cut link*, *par link* and *times link*. Moreover, (i) and (ii) must satisfy the following requirements:

- every formula-occurrence in the structure is the conclusion of one and only one link;
- every formula-occurrence is the premise of at most one link.

The correspondence between a proof Γ in linear sequent calculus and a proof-structure with as conclusions the formulas of Γ is given by the following procedure [44]:

1 The *identity axiom* is associated to an *axiom link*:

$$\vdash A^{\perp}, A \quad \rightsquigarrow \quad \overline{A \quad A^{\perp}}$$

2 The *par rule* and the *par link*:

$$\frac{\begin{array}{c}\Pi \\ \vdots \\ \vdash \Gamma, A, B\end{array}}{\vdash \Gamma, A \wp B} \quad \leadsto \quad \begin{array}{c}PS \\ \Gamma \\ \dfrac{A \quad B}{A \wp B}\end{array}$$

where PS is a proof-structure correspondent to the proof Π.

3 The *times rule* and the *times link*:

$$\frac{\begin{array}{c}\Pi_1 \\ \vdots \\ \vdash \Gamma, A\end{array} \qquad \begin{array}{c}\Pi_2 \\ \vdots \\ \vdash B, \Delta\end{array}}{\vdash \Gamma, \Delta, A \otimes B} \quad \leadsto \quad \begin{array}{ccc}PS_1 & & PS_2 \\ \Gamma & & \Delta \\ & \dfrac{A \quad B}{A \otimes B} & \end{array}$$

where PS_1 is the proof-structure associated to Π_1 and PS_2 to Π_2.

3.4.1 Examples.

Some examples of proof-structures are given here.

Example 20 The proof-structure shown in Figure 1.4 is built from the deductions in linear sequent calculus of Example 18.

Note (Example 18) that two different proofs in linear sequent calculus correspond to a single proof-structure (the one in Figure 1.4).

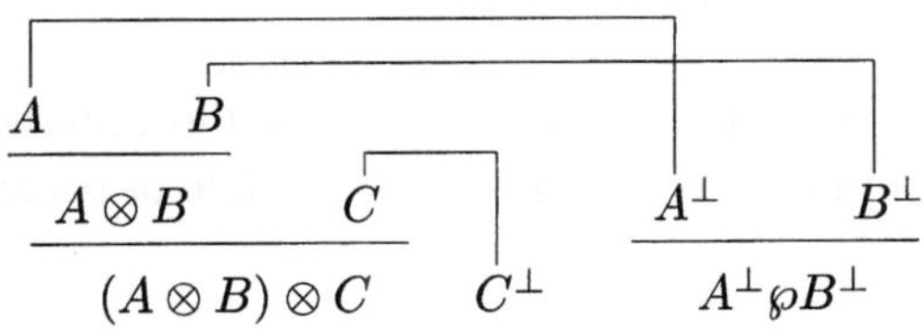

Figure 1.4. Proof-structure

The proof-structure shown in this first example is a proof-net, since it can be verified that it is logically correct. However, that one shown in Figure 1.5 is not a proof-net.

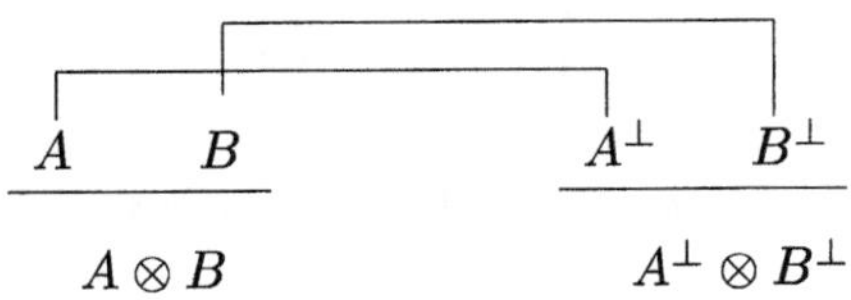

Figure 1.5. Proof-structure

Example 21 Another example of proof-structure which is a proof-net is illustrated in Figure 1.6.

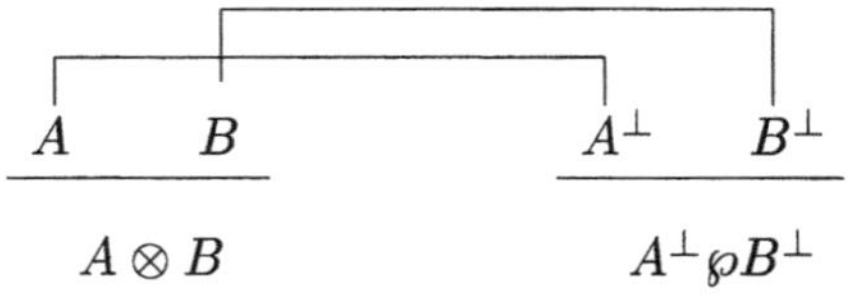

Figure 1.6. Proof-structure

3.5 The soundness criterion

The most interesting feature of the theory of proof-nets is the definition of soundness criteria based on a purely geometrical characterisation of which proof-structures represent a sound proof.

There are various criteria of soundness, here we emphasise the following three criteria:

1 *No short trip condition*: given by Girard in his seminal paper [38];

2 Danos–Regnier criterion: a simplification of Girard's original criterion [28];

3 Asperti criterion: a generalisation of Girard's *no-short-trip* condition, where trips are distributed processes, as opposed to sequential processes like in the original formulation [6].

3.5.1 No short trip condition.

In this criterion each formula is viewed as a box, as in Figure 1.7, in which a particle or some information may travel.

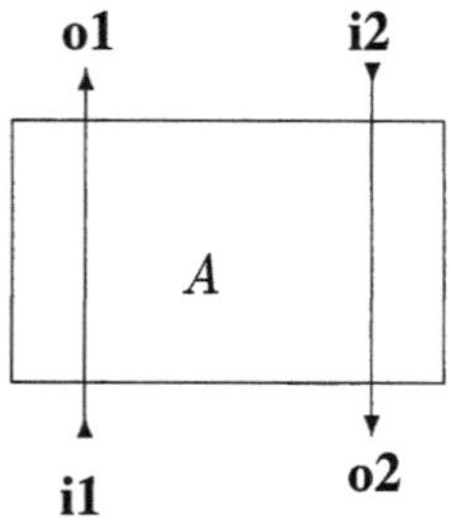

Figure 1.7. Formula as a box

The concept of a cyclic, discrete and finite time is used to control the particle travelling through the proof-structure.

The *trip* of the particle through a given proof-structure is as follows:

- An arbitrary formula is chosen to begin the trip;
- there are two gates where the particle will enter and two out gates, as shown in Figure 1.7;
- the *dual enter gate* of the *exit gates* **o1** and **o2** are **i1** and **i2**, respectively;
- at the unit of time $t^{\wedge}$ the particle will enter A by the gate called "**i1**" and will immediately go out through the gate "**o1**". This movement is denoted by $A^{\wedge}$;
- at the moment $t_{\vee}$ the particle will re-enter A by "**i2**" and will immediately go out through "**o2**". This movement is denoted by $A_{\vee}$;
- the particle will enter another formula through the links of the proof-structure, following the pattern of the trip for the link as defined below;
- if every formula in the proof-structure is visited twice the trip is *long* and the proof-structure is a proof-net. Otherwise the trip is *short* and the proof-structure is not a sound proof.

Axiom link The pattern of a trip in the *axiom link* is shown in Figure 1.8. The trip is performed as following:

- $t(A^{\perp}_{\vee}) = t(A^{\wedge}) + 1$: the particle enters $A^{\perp}$ through gate $i2$ immediately after exiting A through $o1$.
- $t(A_{\vee}) = t(A^{\perp\wedge}) + 1$: immediately after exiting $A^{\perp}$ from $o1$ the particle enters A through $i2$.

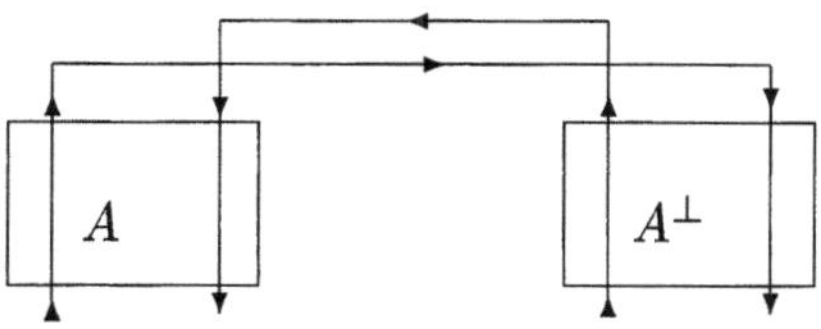

Figure 1.8. Axiom link

Terminal formula The corresponding picture is illustrated in Figure 1.9.

- $t(A^{\wedge}) = t(A_{\vee}) + 1$: immediately after exiting A from $o2$, the particle re-enters A through $i1$.

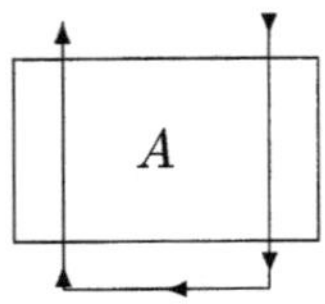

Figure 1.9. Terminal formula

Times link There are two pictures associated with this link:

1 In the pattern shown in Figure 1.10 the link is *switched* on "L", *i.e.*, the conclusion is reached through the left premise (A). In this case:

- $t(B^\wedge) \;=\; t(A \otimes B^\wedge) + 1;$
- $t(A^\wedge) \;=\; t(B_\vee) + 1;$
- $t(A \otimes B_\vee) \;=\; t(A_\vee) + 1.$

2 Respectively, in the link *switched* on "R", the conclusion is reached by the right premise (B), see Figure 1.11:

- $t(A^\wedge) \;=\; t(A \otimes B^\wedge) + 1;$
- $t(B^\wedge) \;=\; t(A_\vee) + 1;$
- $t(A \otimes B_\vee) \;=\; t(B_\vee) + 1.$

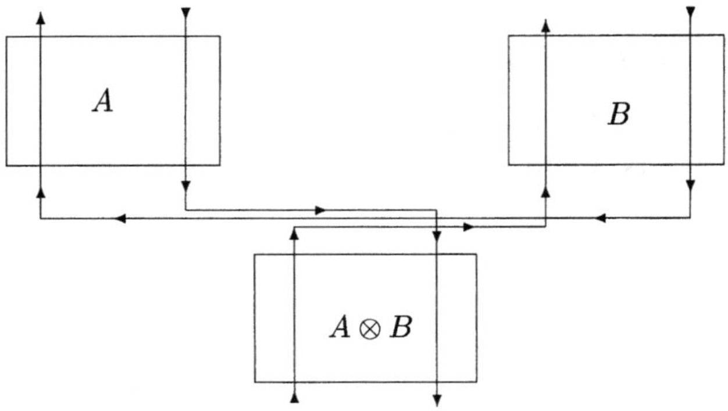

Figure 1.10. Times link switched on "L"

Par link As in the case of *times link*, associated with a *par link* there are two patterns of trip:

1 The *switch* "L" is shown in Figure 1.12. The trip is as follows:

- $t(A^\wedge) \;=\; t(A \wp B^\wedge) + 1;$

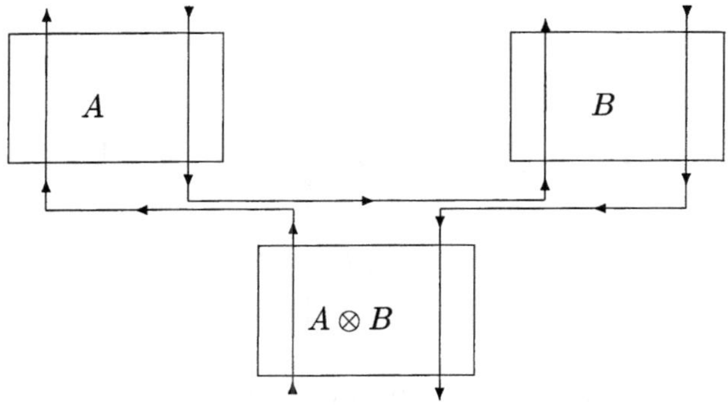

Figure 1.11. Times link switched on "R"

- $t(A\wp B_\vee) = t(A_\vee) + 1;$
- $t(B^\wedge) = t(B_\vee) + 1.$

2 Figure 1.13 illustrates the *switch* "R". The trip is as follows:

- $t(B^\wedge) = t(A\wp B^\wedge) + 1;$
- $t(A\wp B_\vee) = t(B_\vee) + 1;$
- $t(A^\wedge) = t(A_\vee) + 1.$

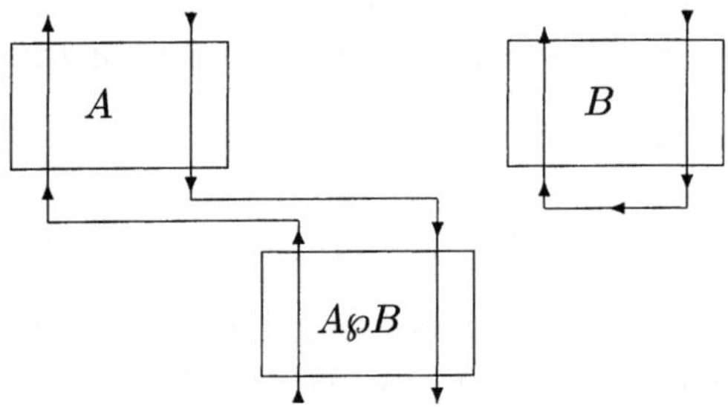

Figure 1.12. Par link switched on "L"

Proof-nets After defining the different patterns of trip, the steps used to define the no-shorttrip condition is as follows:

1 Set the switches of all *par* and *times link* on arbitrary positions;

2 select one of the exit gates of an arbitrary formula A at time 0;

3 define "p" as a number of formulas of the structure, and "k" as the moment when the particle re-enters in the formula A through the dual enter gate of the initial exit gate;

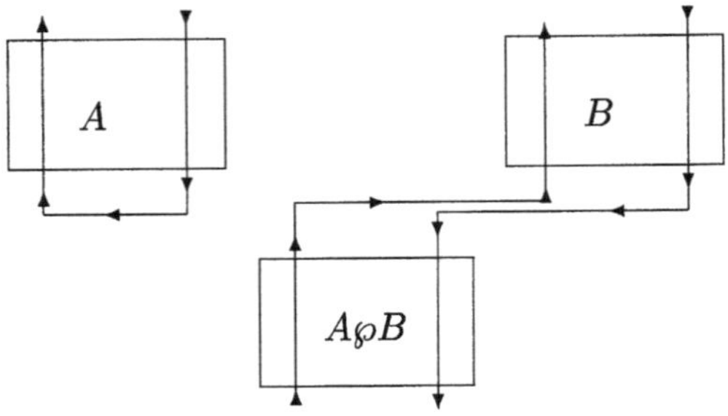

Figure 1.13. Par link switched on "R"

4 if $k < 2p$ then the trip is called a *shorttrip*;

5 if $k = 2p$ there is a *longtrip*.

Definition 22 (proof-nets [38]) A proof-structure is said to be a proof-net when it admits no short trip.

Cost of the criterion One can easily verify that the cost of the criterion is exponential, *i.e.* 2^n, where n is the number of *par links* of the proof-structure.

Examples

Example 23 In this example the proof-structure shown in Figure 1.6 is checked. Setting all switches on "L" we obtain the following longtrip, where $p = 6$ and $k = 12$:

$$A^{\wedge}, A^{\perp}_{\vee}, A^{\perp}\wp B^{\perp}_{\vee}, A^{\perp}\wp B^{\perp\wedge}, A^{\perp\wedge}, A_{\vee}, A\otimes B_{\vee}, A\otimes B^{\wedge}, B^{\wedge}, B^{\perp}_{\vee}, B^{\perp\wedge}, B_{\vee}, A^{\wedge}$$

Example 24 In this example the proof-structure shown in Figure 1.5 is checked and a shortrip is obtained, where $p = 6$ and $k = 8$:

$$A^{\wedge},\ A^{\perp}_{\vee},\ A^{\perp}\otimes B^{\perp}_{\vee},\ A^{\perp}\otimes B^{\perp\wedge},\ B^{\perp\wedge},\ B_{\vee},\ A\otimes B_{\vee},\ A\otimes B^{\wedge},\ A^{\wedge}$$

3.5.2 Danos–Regnier criterion. This correctness criterion, proposed by Danos and Regnier in [28], is a refinement of the no-shorttrip condition [38] in the sense that it is simpler and more manageable. Moreover, with this criterion Gallier in [35] gives a quadratic algorithm for testing whether a proof-structure is a proof-net, while the previous algorithms are of exponential time on the number of *par links* (2^n).

The idea is to associate a set of graphs to a proof-structure. If each graph of the associated collection is connected and acyclic, *i.e.* a tree, the proof-structure is a proof-net.

Associating graphs to proof-structures Proof-structures can be naturally defined in terms of unoriented connected graphs whose vertices are labelled with formula-occurrences and whose edges are defined from the links as follows:

- *identity links*: there is an edge between the two formulas in the link;
- *par/times link*: there is an edge from each premise to the conclusion of the link.

From this definition of proof-structures in terms of graphs we introduce the notion of *D-R-graphs* as follows:

Definition 25 (D-R graph) Given a proof-structure P defined in terms of graphs, a D-R graph G associated with P is any spanning subgraph of P obtained by removing exactly one of the two edges (*i.e.* $(A, A\wp B)$, $(B, A\wp B)$) of every *par link* in P.

The choice of the edges of a *par link* corresponds to the switches ("L" or "R") on Girard's trips.

Proof-nets

Definition 26 (proof-net) A proof-structure P is a proof-net iff every D-R graph associated with P is acyclic and connected.

Examples

Example 27 The D-R graphs associated with the proof-structure of Figure 1.6 are as in Figure 1.14. They are acyclic and connected thus proving that the proof-structure is a proof-net.

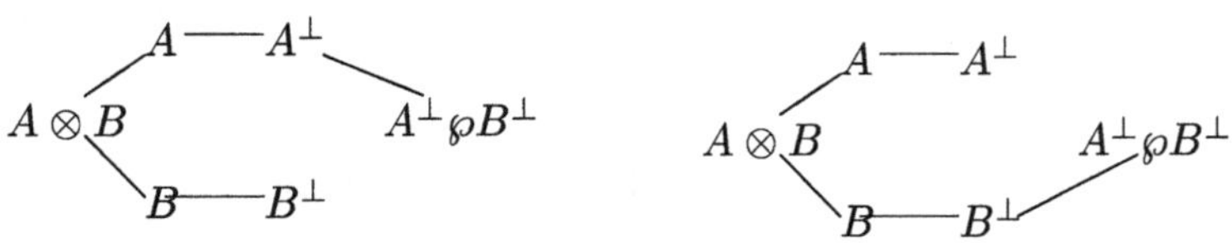

Figure 1.14. D-R graphs

Example 28 The only D-R graph associated with the proof-structure of Figure1.5 is cyclic as shown in Figure 1.15.

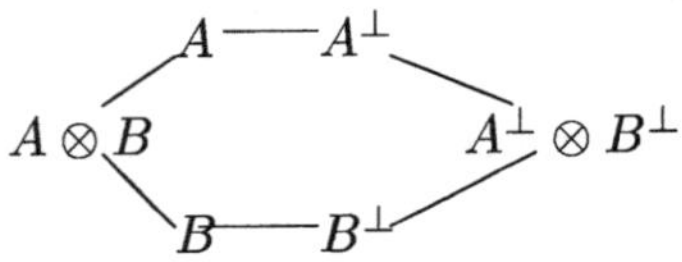

Figure 1.15. D-R graph

3.5.3 Asperti criterion.

Asperti's correctness criterion [6, 5] is based on the interpretation of proof-structures as a distributed system. In this approach, logical formulas are seen as distributed processes, the flow of information inside the proof structure is the computation of the system and proof-nets are deadlock-free proof-structures.

A. Asperti gives this criterion for the multiplicative fragment of Linear Logic (MLL) with mix rule (this fragment is called direct logic). He uses the symbol "$||$" (parallel composition) to denote the connective "$\wp$" (par).

The mechanism The mechanism to verify whether a proof structure is deadlock-free is based on the flow of information in the system:

- Initially all final processes are active (final processes corresponding to conclusions);
- the *axiom link* represents synchronisation. The processes (formulas) A and $A^{\perp}$ connected by the *axiom link* terminate if both are active;
- the *par link* interprets parallelism. The activation of $A||B$ implies in the activation of A and B. The process $A||B$ terminates when A and B terminate;
- the *times link* represents mutual exclusion. The activation of $A \otimes B$ produces first the activation of A or B. After the termination of the first subprocess the other one is activated. The choice of which process will be first activated is non-deterministic. The process $A \otimes B$ terminates when the second subprocess terminates.

3.6 The soundness of the criteria

In order to guarantee the soundness of the criterion established the following two theorems must be proved:

Theorem 29 (Th. 2.7 [38], lemma 29 [35]) *Given a deduction Π of a multiplicative sequent $\vdash A_1, \ldots, A_n$, then we can naturally associate with Π a proof-net P whose terminal nodes are in one-to-one correspondence with the formula-occurrences $A_1, \ldots, A_n$.*

Theorem 30 (sequentialisation) *If P is a proof-net whose terminal nodes are the formula-occurrences $A_1, \ldots, A_n$ one can find a deduction Π of a multiplicative sequent $\vdash A_1, \ldots, A_n$.*

Bellin and van de Wiele in [11] give nice and simple proofs of these results using the notion of *subnets*.

The proof of Theorem 29 is simple: one just shows a procedure to map a sequent calculus proof to a proof-structure and then apply some global soundness criterion to check its correctness. This procedure was already sketched in Section 3.4. However the proof of Theorem 30 is more complicated. It is proved by induction on the number of links in the proof-net. The original proof is given by Girard in [38]. With this proof he proves the soundness of the *no-shorttrip* condition. In [28], Danos and Regnier give a proof of the correspondence between their method and Girard's *no-shorttrip* condition. In [35], Gallier presents an elegant proof based on D-R-graphs. This proof yields a quadratic algorithm for testing whether a proof-structure corresponds to a sequent calculus derivation. In [44] Girard also presents a quadratic proof based on D-R graphs.

Before we proceed to a discussion of the difficulties of the proof of theorem 30, let us first present a sketch of its proof based on [44]. The proof proceeds by induction on the number of links of a given proof-structure P whose all D-R graphs are acyclic and connected:

1 one link: P consists of an *axiom link*; the result is immediate:

$$\overline{A \quad A^\perp} \quad \rightsquigarrow \quad \vdash A, A^\perp$$

2 if there is more than one link, there must be a link which is not an axiom (otherwise P cannot be connected); such a link can be chosen *terminal*, *i.e.* the conclusion of the link (if any: remember that a *cut link* has no conclusion) is a conclusion of the proof-net. Then:

2A there is a terminal $\wp$-link: remove it, and note that the structure one gets is still connected and acyclic, and then apply the induction hypothesis;

2B there is no terminal $\wp$-link as illustrated in Figure 1.16.This case is not easy to prove. For example, if the $\otimes$ link whose conclusion is $A^\perp \otimes B^\perp$ is chosen to be deleted, the result is not two disjoint proof-nets. However, at the formula $C \otimes (A \wp B)$ the proof-net can be split into two disjoint proof-nets. Through another theorem called *Splitting theorem* (Theorem 2.9.7, [38, p39]), Girard proves the existence of a *times link* with such feature, *i.e.* with *split property*. Since the two components are immediately acyclic and connected, the induction hypothesis can be easily applied. But the link with the *split property* is not easy to find and the solution is not unique as we will see below.

In order to prove the existence of a terminal $\otimes$ link with *split property* Girard needs additional notions of *empire* of a given *formula*:

Definition 31 (empire [11]) The empire of a formula-occurrence A in a proof-net P, denoted by eA, is the largest subnet of P that has A among their conclusions.

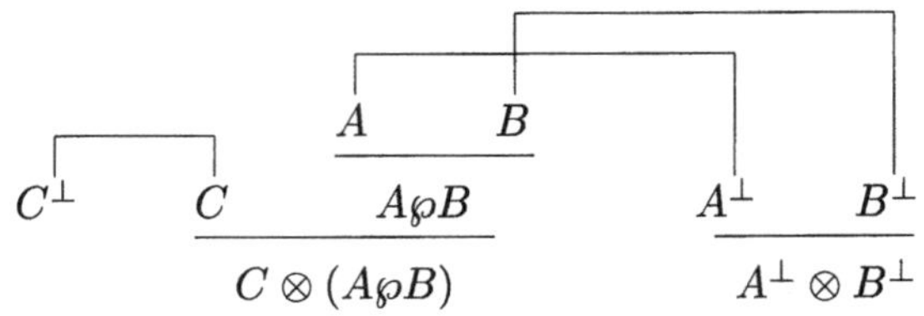

Figure 1.16. Proof-net of $\vdash C^{\perp}, C \otimes (A\wp B), A^{\perp} \otimes B^{\perp}$

Through the *Trip Theorem* (Theorem 2.9.5, [38, p. 39]) Girard defines how to construct an empire of a formula. The notion of empire allows one to recover the moment of the inclusion of a given *times link* during the construction of a proof-net. As we will see in the section after the next one, in a multiple-conclusion calculus one must define conditions to deal with two-premise rules, such as $\wedge$-introduction, that in the context of MLL^{-} corresponds to addition of a *times link*. The problem of such kind of rules is that they combine two separate derivations (proof-nets), while for example the *par link* operates in only one proof-net. The criterion proposed by Girard is very interesting and innovative because the logical dependencies are intrinsic in his method. Alternatively, one could also give a definition of a set operations on proof-structures which represent a deduction step in the proof and then give an inductive definition of proof-nets that naturally correspond to sequent derivations. In fact, Bellin in [7] gives such presentation of proof-nets. However, as himself analyses "this is hardly a surprise ... the non trivial task is to provide (algebraic or geometric) conditions that allow us to recognise whether a given proof-structure can be generated inductively. Moreover, such conditions should be relatively simple when compared to the formalism of sequent calculus and its allegedly excessive 'bureaucracy' " [7, p. 22]. Let us look at an example of construction of the proof-net shown in Figure 1.16.

Example 32 Let us take the two proof-nets as in Figure 1.17. Then we can connect these two proof-nets by adding a *times link*. The result is the proof-net as in Figure 1.18. Alternatively, if we would try to add a *par link* to connect the two proof-nets of Figure 1.17, the result shown in Figure 1.19 would not be a proof-net. Thus, we see that there is a priority between a *times link* over the *par link*, because the latter operates on only one structure while the former operates on two structures of proofs. From the proof-net in Figure 1.18 we add a *par link*, resulting in the proof-net already shown in Figure 1.6. By connecting this proof-net with an *axiom link* we finish the construction of the proof-net illustrated in Figure 1.16.

Figure 1.17. Proof-nets P_1 and P_2, respectively

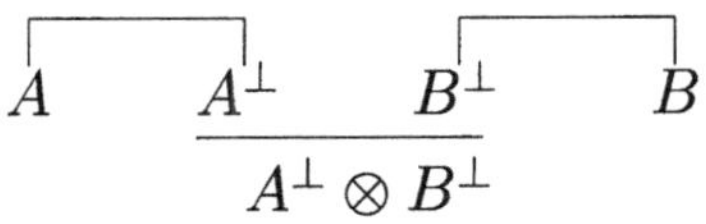

Figure 1.18. Connecting P_1 and P_2 through a times link

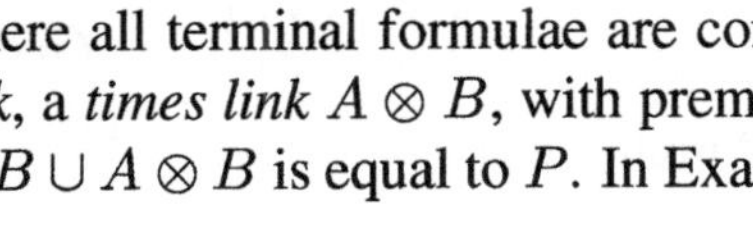

Figure 1.19. Connecting P_1 and P_2 through a par link

In a proof-net P where all terminal formulae are conclusions either of an axiom or of a *times link*, a *times link* $A \otimes B$, with premises A and B, has the *split property* if $eA \cup eB \cup A \otimes B$ is equal to P. In Example 32 we have that:

- $eA^{\perp}$ and $eB^{\perp}$ are the proof-nets P_1 and P_2, respectively, of Figure 1.17;
- $eA\wp B$ is the proof-net of Figure 1.6;
- and eC is the proof-net with only one link, *i.e.* the *axiom link* with the edge $(C, C^{\perp})$.

Thus we can see that $eA^{\perp} \cup eB^{\perp} \cup A^{\perp} \otimes B^{\perp}$ is not maximal, *i.e.* is not equal to the proof-net of Figure 1.16, therefore the *times link* $A^{\perp} \otimes B^{\perp}$ does not have the *split property*. However, $eC \cup eA\wp B \cup C \otimes A\wp B$ is equal to the proof-net of Figure 1.16, hence the link $C \otimes A\wp B$ has the *split property*.

3.7 Cut-elimination

An important result for the theory of proof-nets is the Strong Normalisation Theorem given by Girard [38] which says:

Theorem 33 (Strong normalisation) *A proof-net of size n normalises to a cut-free proof-net in less than n steps; the result is called the normal form of our proof-net.*

The proof of the theorem is based on the following basic reductions:

Axiom reductions

$$\begin{array}{ccc} \vdots & & \\ A & \overline{A^{\perp} \quad A} & \\ \overline{CUT} & & \vdots \end{array}$$

$$\textit{reduces to} \quad \begin{array}{c} \vdots \\ A \\ \vdots \end{array}$$

Symmetric reductions

$$\dfrac{\dfrac{\begin{array}{cc}\vdots & \vdots \\ B & C\end{array}}{B\, m\, C} \qquad \dfrac{\begin{array}{cc}\vdots & \vdots \\ B^{\perp} & C^{\perp}\end{array}}{B^{\perp}\, m^{\perp}\, C^{\perp}}}{CUT} \quad \text{reduces to} \quad \dfrac{\begin{array}{cc}\vdots & \vdots \\ B & B^{\perp}\end{array}}{CUT} \qquad \dfrac{\begin{array}{cc}\vdots & \vdots \\ C & C^{\perp}\end{array}}{CUT}$$

where $m,\ m^{\perp}$ are dual multiplicatives.

4. Logical flow graphs

The concept of *logical flow graph* was introduced by S. Buss [13] with the intention of studying the complexity of certain specific proofs in the sequent calculus for First-Order Logic. He used this notion to prove *the undecidability of k-provability, i.e.* given a formula A and an integer k, to determine if A has a proof with k or fewer lines.

Again with the aim of studying complexity aspects of proofs in the sequent calculus, though in a different perspective, Carbone [14, 16, 17, 18] also uses the notion of logical flow graphs to analyze the complexity of proofs. She has done an extensive work using the notion of *logical flow graph* (*LFG*), such as, for example providing an explanation for the exponential blow up of proof sizes caused by the cut-elimination process. In [14] she proves the acyclicity of cut-free proofs and the acyclicity of contraction-free proofs (possibly containing cuts). In a recent work Carbone and Semmes [18, 22] develop a combinatorial model to study the evolution of graph underlying proofs during the process of cut-elimination. The model has not been completely elaborated in [22], and it tooks its co mplete form in [18].

Here focus our attention on the study of structural properties of Sequent Calculus via *logical flow graphs* (*LFG*)[8] proposed by Carbone.

4.1 Motivation

As we have previously said, Statman was the one who first noticed the importance of studying the structural properties of proofs through a geometrical perspective. On the other hand, there are some interesting questions which involve mathematical logic and computational complexity and represent a challenge for researchers in this area. A deeper understanding of structural properties of proofs might help clarify several of those questions.

A fundamental result which links mathematical logic and computational complexity was pointed out by Cook and Reckhow in [27], where they proved that NP is closed under complementation if and only if there is a propositional proof system where all tautologies can be proved by polynomial size proofs in the length of the tautology.

Since then, a lot of work has been done with the intention of classifying the relative complexity of various proof systems by proving lower bounds for the well known proof systems. For instance, in [26] Cook and Reckhow already define a hierarchy of various propositional proof systems. They placed the resolution into the class of "Known not to be *super*" (a proof system is said to be "*super*" if it admits polynomial length proofs of all tautologies). Many years later, Haken [46] showed that resolution requires exponential time on the pigeonhole formula[9]. A recent classification of various proof systems can be found in [61, 68].

For some proof systems the task of finding hard tautologies is very difficult. Gentzen sequent calculus with cut is an example of such a system. There are families of tautologies in propositional logic which have proofs with polynomial size if the system has the cut rule, but with exponential size with a proof without cuts. The pigeonhole principle is an example, and another one is given by Carbone in [18, 20] called "a concrete example" explained here in subsection 4.3.1. Carbone concentrates her studies in understanding *why* a tautology might be 'hard to prove' [16]. She then analyses the structure of proofs in sequent calculus studying the procedure of cut-elimination and the Interpolation theorem in combinatorial terms. For instance, in [16] is given another proof of the interpolation theorem through *LFG*. Here, we present some of her results about the cut-elimination procedure.

4.2 Defining logical flow graphs

Logical flow graphs are constructed by tracing the flow of formula-occurrences in proofs in the sequent calculus. Before proceeding to the formal definition, let us first present a summary of the way in which Carbone uses the sequent calculus, and then give an informal definition. Once we have done that, we will then show the formal definition as presented by Carbone [16] and Buss [13].

4.2.1 The sequent calculus.

In the context of Carbone's work [16] the axioms in the sequent calculus assume the form $A, \Gamma \vdash \Delta, A$, where A is atomic and formulas occurring in Γ, Δ are referred to as **weak formulas.** The only structural rule used is the contraction and the cut rule. The sets Γ and Δ are *multisets, i.e.* finite sets of formulas which allow repetitions.

Other definitions are given [13, 16] next.

Definition 34 (principal or main formula) The principal formula of an inference is the formula in the lower sequent of the inference upon which the inference acted.

Definition 35 (auxiliary formula(s)) The auxiliary formula(s) of an inference are those one in the upper sequent which are used by the inference.

Definition 36 (side formulas) The side formulas are those ones which do not participate directly in the inference.

Example 37 In the $\wedge$-*right* rule the formula $A \wedge B$ is the principal formula, the formulas A and B in the upper sequent are auxiliaries and the other formulas of $\Gamma, \Gamma', \Delta, \Delta'$ are side formulas:

$$\frac{\Gamma \vdash \Delta, A \qquad \Gamma' \vdash \Delta', B}{\Gamma, \Gamma' \vdash A \wedge B, \Delta, \Delta'}$$

Definition 38 (cut-formulas) The auxiliary formulas of the cut rule are called cut-formulas.

Definition 39 (contraction formulas) The contraction formulas are the auxiliary formulas of the contraction rule.

4.2.2 Building a logical flow graph.

Through *logical flow graphs* we can trace the flow of information in a proof, *i.e.* the formulas. The different occurrences of a formula in a proof are linked by the edges of the graph. Two concepts are defined in [16, 13]:

Definition 40 (s-formula) An s-formula is an occurrence of a subformula of a formula occurring in an inference, where "s" stands for "semi" or "sub".

The difference between the subformula which may occur many times in a proof and an *occurrence* of a subformula is given by the notion of variants.

Definition 41 (variants) Two distinct occurrences of the same formula A are called variants.

The orientation of the edges in the *LFG* is determined by the sign of the s-formula as illustrated in Figure 1.20.[10]

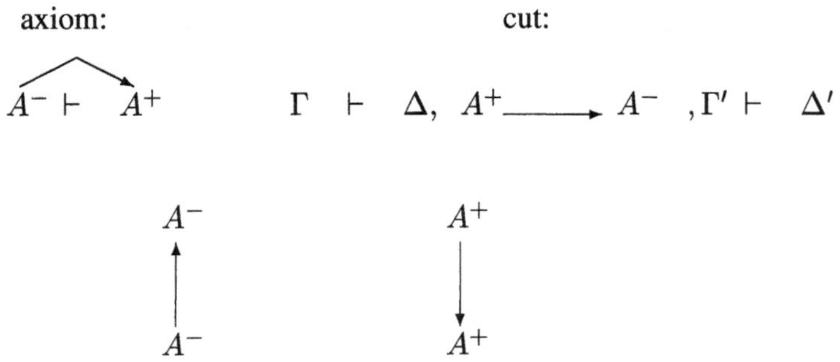

Figure 1.20. The orientation of the edges in a LFG

A simple example that shows how the *LFG* is a helpful tool in the study of the structure of proofs is given in [14].

Example 42 In the tautology $p \vee \neg p \rightarrow q \vee \neg q$ the two q's must have the same name and therefore must be connected inside the proof. The two p's do not need to be linked, as shown in Figure 1.21.

This example shows that for the propositional logic when in the *LFG* the occurrences of some variable come in distinct connected components of the graph, as the occurrences of p, we can replace these occurrences by another formula and the proof continue to be valid.

Through this example we can illustrate other definitions given in [16].

Definition 43 (logical path) Any sequence of consecutive edges in a logical flow graph is called logical path or simply path.

Definition 44 (bridge) Any logical path starting and ending with two (distinct) s-formulas occurring in the end-sequent is called a bridge.

Definition 45 (direct path) A logical path passing through either positive or negative occurrences only is called a direct path.

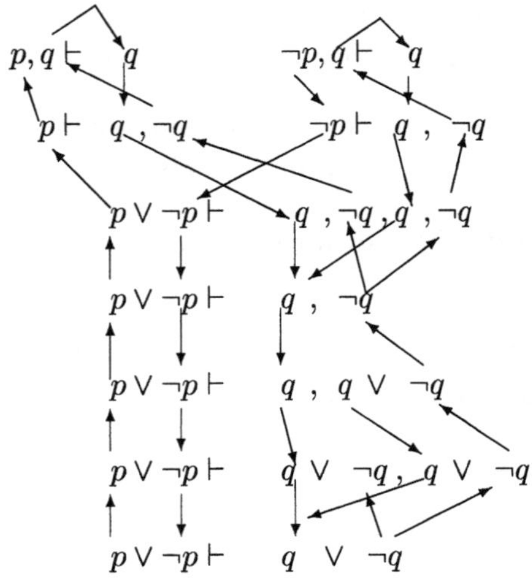

Figure 1.21. Logical flow graph to the formulas p and q

Note how the *logical flow graphs* allow us control the flow of information in proofs of sequent calculus, *i.e.* they tell us where the information (formula) came from and how it will be used later.

4.2.3 A more precise definition. There are some differences between the definition of *LFG* given by Carbone and S. Buss [13]. Carbone applies the concept of *LFG* only to atomic formulas, although in [19] she considers links between non-atomic formulas.

Even though, our analysis focus on Carbone's work, here we also include Buss definition.

Logical flow graphs are defined by specifying the edges [14, 16, 20, 13]:

1 In an axiom there are two cases:

- $A, \Gamma \vdash \Delta, A$: there is an edge directed from the left-hand A to the right-hand A;
- equality axiom:
  -
    $$s_1 = t_1, \ldots, s_k = t_k, P(s_1, \ldots, s_k) \vdash P(t_1, \ldots, t_k)$$
    there is an edge directed from the $P(\mathbf{s})$ to the $P(\mathbf{t})$;
  - In all other equality axioms, there is an edge from each formula in the antecedent to the formula in the succeedent.

2 For side-formulas (*i.e.* those formulas which belong to the contexts $\Gamma, \Delta, \Gamma', \Delta'$) in any logical and structural inferences of the sequent calculus:

- there is an edge directed from each side-formula in the antecedent in the *lower* sequent to the corresponding side formula in the antecedent in the *upper* sequent(s);
- there is and edge directed from each side-formula in the succeedent in the *upper* sequent to the corresponding formula in the *lower* sequent.

3 For auxiliary formulas in any inference:

- if A (or B) is an auxiliary formula which appears in the succeedent of an upper sequent of the inference then there is an edge directed from A (or B) to the corresponding s-formula in the lower sequent;
- if A (or B) is an auxiliary formula which appears in the antecedent of an upper sequent of the inference then there is an edge directed towards that A (or B) from the corresponding s-formula in the lower sequent.

4 In a cut inference there is an edge directed from the cut-formula A in the succeedent of the left-hand upper sequent to the occurrence of A in the antecedent of the right-hand upper sequent.

5 Finally, suppose there is a directed edge from an s-formula A_1 to A_2 and suppose B_1 is a subformula of A_1. Since A_1 and A_2 are variants there is a subformula B_2 of A_2 which corresponds to the subformula B_1 of A_1; clearly B_1 and B_2 are variants. If B_1 occurs positively in A_1 then there is an edge from B_1 to B_2. If B_1 occurs negatively in A_1 then there is an edge from B_2 to B_1.

The orientation of the edges is given by the *sign* of the s-formula, which is defined as follows:

Definition 46 (sign of an s-formula) An s-formula occurs positively if and only if it is in a sequent $\Gamma \vdash \Delta$ and either occurs positively in a formula in Δ or negatively in a formula in Γ.

From this notion, the orientation of the edges is easily defined: in a proof every *downward* edge connects two s-formulas which occurs positively and every *upward* edge connects s-formulas which occur negatively. *Lateral* edges occur in an axiom and in a cut rule. In axiom, the edge is oriented from the *negative* occurrence to the positive one. In the case of a cut rule, the orientation is from positive occurrences to negative occurrences ([13, 18]). In Figure 1.20 we have already seen the orientation of the edges in a *LFG*.

4.2.4 Examples.

Some examples are listed here:

Example 47 This example is shown in [16]. The following proof has the logical flow graph restricted to the formula A illustrated in Figure 1.22.

$$\dfrac{\dfrac{\dfrac{\dfrac{\dfrac{A \vdash A \quad B \vdash B}{A, B \vdash A \wedge B} \quad C \vdash C}{A \vee C, B \vdash A \wedge B, C}}{(A \vee C) \wedge B \vdash A \wedge B, C}}{\vdots}}{(A \vee C) \wedge B \vdash (A \wedge B) \vee C}$$

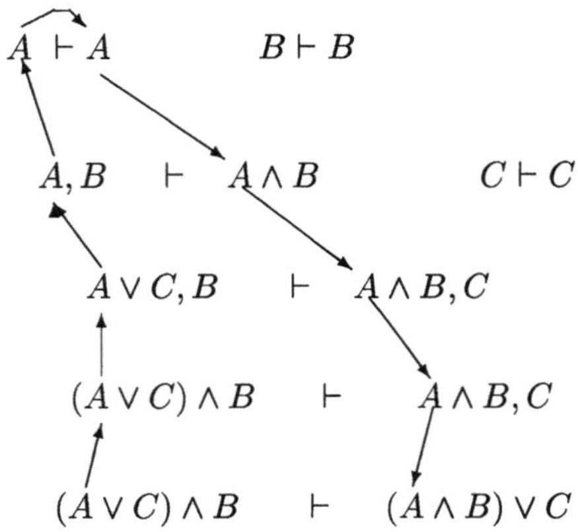

Figure 1.22. Logical flow graph to the formula A

Example 48 The following proof has the logical flow graph restricted to the formula A and formula B shown in 1.23 [13].

$$\dfrac{\dfrac{\dfrac{\dfrac{A \vdash A}{\neg A, A \vdash}}{\neg A, A \vdash B}}{A \vdash (\neg A) \to B} \qquad \dfrac{\dfrac{B \vdash B}{\neg A, B \vdash B}}{B \vdash (\neg A) \to B}}{A \vee B \vdash (\neg A) \to B}$$

4.3 Analyzing the cut-elimination process

The cut-elimination theorem, also known as *Hauptsatz*, says that any proof in sequent calculus with cut can be effectively transformed into a cut-free proof. From the proof theory view point it has nice consequences such as the **subformula property**, *i.e.* every formula which appears in a cut-free proof

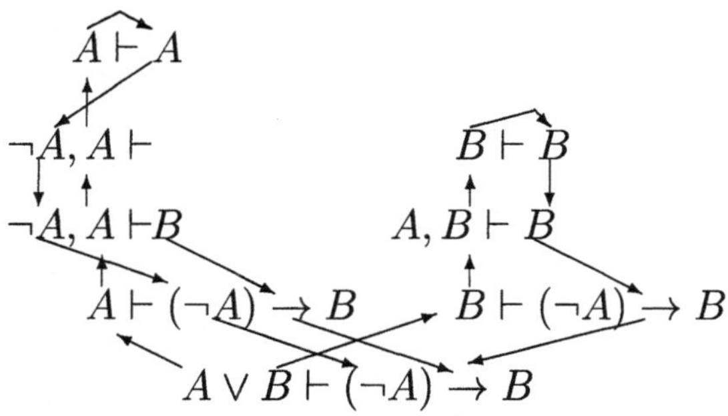

Figure 1.23. Logical flow graph to the formulas A and B

also occurs as subformula of a formula in the end-sequent. However in terms of computational complexity it has a cost. In general, proofs with cuts are smaller than cut-free proofs. There are many examples of propositional tautologies, including the Pigeonhole Principle, for which proofs without cuts must be exponentially larger than proofs with cuts. In [18] Carbone lists various references where the reader can have such examples. Here in Section 4.3.1 we show one of these examples.

Carbone wants to explain the expansion of proofs after cut-elimination by a geometrical analysis. In order to do this, she firstly examines the procedure of cut-elimination looking for the cases which cause expansion. In [15] she explains how this happens. Here we report the example used to illustrate this situation:

Example 49 Consider the following case:

$$\dfrac{\begin{array}{c}\Pi_1\\ \Gamma\vdash A\end{array} \qquad \dfrac{\dfrac{\begin{array}{cc}\Pi_2 & \Pi_3\\ A\vdash B & A\vdash C\end{array}}{A,A\vdash B\wedge C}}{A\vdash B\wedge C}}{\Gamma\vdash B\wedge C}$$

after the cut is pushed up over the contraction, the following proof is obtained:

$$\dfrac{\dfrac{\dfrac{\begin{array}{cc}\Pi_1 & \Pi_2\\ \Gamma\vdash A & A\vdash B\end{array}}{\Gamma\vdash B} \qquad \dfrac{\begin{array}{cc}\Pi_1 & \Pi_3\\ \Gamma\vdash A & A\vdash C\end{array}}{\Gamma\vdash C}}{\Gamma,\Gamma\vdash B\wedge C}}{\Gamma\vdash B\wedge C}$$

as we can see the subproof Π_1 is duplicated.

As we know the usual Gentzen's cut-elimination procedure is performed by pushing the cuts up until one can eliminate it.[11] Roughly speaking, this is done systematically in such a way that the degree of the cut-formula is reduced in each stage of the procedure [45, 36]. The proof of the theorem is made by the verification of the various cases.[12]

The general scheme which illustrates the effect of pushing a cut up across a contraction is as follows:

$$
\frac{\dfrac{\begin{array}{c}\Pi_1\\ \Gamma_1 \vdash \Delta_1, A, A\end{array}}{\Gamma_1 \vdash \Delta_1, A} \quad \begin{array}{c}\Pi_2\\ A, \Gamma_2 \vdash \Delta_2\end{array}}{\Gamma_1, \Gamma_2 \vdash \Delta_1, \Delta_2}
\quad \rightsquigarrow \quad
\begin{array}{c}
\dfrac{\dfrac{\begin{array}{c}\Pi_1\\ \Gamma_1 \vdash \Delta_1, A, A\end{array} \quad \begin{array}{c}\Pi_2\\ A, \Gamma_2 \vdash \Delta_2\end{array}}{\Gamma_1, \Gamma_2 \vdash \Delta_1, \Delta_2, A} \quad \begin{array}{c}\Pi_2\\ A, \Gamma_2 \vdash \Delta_2\end{array}}{\Gamma_1, \Gamma_2, \Gamma_2 \vdash \Delta_1, \Delta_2, \Delta_2}\\
\vdots \; contractions\\
\Gamma_1, \Gamma_2 \vdash \Delta_1, \Delta_2
\end{array}
$$

A concrete example of expansion of a proof after cut-elimination is shown in the next subsection.

4.3.1 An example of expansion after cut-elimination.

In [18] the predicate sequent calculus is used with the following extension rule:

$$
\frac{\Gamma_1 \vdash \Delta_1, F(s) \quad \Gamma_2 \vdash \Delta_2, F(t)}{\Gamma_1, \Gamma_2 \vdash \Delta_1, \Delta_2, F(s * t)} F : times
$$

where F is a unary predicate and $*$ a binary function symbol. The constant symbol 2 is also added to the language. The symbol 2 can be interpreted by the number 2 and * by the multiplication operation. The exponential function is also freely used to denote a term written with the symbols 2, $*$ just to abbreviate it because it becomes very large in most cases.

The proof of $F(2) \vdash F(2^{2^n})$ in the propositional fragment of the sequent calculus is analyzed. Using cuts the proof has $O(n)$ lines, however any cut-free proof has $2^{O(n)}$ lines.

The logical flow graph of the proof looks as shown in Figure 1.24. Where B_i $(1 \leq i \leq n)$ are "building blocks" proofs of $F(2^{i-1}) \vdash F(2^{2^i})$, illustrated in Figure 1.25.

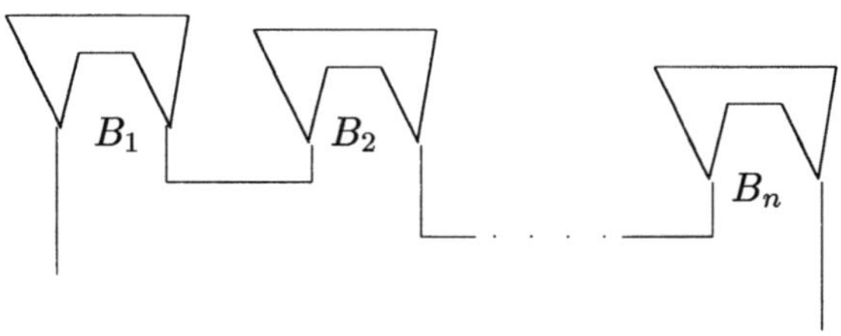

Figure 1.24. An example of expansion

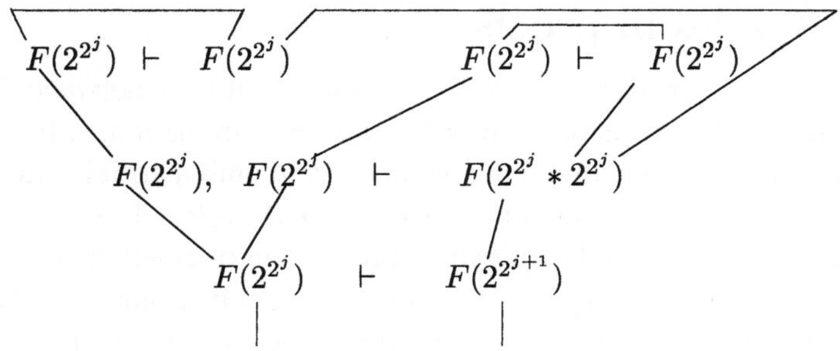

Figure 1.25. The building block B_{j+1}

Each building block B_i contains two branches: the left for the contraction and the right for the F : *times* rule. In Figure 1.26 we can see how the building blocks are linked. In the whole proof there is a *chain* of n pairs of branches which are eliminated during the process of cut-elimination, where the following expansion occurs:

$$\dfrac{\dfrac{\begin{array}{c}B_1\\ \vdots\\ F(2)\vdash F(2^2)\end{array} \qquad \dfrac{\begin{array}{c}B_1\\ \vdots\\ F(2)\vdash F(2^2)\end{array} \qquad \dfrac{F(2^2)\vdash F(2^2) \quad F(2^2)\vdash F(2^2)}{F(2^2),F(2^2)\vdash F(2^2 * 2^2)}}{F(2^2),F(2)\vdash F(2^{2^2})}}{F(2),F(2)\vdash F(2^{2^2})}}{F(2)\vdash F(2^{2^2})}$$

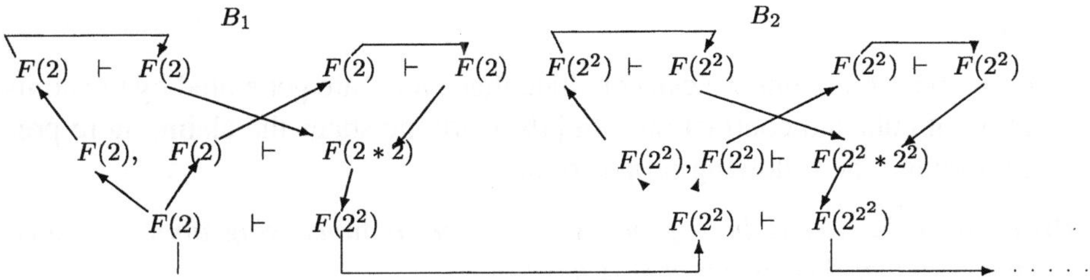

Figure 1.26. Building blocks linked

As in Example 49, this "concrete example" shows the duplication of subproofs which occurs during the process of cut-elimination. Based on this pattern of expansion (*i.e.* caused when a cut is pushed up over a contraction), a combinatorial model is developed as an attempt to explain the expansion of proofs by a geometrical perspective.

4.4 Cycles in proofs

The examples shown in Section 4.3 are important to understand Carbone's analysis about the expansion of proofs after cut-elimination. She notes that cut-free proofs have much simpler geometrical/combinatorial structure than the ones with cuts, *i.e.* they do not have *oriented cycles*. In other words, the logical flow graphs of proofs with cuts may contain oriented cycles.

There are two kinds of cycles in proofs: *non-oriented* and *oriented* cycles. The former may appear in cut-free proofs. The pigeonhole principle is an example which has an exponential number of non-oriented cycles in its cut-free proof, as noticed by Carbone in [19]. Moreover, the logical paths which form these unoriented cycles cannot be eliminated.

In her studies, Carbone considers *oriented* cycles, which can be eliminated by the cut-elimination process. In [16] two important theorems about acyclicity of proofs are given. One deals with the acyclicity of cut-free proofs and the other refers to acyclicity of contraction-free proofs possibly containing cuts. They are reported here as follows:

Theorem 50 (acyclicity of cut-free proofs) *Let* $\Pi : S$ *be a cut-free proof. The logical flow graph of* Π *is acyclic.*

Since in a cut-free proof there are only three kinds of edges, namely upwards (linking negative occurrences of formulas), downwards (linking positive occurrences of formulas) and from negative to positive (axiom), there is no way to link a positive occurrence to a negative one and then close the cycle. The only way to have such a kind of connection is through the lateral edge which links cut-formulas.

Theorem 51 (acyclicity of contraction-free proofs) *Let* $\Pi : S$ *be a contraction-free proof (possibly containing cuts). The logical flow graph of* Π *is acyclic.*

From these theorems we can conclude that cycles are generated by the combination of cuts and contractions. In [19] Carbone states the claim, more precisely through the following proposition:

Proposition 52 *Let* Π *be a proof with no contractions lying above its cuts. Then the logical flow graph of* Π *is acyclic.*

Moreover, proofs with only a certain kind of cut-formulas contain cycles, *i.e.* those one which have two distinguished occurrences, one positive and the other negative. Notice that atomic cuts do not permit cycles to appear. Figure 1.27 shows a general scheme of a cycle in a proof, where P^- and P^+ are distinguished occurrences of P, a subformula of A.

Cycles are eliminated by the cut-elimination procedure. The general scheme of the split of the cycles during the cut-elimination is shown in Figure 1.28.

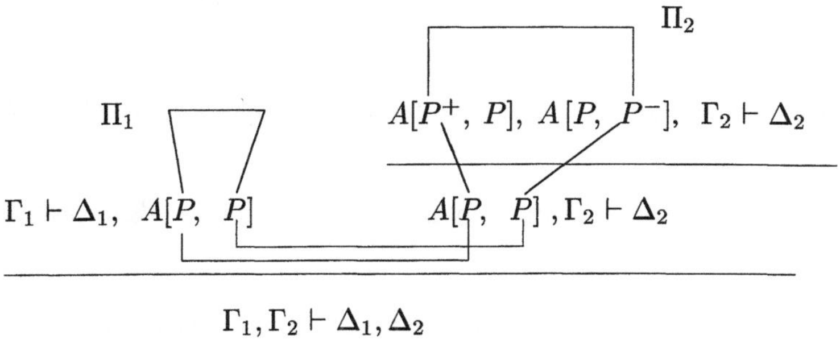

Figure 1.27. General scheme of a cycle in a proof

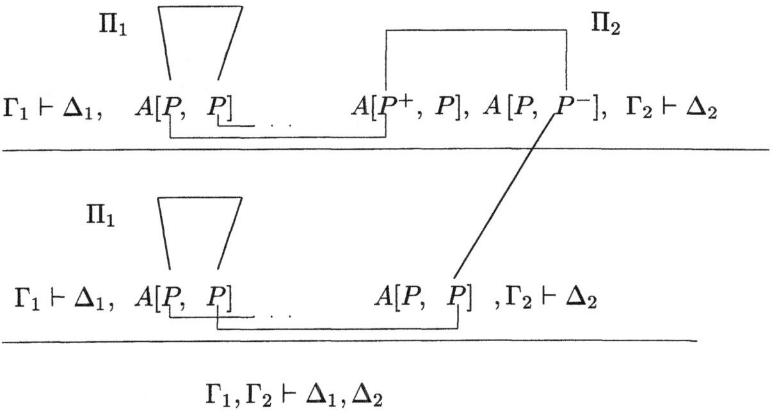

Figure 1.28. Split of a cycle in a proof

4.5 A combinatorial model

In [22] a combinatorial model is introduced with the intention of relating the expansion of a proof after cut-elimination to the geometry of its underlying graph.[13] The model tooks its complete form in [18], where the *operation of duplication* on graphs, *i.e. optical graphs*, (defined as follows) is given.

4.5.1 Optical graphs.

Through the notion of an optical graph, a class of graphs that includes graphs of proofs is defined [18].

Definition 53 (optical graph) An optical graph is an oriented graph in which each vertex v has at most three edges attached to it and no more than two edges are oriented away from v and no more than two are oriented towards v.

From this definition the notions of *branch point, focussing branch*, and *defocussing branch* are given:

- Branch point: is a vertex in a optical graph with three edges attached to it;
- focussing branch point: is a branch point with two edges oriented towards it;
- defocussing branch point: a branch point with two edges oriented away from it.

An *input vertex* (*output vertex*) is a vertex with no edges in the graph which are oriented towards it (away from it).

The notion of labelled optical graphs is also given and Figure 1.29 illustrates the definitions presented here.

Definition 54 (labelled optical graph) A labelled optical graph is an optical graph with the property that the edges oriented away from a defocussing point and the edges oriented towards a focussing point are labelled by the number 1 and 2 respectively.

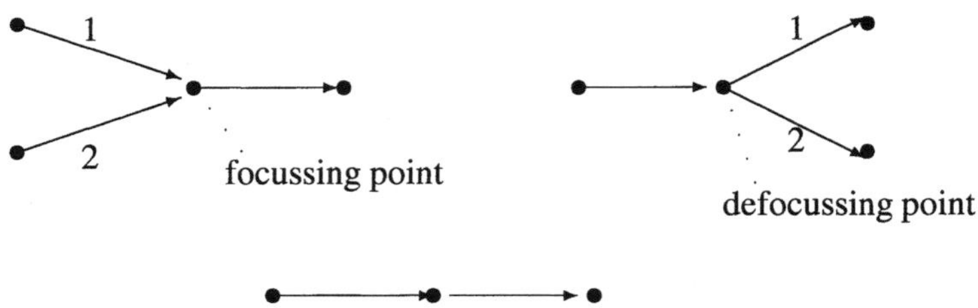

Figure 1.29. Optical graphs

4.5.2 The duplication operation.

This operation is defined to describe the changes (*i.e.* "topological" changes) which occur on the graphs of a proof during the procedure of cut-elimination. We report here the definition given in [18]:

Definition 55 (duplication operation) The duplication operation D is a binary operation applied to a labelled graph G and a subgraph G' of G with the following properties:

1 if a vertex of G' is a focussing point in G then either its immediate predecessor vertices both lie in G' or none of them does, and

2 if a vertex of G' is a defocussing point in G then either its immediate successor vertices both lie in G' or none of them does, and

3 at least one input vertex in G' is a focussing point, or at least one output vertex in G' is a defocussing point.

Let $v_1, \ldots, v_n$ be the input vertices of G' which are not inputs in G, and let $w_1, \ldots, w_n$ be the output vertices of G' which are not outputs in G. The result of duplication applied to G, G' is a graph $D(G, G')$ which is defined as G except on the subgraph G' which will be substituted by two copies of it.

Namely all vertices of G which are not in G' lie in $D(G', G)$ as well as those edges in G which connect any two of these vertices. Moreover, $D(G', G)$ contains two copies G'_1, G'_2 of G' which are attached to the 'rest' of $D(G', G)$ as follows:

- let v_i be an input vertex of G' which is not focussing in G and let v_i^1, v_i^2 be the two copies of it in G'_1, G'_2 of $D(G', G)$ respectively. Add a new vertex u_i in $D(G', G)$ and two edges from u_i to v_i^1, v_i^2 with labels 1 and 2 respectively. If s_i is the vertex in G with an edge to v_i, add an edge from s_i to u_i in $D(G', G)$;
- let w_i be an output vertex of G' which is not defocussing in G and let w_i^1, w_i^2 be the two copies of it in G'_1, G'_2 of $D(G', G)$ respectively. Add a new t_i in $D(G', G)$ and two edges from w_i^1, w_i^2 to t_i with labels 1 and 2 respectively. If r_i is the vertex in G with an edge from w_i to it, add an edge from t_i to r_i in $D(G', G)$;
- let v_i be an input vertex of G' which is focussing in G and let v_i^1, v_i^2 be the two copies of it in G'_1, G'_2 of $D(G', G)$ respectively. If s_i^1, s_i^2 are the vertices in G with edges to v_i labelled 1 and 2 respectively, add an edge from s_i^1 to v_i^1 in $D(G', G)$, and add an edge from s_i^2 to v_i^2 in $D(G', G)$.
- let w_i be an output vertex of G' which is defocussing in G and let w_i^1, w_i^2 be the two copies of it in G'_1, G'_2 of $D(G', G)$ respectively. If s_i^1, s_i^2 are the vertices in G with an edge from w_i, add an edge from w_i^1 to s_i^1 in $D(G', G)$, and add an edge from w_i^2 to s_i^2 in $D(G', G)$.

This concludes the definition of the graph $D(G', G)$.

The application of repeated operation of duplication to an optical graph will transform it into another graph which can be linearly, polynomially or exponentially larger, depending on the way in which the transformation is performed, *i.e.* to which subgraphs G' of a given graph G one will apply the operation of duplication. In Figure 1.31, we show an example from [18] that illustrates this kind of transformation which has an exponential growth.

Carbone notices that the exponential growth depends on the existence of chains of alternating branching points in the graphs. This observation is impor-

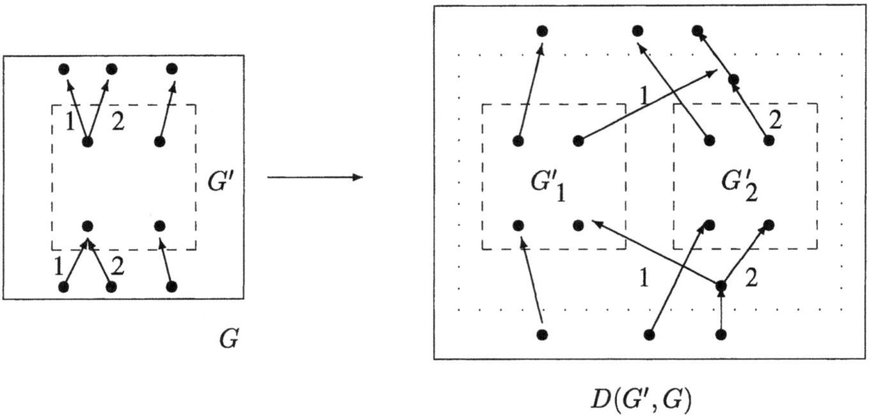

Figure 1.30. Duplication operation

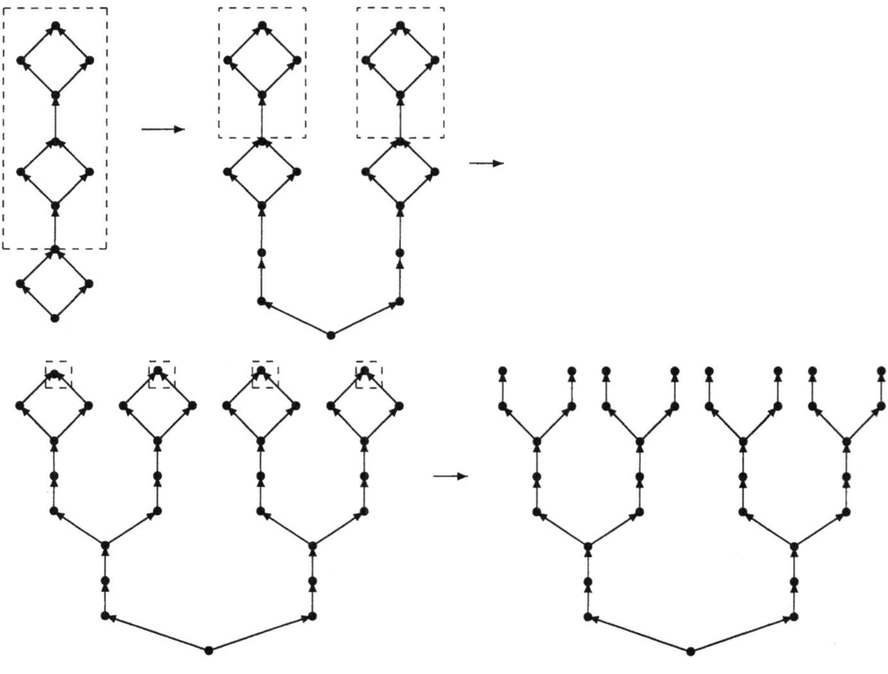

Figure 1.31. An exponential growth

tant to understand the expansion of proofs after cut-elimination by a geometric analysis.

4.5.3 Connections with cut-elimination.

The cut-elimination process can be interpreted as transformations on optical graphs. In particular, the duplication operation is the combinatorial counterpart of the step of cut-elimination which eliminates contractions. For example, let G be the optical graph of the following proof:

$$\dfrac{\begin{array}{c}\Pi_1\\ \Gamma_1 \vdash \Delta_1, A\end{array} \qquad \dfrac{\begin{array}{c}\Pi_2\\ A, A, \Gamma_2 \vdash \Delta_2\end{array}}{A, \Gamma_2 \vdash \Delta_2}}{\Gamma_1, \Gamma_2 \vdash \Delta_1, \Delta_2}$$

The subgraph G' contains the *LFG* of the subproof Π_1 as well as the cut edges, as illustrated illustrated in Figure 1.32. The input and output vertices of G' which are focussing and defocussing points correspond to negative and positive occurrences of contracted cut-formulas. All of other input and output vertices of G' are weak formulas or formulas passing through the side formulas Γ_1, Δ_1. For instance, if A is the formula $\neg p \vee p$, as in Figure 1.32, G' has as an input vertex, which is a focussing point, the negated occurrence of p in $\neg p \vee p$ and as output vertex, which is a defocussing point, the positive occurrence of p in $\neg p \vee p$.

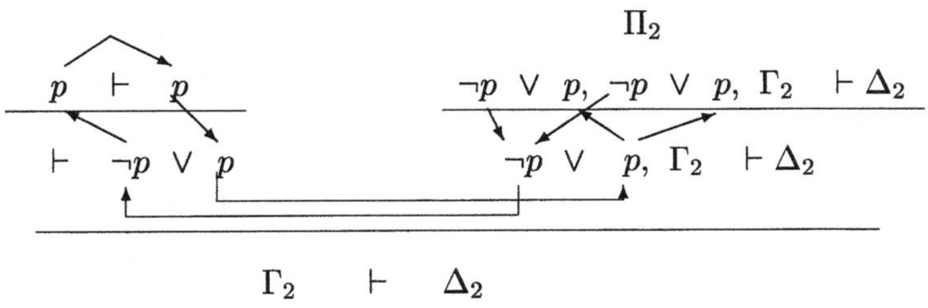

Figure 1.32. Focussing point followed by a defocussing one in the *LFG* of a proof

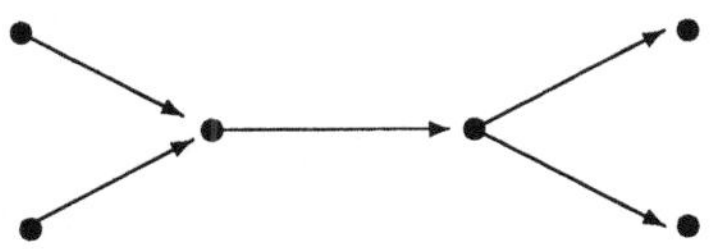

Figure 1.33. Focussing point followed by a defocussing one

As we have seen, this kind of combination of cut and contraction causes cycles in proofs. The cut-elimination procedure drops theses cycles. From the geometric perspective, the configurations shown in Figure 1.33 are eliminated. They represent a focussing point followed by a defocussing one. The logical flow graphs without these configurations correspond to cut-free proofs, they are called H-graphs and are defined as follows [18].

Definition 56 (H-graph) An optical graph is an H-graph if none of its focussing points is followed by a defocussing one.

Yet as a basic property of the duplication operation which is related with cut-elimination we have that its application to a given graph can disrupt a cycle, as shown in Figure 1.34.

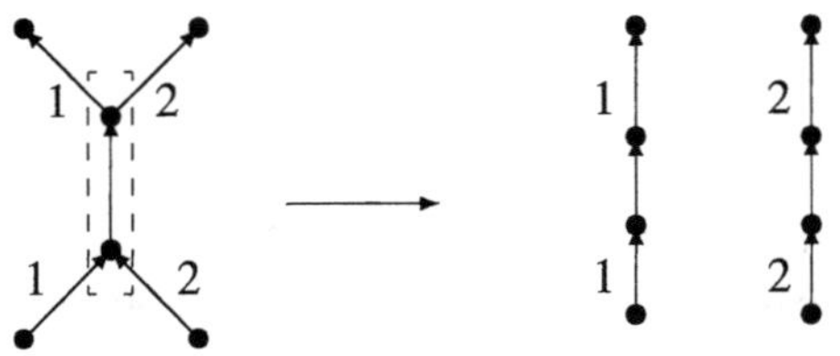

Figure 1.34. An operation of duplication breaking a cycle

In order to complete her combinatorial model to study cut-elimination, Carbone defines a specific strategy to transform any acyclic optical graph G into an H-graph with no focussing vertices. This strategy is called *positive resolution*. She also gives examples of proofs where this strategy is applied. The notions of *focal pairs* and *visibility* graph are introduced to compute lower and upper bounds of the size of the expansion [22, 18].

5. Multiple-conclusion classical calculi

In a previous section we have studied the well-known multiple-conclusion calculus for linear logic, *i.e.* the so-called *proof-nets*. Here we look at two important multiple-conclusion classical calculi: Kneale's tables of development [49, 50] as well as the improvements proposed by Shoesmith–Smiley in [63]; and Ungar's ND system [67]. Both systems are motivated from the difficulties of ND classical systems.

Kneale pioneered this kind of calculus. His intention in proposing the "tables of development" was to improve Gentzen's results in order to obtain a symmetrical ND system. In his analysis (which first appeared in [49] and subsequently in [50]) he notes that the rules which discharge assumptions (*i.e.* $\rightarrow$ *-I*, $\neg$*-I* , $\vee$*-E* and $\exists$*-E*) cause the problem of non-symmetry of ND systems. Looking for a better way to get a good sense of duality for those rules, he proposes what he calls *tables of development*, which has the following general pattern:

$$\frac{P \qquad Q}{R \qquad S}$$

where P and Q are premises and R and S the *limits of the development* (*i.e.* conclusions).

On the other hand, the aim of Ungar's work [67] is to study the relationship between normalisation and cut-elimination as did Zucker in [69]. He wants to give more positive results than Zucker's rather negative ones. For this he proposes a representation of proofs in natural deduction by graphs, instead of the traditional tree structure, where derivations can have more than one conclusion.

Multiple-conclusion calculi can be considered as one alternative to deal with the problem of lack of symmetry in ND classical systems. Before we present these two proposals we shall first look more carefully at this difficulty of ND systems as well as some extensions to Gentzen's original ND system which do not change the structure of ND derivations, *i.e.* the tree-like structure.

5.1 Lack of symmetry in natural deduction

As we have pointed out above, it seems to be the case that the lack of symmetry in natural deduction systems turns out to be its "Achilles' heel" since its formulation by Gentzen [36]. Besides not having invertible rules, the NK calculus (natural deduction for classical logic) presented by Gentzen falls outside the framework proposed for the NJ calculus (intuitionist fragment) because it cannot be integrated to the pattern of *introduction* and *elimination* of the different logical constants: NK includes either the *"law of the excluded middle"* ($A \vee \neg A$, where A stands for any arbitrary formula) as an axiom schema or a new elimination of the negation connective, say $\frac{\neg\neg A}{A}$.

Due to the mismatch between a rather symmetrical system (classical logic) and the system of natural deduction proof rules, Gentzen then abandoned ND in order to be able to prove a fundamental result about the general structure of proofs: the *Hauptsatz*. He went on to formulate a rather symmetrical alternative: the sequent calculus which turned out to be suitable for both intuitionistic and classical logic. Here we quote:

> "In order to be able to enunciate and prove the Hauptsatz in a convenient form, I had to provide a logical calculus especially suited to the purpose. For this the natural calculus proved unsuitable. For, although it already contains the properties essential to the validity of the Hauptsatz, it does so only with respect to its intuitionist form..."

As is well known, Prawitz in [58, 59] proved the corresponding *"Hauptsatz"* for natural deduction derivations, the *normal form theorems*. This represented an important progress for natural deduction systems. In his classical ND system, instead of the law of excluded middle, the classical absurdity rule (Λ_C), $\frac{\begin{matrix}[\neg A]\\ \Lambda\end{matrix}}{A}$, is added to the system. Prawitz then followed the second option given by Gentzen to formulate classical ND systems. Since $\neg\neg A$ is equivalent to A, if you begin a proof with $\neg A$ as assumption and find Λ (*'false'*), then you have a proof of A, because of the implication rule and the correspondence between

$\neg A$ and $A \to \Lambda$, as we can see in the scheme below:

$$\dfrac{\begin{array}{c}[\neg A]\\ \vdots \\ \Lambda\end{array}}{\neg A \to \Lambda}\to\text{-}I \quad\equiv\quad \dfrac{\begin{array}{c}[\neg A]\\ \vdots \\ \Lambda\end{array}}{\neg\neg A} \quad\equiv\quad \dfrac{\begin{array}{c}[\neg A]\\ \vdots \\ \Lambda\end{array}}{A}$$

Prawitz' work represents a significant improvement to ND systems, however when he considers the normal form for ND classical derivations, the logical constants $\vee$ and $\exists$ are not included. This restriction has motivated various studies for a presentation of a proof of normal form theorems for full classical logic, using a formulation including the connectives $\vee$ and $\exists$ as primitive. Statman's work [65] was the first presentation of the normalisation theorem for full first order classical ND systems. He was interested in establishing a connection between structural and semantic properties, in particular, the length of normal proofs in ND classical systems and their semantic properties. He defined various systems, such as for example the second order systems with the choice axiom, etc., and proved the strong normalisation theorem through the definition of homomorphisms between the systems previously defined. Although with different and simpler methods to prove the normal form theorems for full first order classical logic, Stalmarck [64] as well as L. Pereira and C. Massi [56] follow Statman in some technical details and present an alternative solution to the problem of proving normalisation for ND with full classical logic. These works together with the proposal of Andou [4] solve the technical problem, even if a philosophical question concerning constructive proofs remains open. In his proposal, Prawitz restricts the application of the Λ_C-rule to atomic formulas because he wanted to analyze the classical inferences constructively. Thus, the classical operations are composed by constructive ones plus the principle of indirect proof (Λ_C) for atomic sentences (see [59, pp. 244–245]).[14] Thus Prawitz defines a set of reduction rules which replaces each application of the classical absurdity rule (Λ_C) by one or more applications in which the conclusion is atomic. For example, for the connective $\wedge$ the reduction rule is framed as follows:

$$\dfrac{\dfrac{\begin{array}{c}[\neg(B \wedge C)]\\ \Sigma\end{array}}{\Lambda}}{B \wedge C} \quad\rightsquigarrow\quad \dfrac{\dfrac{\dfrac{\begin{array}{c}\dfrac{\dfrac{\dfrac{B \wedge C}{B} \quad \neg B}{\Lambda}}{[\neg(B \wedge C)]}\\ \Sigma\end{array}}{\Lambda}}{B} \qquad \dfrac{\dfrac{\begin{array}{c}\dfrac{\dfrac{\dfrac{B \wedge C}{B} \quad \neg C}{\Lambda}}{[\neg(B \wedge C)]}\\ \Sigma\end{array}}{\Lambda}}{C}}{B \wedge C}$$

It is at this point that the proof fails in the presence of $\vee$ and $\exists$. It is not possible to "atomicise" the applications of the Λ_C rule for these connectives.

Statman [65], L.C. Pereira and C. Massi [56], G. Stalmarck [64] and Andou [4] define new reduction rules that remove a compound formula which is a conclusion of Λ_C rule **only** if it is at the same time a major premise in an application of an elimination rule. Hence, a *closed normal derivation* can end with an application of the Λ_C rule applied to a compound formula.

Another alternative to deal with a full classical ND is to approach it through the inclusion of the law of excluded middle. Tennant [66], as well as von Plato [57], give the following rule of natural deduction excluded middle:

$$\frac{\begin{array}{cc}[A]^1 & [\neg A]^2 \\ C & C\end{array}}{C}\,(1,2)$$

where 1 and 2 are discharged.

In [62], Seldin uses the following rule to obtain classical logic from intuitionistic logic:

$$\neg D\text{-rule} \qquad \frac{\begin{array}{c}[\neg A] \\ \vdots \\ A\end{array}}{A}$$

As noted by Seldin, this rule was previously proposed by Curry,[15] and it is a version of a rule which corresponds to the Peirce's law:

$$\to K\text{-rule} \qquad \frac{\begin{array}{c}[A \to B] \\ \vdots \\ A\end{array}}{A}$$

For our purposes, we believe that the rule for the *law of excluded middle* proposed by Tennant and von Plato ([66, 57]) is a good alternative for the classical case. Although we did not look too deeply at the details of the normalisation procedure proposed by von Plato, it looks rather promising. The rule proposed by Seldin is very similar to the Λ_C and the normalisation procedure is not as simple as Prawitz', although technically it works well.

5.2 Kneale's tables of development

The *tables of development*, has the following general pattern:

$$\frac{P \quad Q}{R \quad S}$$

where P and Q are premises and R and S the *limits of the development*(*i.e.* conclusions). Between the initial formulas and the end formulas there may be

(1) $\dfrac{P \quad Q}{P \wedge Q}$		(2a) $\dfrac{P \wedge Q}{P}$	(2b) $\dfrac{P \wedge Q}{Q}$
(3a) $\dfrac{P}{P \vee Q}$	(3b) $\dfrac{Q}{P \vee Q}$	(4) $\dfrac{P \vee Q}{P \quad Q}$	
(5) $\dfrac{*}{P \quad \neg P}$		(6) $\dfrac{P \quad \neg P}{*}$	
(7a) $\dfrac{*}{P \quad P \rightarrow Q}$	(7b) $\dfrac{Q}{P \rightarrow Q}$	(8) $\dfrac{P \quad P \rightarrow Q}{Q}$	

Table 1.1. Basic development schemata for propositional fragment

many intermediate expressions obtained from the basic development schemata, shown in Table 1.1 for the propositional fragment.

Unlike Gentzen ND system, in which an inference cannot have more than one conclusion, Kneale's "tables of development" allows the construction of multiple-conclusion proofs. In the propositional fragment in rules (4), (5) and (7a) there are two formulas below the line.

Another novelty in Kneale's development is the presence of the asterisk in rules (5), (6) and (7a). It indicates a place where one may put any propositional expression. Rule (5) is a formulation of the law of excluded middle while rule (6) represents the principle of non-contradiction. In [49] Kneale puts a propositional expression in the asterisk place, but in [50] he keeps the asterisk to mean a 'hole' which can be filled in by any proposition. Through some examples we can see how this works out in practice.

Kneale explains the rather innovative rule (7a) by showing its equivalence to the principle that a material implication is entailed by the negation of its antecedent:

Example 57 The development of $\neg P/P \rightarrow Q$ (the notation P/Q indicates a entailment statement - where P is the premise and Q is the conclusion):

In [50] version: Figure 1.35;

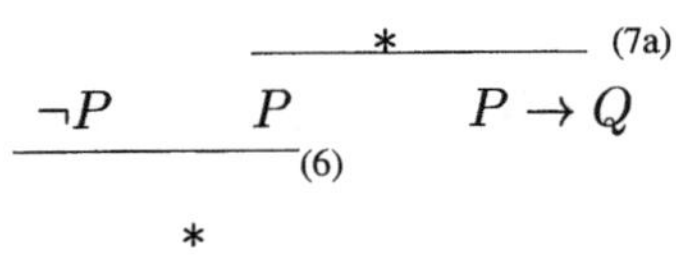

Figure 1.35. Development of $\neg P/P \rightarrow Q$

in [49] : Figure 1.36.

Here we will adopt the [50] version.

$$\begin{array}{ccc} & \dfrac{\neg P}{\quad\quad\quad\quad\quad P \to Q}\,(7a) \\ \dfrac{\neg P \qquad P}{P \to Q}\,(6) & \end{array}$$

Figure 1.36. Development of $\neg P/P \to Q$

$$\dfrac{\dfrac{A \vee B}{A \qquad B}}{A \wedge B}$$

Figure 1.37. A fallacy

In order to avoid fallacious proofs such as the one illustrated in Figure 1.37 Kneale defines the condition that formulas which are already connected, either directly by one single horizontal line or indirectly through several horizontal lines, should not be connected again in any way.

The global rule of $\to$ *-I* in ND Gentzen system can be derived in Kneale's tables of development, as he explains in [50, p. 248], by showing that $P \to Q$ is provable if P entails Q (if P/Q then $*/P \to Q$):

$$\dfrac{*}{\dfrac{\dfrac{P}{Q}(Hyp.)}{P \to Q}(7b) \qquad\qquad P \to Q}$$

5.2.1 Classical and intuitionistic logic. As explained by Kneale [49, p. 253], unlike Gentzen ND system, the rules of development is a version of classical logic. In order to obtain intuitionistic logic, rule (5) must be dropped and the corresponding ND system rule restored in its place. This is a disadvantage of Kneale's proposal when dealing with intuitionistic logic, since the device of assumption discharge falls out of the pattern of the other rules of the system.

5.2.2 Some examples.

Example 58 $P \to Q/\neg P \vee Q$:

$$\begin{array}{cc} & \dfrac{*}{\quad\quad\quad\quad\quad \neg P}(5) \\ \dfrac{\dfrac{P \to Q \quad P}{Q}(8)}{\neg P \vee Q}(3b) & \dfrac{\neg P}{\neg P \vee Q}(3a) \end{array}$$

Example 59 $*/(P \vee Q) \to (Q \vee P)$:

$$\cfrac{*}{\cfrac{P\vee Q}{\cfrac{\cfrac{P}{Q\vee P}(3b)}{(P\vee Q)\to(Q\vee P)}(7b)\qquad \cfrac{\cfrac{Q}{Q\vee P}(3a)}{(P\vee Q)\to(Q\vee P)}(7b)}(4)\qquad (P\vee Q)\to(Q\vee P)}(7a)$$

Example 60 $\neg Q/P, \neg(P \vee Q)$:

$$\cfrac{*}{\cfrac{P\vee Q}{P\quad \cfrac{Q\qquad \neg Q}{*}(6)}(4)\qquad \neg(P\vee Q)}(5)$$

5.2.3 The inadequacy. In [63] Shoesmith and Smiley point out the inadequacy[16] of Kneale's tables of development. It is an inherent problem of multiple-conclusion systems of proofs. Ungar in [67], Chapter 4, calls this difficulty "the problem of substitution" in the $\vee$-*elimination*. In a few words, Shoesmith and Smiley show that there may be developments of S from R and of R from Q, without there being any development of S from Q. This kind of problem does not arise in single conclusion calculi, such as ND, because there is no repetition of formulas in the conclusion and the assumptions are grouped into equivalence classes. In order to illustrate this difficulty of multiple-conclusion calculi, let us look at a derivation Π of C from R in ND:

$$\Pi: \quad \cfrac{\begin{matrix}[R]\\ \Pi_0\\ A\vee B\end{matrix}\qquad \begin{matrix}[A]\\ \Pi_1\\ C\end{matrix}\quad \begin{matrix}[B]\\ \Pi_2\\ C\end{matrix}}{C}$$

Thus, if we have a derivation Π' of S with various occurrences of C as open assumptions:

$$\Pi': \quad \cfrac{\begin{matrix}[C]\ldots[C]\\ \Pi_3\end{matrix}}{S}$$

then we can easily construct a derivation of S from R substituting a derivation Π of C for each occurrence of C in Π'. However, if we are in a context of a multiple-conclusion calculus, such as the tables of development, the derivation Π should be as follows:

$$\Pi_1 \qquad\qquad\qquad\qquad \Pi_2$$

$$\dfrac{\dfrac{}{A}}{A \to A} \qquad A \to A \qquad \dfrac{A \to A \qquad \dfrac{A \to A}{A \vee (A \to A)}}{(A \to A) \wedge (A \vee (A \to A))}$$

$$\Pi$$

$$\begin{array}{c} \\ \dfrac{\dfrac{}{A \qquad A \to A}}{} \end{array} \qquad \dfrac{A \to A}{A \vee (A \to A)}$$

$$\dfrac{A}{A \to A} \qquad \dfrac{A \to A \qquad A \vee (A \to A)}{(A \to A) \wedge (A \vee (A \to A))}$$

Table 1.2. Inadequacy of Kneale's developments

$$\begin{array}{c} R \\ \Pi_0 \\ \dfrac{A \vee B}{A \qquad B} \\ \vdots \qquad \vdots \\ C \qquad C \end{array}$$

It is not obvious how to construct a derivation of S from R from the derivations Π and Π'.

Shoesmith and Smiley provide an example of a tautology, namely $(A \to A) \wedge (A \vee (A \to A))$, in which there is not a Kneale development for it. They show that there is a development Π_1 of $A \to A$, a development Π_2 from $A \to A$ to $(A \to A) \wedge (A \vee (A \to A))$, but we cannot construct a development Π from Π_1 and Π_2, as illustrated in Table 1.2.

5.3 Shoesmith and Smiley refinements

5.3.1 Developments as graph arguments. In order to solve the inadequacy of Kneale's proposal, Shoesmith and Smiley [63] refine Kneale's development by presenting it through a graph, called *graph argument*, where vertices are labelled with occurrences of formulae and inference strokes; and premises and conclusions have different graphical representation.

The inference strokes which label the vertices represent the horizontal line of the developments, as in Figure 1.38. Conclusions are represented by triangles ($\triangle$), premises by inverted triangles ($\bigtriangledown$), and other formula vertices by circles ($\bigcirc$). An isolated vertex being both premise and conclusion is represented by the combination of the two notations (✡).

From graph theory, Shoesmith and Smiley use the notions of initial/final/intermediate vertex and bipartite graph to formalise the definition of *graph argument*.

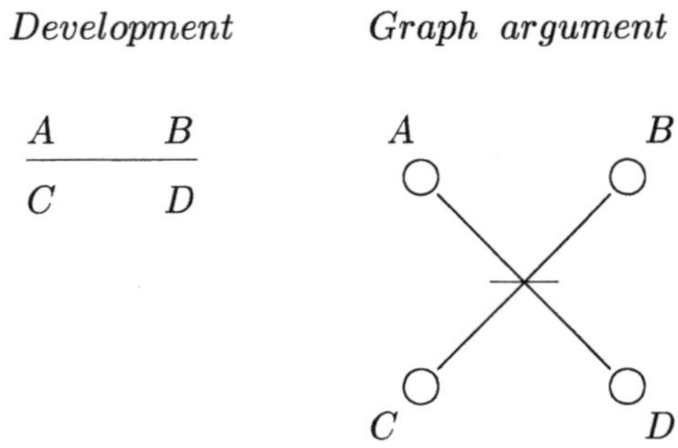

Figure 1.38. Development versus Graph argument

Definition 61 (graph argument) A graph argument is a finite bipartite graph of formulas and strokes in which a subset of the initial and final vertices respectively are specified as the premises and conclusions of the argument.

Definition 62 (standard graph argument) A standard graph argument (or standard argument for short) is a graph argument where all the initial formulas are premises and all the final ones are conclusions.

Definition 63 (non-standard vertex) Any initial formula vertex that is not a premise of the argument, and any final one that is not a conclusion, is called non-standard.

In the case of a standard graph argument, there is no need for a special notation for premises and conclusions.

The graph argument which is nonempty, connected, circuit-free and corner-free is called *Kneale argument.* If it is standard, then it is called *standard Kneale argument.* The developments are represented by standard Kneale arguments.

The requirement of a circuit-free graph argument avoids proofs such as that one illustrated in Figure 1.37 of Section 5.2.

In order to avoid representations of developments in which a formula appears as premise (or as conclusion) in more than one step, as illustrated in Figure 1.39, a property called "corner-free graph argument" is required. This property is defined as follows.

Definition 64 (corner) A corner is a set $\{E_1, b, E_2\}$ in which b is a formula vertex and E_1 and E_2 are distinct edges both going to b or both from it.

Definition 65 (corner-free graph) A corner-free graph is a graph without corners.

In Figure 1.40 we can see a graph with a corner at B[17]

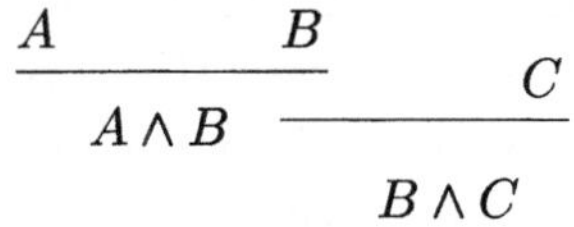

Figure 1.39. Development with a corner

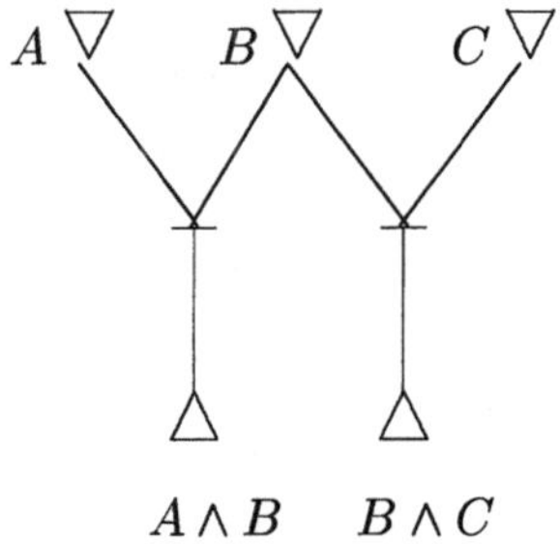

Figure 1.40. Corner at B

5.3.2 Solving the inadequacy.

With the definition of developments as Kneale arguments as presented above, Shoesmith and Smiley show the situations where the problem arises and give an alternative solution to deal with them. The notion of an operation on graphs called *junction* is needed.

Definition 66 (junction) Let Π_1 and Π_2 be graph arguments which have just one vertex labelled by A in common, A being a conclusion of Π_1 and a premise of Π_2. The graph Π with premises from Π_2 (other than A) and with conclusions from Π_1 (other than A) is the junction of Π_1 and Π_2, as illustrated in Figure 1.41.

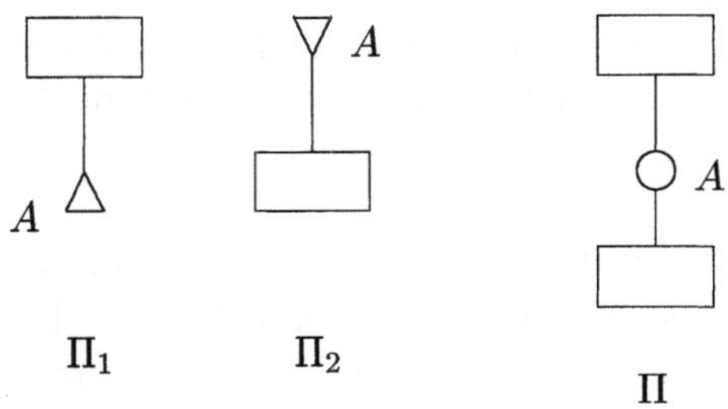

Figure 1.41. Junction

Given a standard Kneale argument Π_1 from R to A and Π_2 from A to Q, the junction of Π_1 and Π_2 does not produce a standard proof from R to Q in the case where A occurs more than once as conclusion in Π_1 and also more than once as premise in Π_2, as illustrated in Figure 1.42. This same situation occurs in the attempt to obtain the development Π of $(A \to A) \land (A \lor (A \to A))$ from the developments Π_1 and Π_2 as shown in Table 1.2: $A \to A$ occurs twice as premise in Π_2 and twice as conclusion in Π_1.

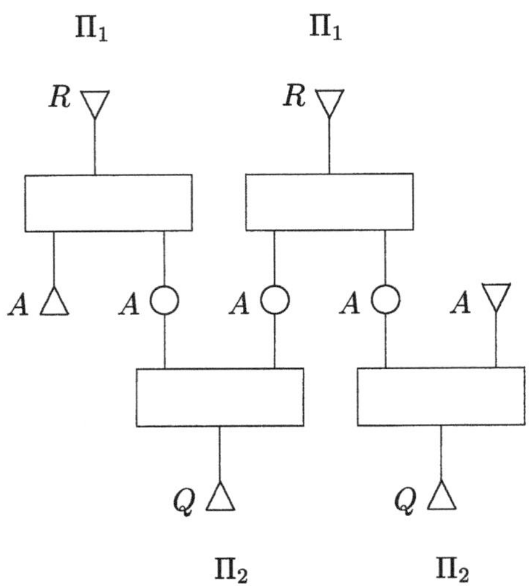

Figure 1.42. Non-standard proof from R to Q

Using the notational feature of graph arguments in which one can indicate which of the initial and final formulas are meant to be premises and conclusions together with an operation called *election* on the graph arguments, Shoesmith and Smiley present a solution to the inadequacy of Kneale's developments.

Definition 67 (election operation) An election operation transforms a graph argument Π with various occurrence of the same formula A as premises into a graph argument Π' which is the same graph as Π with the exception that the number of occurrences of the formula A as premise in Π' is reduced. Similarly the election operation acts over conclusions.

Election is an implicit form of defining the structural rule of *contraction* in the sequent calculus. The inadequacy Kneale's development is then solved by representing it through graph arguments; and by reducing the number of occurrences of a formula as premise or conclusions to just one through an election operation before the application of a junction operation.

Now we can see how to obtain the proof of $(A \to A) \land (A \lor (A \to A))$. The corresponding graph arguments of the developments Π_1 and Π_2 (Table 1.2) are shown in Figure 1.43. By "electing" one of the two occurrences of $A \to A$ in Π_1 and Π_2 the graphs Π_1' and Π_2' are obtained respectively, as in Figure 1.44.

Finally, the intended proof Π is obtained by joining Π'_1 and Π'_2 as in Figure 1.45.

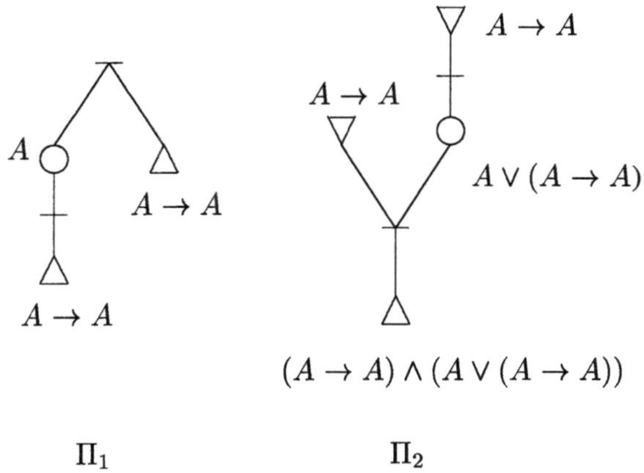

Figure 1.43. Graph arguments Π_1 and Π_2

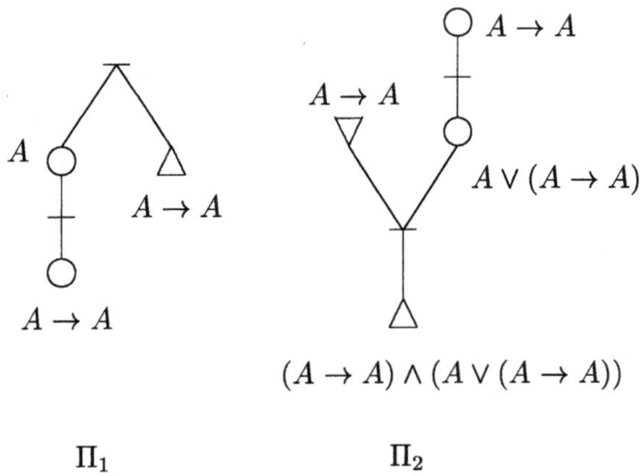

Figure 1.44. Graph arguments Π'_1 and Π'_2

Shoesmith and Smiley (theorem 9.4, [63, p. 152]) prove that the closure of the standard Kneale proofs under junction and election is an adequate class of proofs. However, it is necessary to characterise which class of proofs under election operation is adequate. For example, the arguments shown in Figure 1.46 are not valid. Whenever a vertex u is made non-standard by an election operation on a graph argument Π, there will be a path $u, \ldots, v$ in Π, where u is premise or conclusion that is an occurrence of the same formula as v. If Π is embedded by junction in a larger argument Π', the path $u, \ldots, v$ will also be

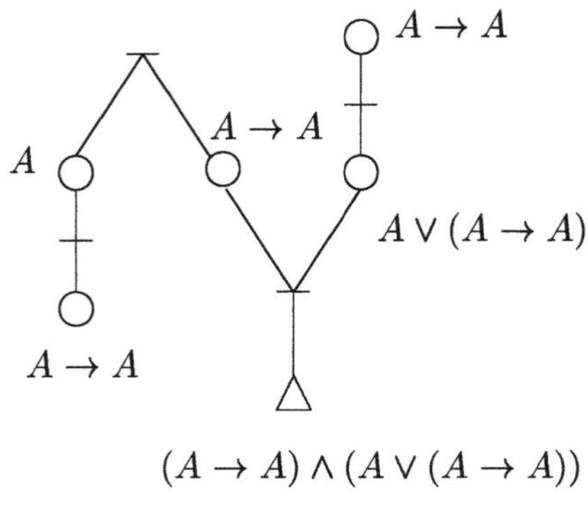

Figure 1.45. Junction of Π'_1 and Π'_2

in Π' as a reminder that u can be justified by the notion of *cross-reference* to v, as following defined.

Definition 68 (cross-reference path) A cross-reference path is a path $u, \ldots, v$ such that u and v are occurrences of the same formula and are both initial and final in the path.

Not all cross-reference paths justify the presence of non-standard vertices in a graph argument. In Figure 1.46 either of the initial occurrences of A in Π_1 could be justified by cross-reference to the intermediate occurrence of the same formula. In order to recognise which cross-reference path justifies an argument, the notion of *cross-referenced Kneale argument* is given.

Definition 69 (cross-referenced Kneale argument) A Kneale argument is cross-referenced if its non-standard vertices can be arranged in a sequence $u_1, \ldots, u_n$ such that there exist cross-reference paths $(u_1, \ldots, v_1), \ldots,$ $(u_n, \ldots, v_n)$, where no v_i appears in $u_j, \ldots, v_j$ for $j > i$ unless $v_i = v_j$. In such a case we say that the sequence of paths justifies the argument.

With this definition we complete the characterisation of Kneale arguments through cross-references paths.

5.3.3 Second solution: multiple junction.

Shoesmith and Smiley relaxed the requirement of a circuit-free graph argument and gave an alternative solution to the inadequacy problem of Kneale's developments. They note that not all circuits in a standard corner-free Kneale argument are problematic. For example, in Figure 1.47 the argument is a fallacy while the one in Figure 1.48 is valid. In the former the formulae A and B are distinct while

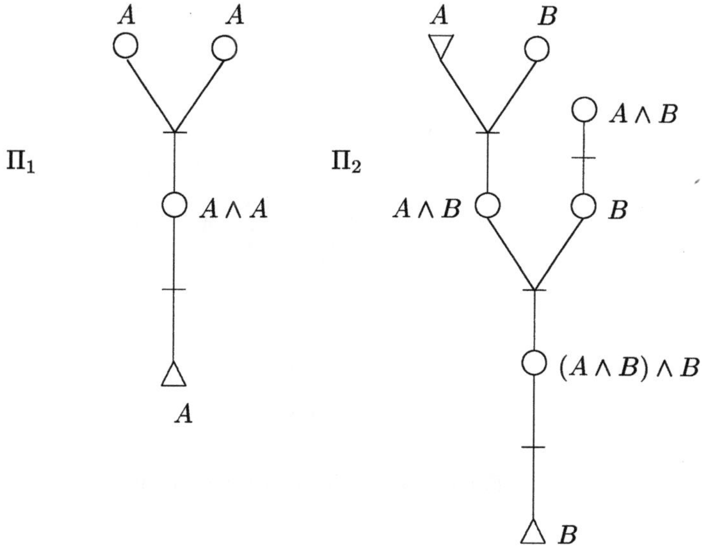

Figure 1.46. Invalid arguments

in the latter there is a recurrence of formulae, namely the formula A, in this case the circuit contains a cross-reference path. In order to characterise these circuits in an argument, the notion of a *cross-referenced-circuit argument* is given:

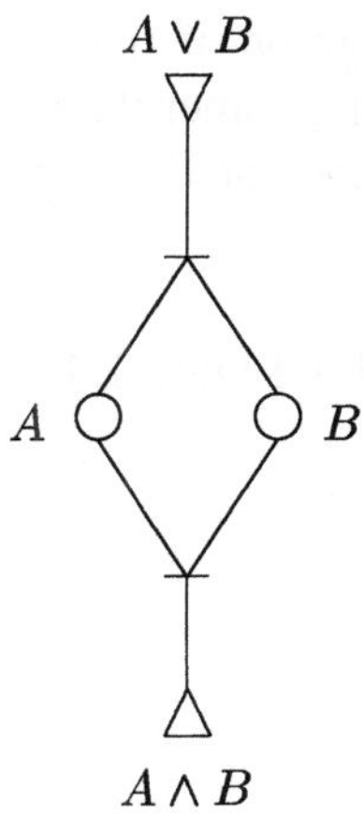

Figure 1.47. Circuit in an invalid argument

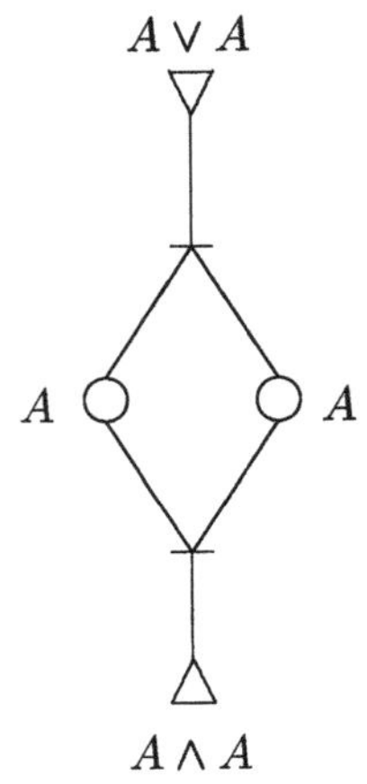

Figure 1.48. Circuit in a valid argument

Definition 70 (cross-referenced-circuit argument) An argument is cross-referenced-circuit if it is connected, nonempty and corner-free, and every circuit contains a cross-reference path.

The problem of adequacy in cross-referenced-circuit arguments is solved through the operation of *multiple-junction*, defined bellow.

Definition 71 (multiple-junction) Let Π_1 and Π_2 be arguments whose common vertices $a_1, \ldots, a_k$ ($k \geq 1$) are conclusions of Π_1 and premises of Π_2, and are all occurrences of the same formula. We say that Π is the *multiple-junction* of Π_1 and Π_2, if the graph of Π is the union of those Π_1 and Π_2, the premises of Π being those of Π_2 (other than $a_1, \ldots, a_k$) plus those of Π_1, and the conclusions of Π being those of Π_1 (other than $a_1, \ldots, a_k$) plus those of Π_2.

This operation is illustrated in Figure 1.49.

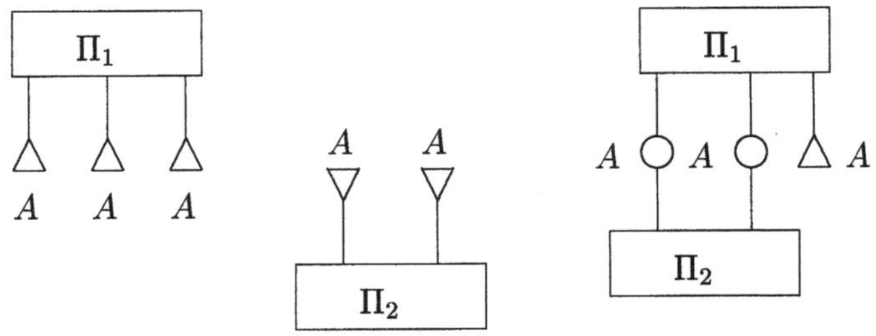

Figure 1.49. Multiple-junction

The example used in the discussion of inadequacy is now with multiple-junction as in Figure 1.50.

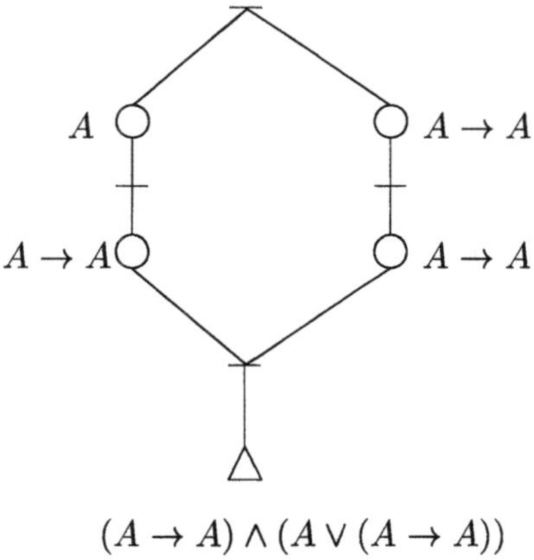

Figure 1.50. "key example" with multiple-junction

There are cases where the multiple-junction operation generates some unwanted occurrences of the relevant formula, as in Figure 1.49, the formula A in the conclusion of Π. One can deal with it by making an appropriate number of copies of Π_1 and Π_2 before the application of the multiple-junction operation. If the same formula occurs m times as a conclusion in Π_1 and n times as a premise in Π_2, one can first bring these up to the same number $m.n$ by considering n copies of Π_1 and m copies of Π_2, and then the multiple-junction is applied, as illustrated in Figure 1.51.

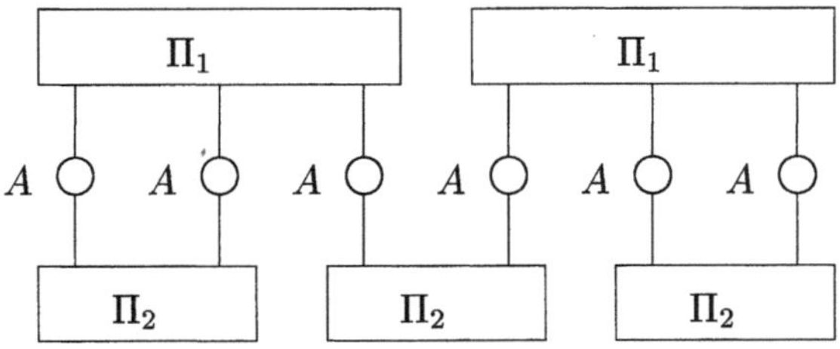

Figure 1.51. Multiple-junction

5.3.4 Third solution: identification of premises and conclusions. A third solution to the problem of inadequacy is given by allowing corners at Kneale arguments and by definition of the operation of *identification* of different occurrences of the same formula as premises (or conclusions) in a proof.

Definition 72 (identification of premises (conclusions)) We say that Π' is obtained by a graph argument Π by identification of premises (conclusions) if it is the result of omitting vertices $a_2, \ldots, a_n$ from Π and replacing every edge from (to) any of them by a corresponding edge from (to) the vertex a_1, where all the a_i are occurrences of the same formula and are premises (conclusions).

Figure 1.52 illustrates this operation. Π_1' is obtained from Π_1 by identification of conclusions and Π_2' is obtained from Π_2 by identification of premises. After these operation Π is obtained by junction of Π_1' and Π_2'. The usual example is illustrated in Figure 1.53.

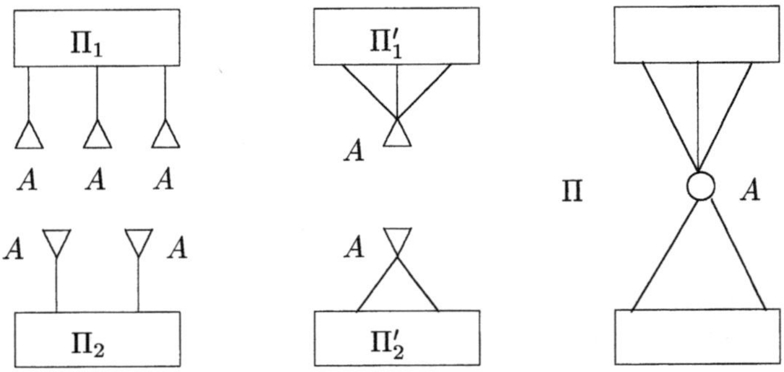

Figure 1.52. Identification of premises and conclusions

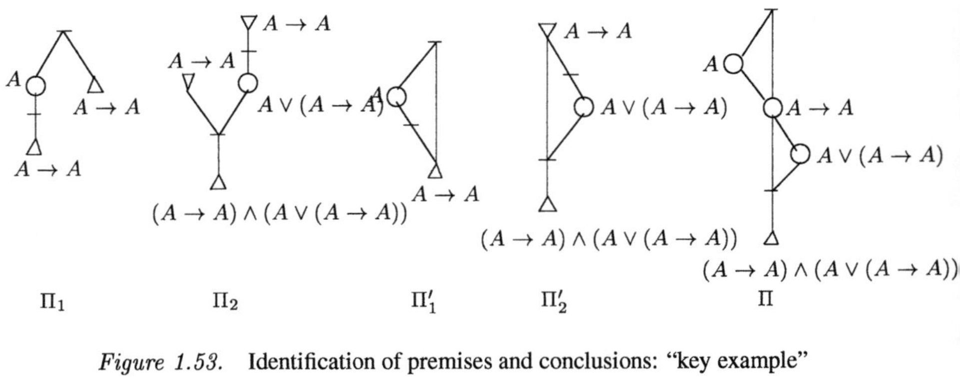

Figure 1.53. Identification of premises and conclusions: "key example"

Any circuit formed as a result of identification of premises or conclusions contain a corner.

Definition 73 (cornered-circuit argument) A nonempty argument in which every circuit is cornered, *i.e.* contain a corner, is called cornered-circuit argument.

Axioms: A_m

Rules:

(1) $\dfrac{A_m \quad B_n}{A \wedge B_p}$ (2) *a.* $\dfrac{A \wedge B_p}{A_m}$ *b.* $\dfrac{A \wedge B_p}{B_n}$

(3) *a.* $\dfrac{A_m}{A \vee B_p}$ *b.* $\dfrac{B_n}{A \vee B_p}$ (4) $\dfrac{A \vee B_p}{A_m \quad B_n}$

(5) $\dfrac{B_m}{A \to B_p \ \{A_n\}}$ (6) $\dfrac{A_n \quad A \to B_p}{B_m}$

(7) $\dfrac{\bot_p}{A_n}$ (8) $\dfrac{A_n}{\top_p}$

Table 1.3. Ungar's system

Through theorems 10.3 and 10.4 [63, p. 170] we have the proof that every standard cornered-circuit argument is valid and adequate.

5.4 Ungar's system

With the intention to study the correspondence between normalisation and cut-elimination as did Zucker in [69], Ungar proposes in his book [67] a revised treatment of ND systems through a multiple-conclusion calculus. He wants to give more positive results than Zucker's. As analyzed in Chapter 1 [67], Zucker gave up the idea of reconciling ND and sequent calculus due to the fact that there may be meaningful properties of proofs which are preserved by all reductions in ND, but not by those of the sequent calculus. Ungar's analysis begins with a discussion about some aspects of ND systems that could be improved in a revised calculus. The main point discussed is that the elimination rule is not a good alternative to represent the actual reasoning involved in the proof. Considering that derivations are formal representations of proofs, the presence of $\vee$-*elimination* rule in derivations of ND systems does not represent the structure of the proofs they are supposed to represent. Thus, a revised treatment for ND system should include a reformulation of this rule to one of the form $\frac{A \vee B}{A \quad B}$ as in Kneale's developments. Before giving a precise notion of derivation in his multiple-conclusion version of ND systems, Ungar analyzes the *problem of substitution* in this kind of calculus, which corresponds to the inadequacy of developments studied by Shoesmith and Smiley. Ungar begins his analysis by listing a set of propositional rules for a multiple-conclusion variant of natural deduction, here shown in Table 1.3. Each formula-occurrence has a subscript, which ranges over natural numbers, to indicate the class to which it belongs.

Some remarks on Ungar's system are enumerated as follows:

- Ungar includes the rule 8 only for aesthetic reasons. In the sequel, he ignores it.

- The notation $\{A_n\}$ in rule 5 indicates that all occurrences of A_n as an assumption are discharged by the inference.
- Derivations in this system are graphs whose vertices are labelled by formulae.
- Each axiom of the form A_m will be represented by one-element whose only vertex is labelled A_m.
- To exclude fallacious derivations such as that one shown in Figure 1.37 (Section 5.2) as did Kneale, Ungar stipulates that the premises of rule (1) and (6) cannot both be conclusions of a single derivation.

5.4.1 The problem of substitution. The problem of substitution in a multiple-conclusion calculus corresponds to the problem of inadequacy of developments, already discussed previously in Table 1.2 and summarised here in Figure 1.54.

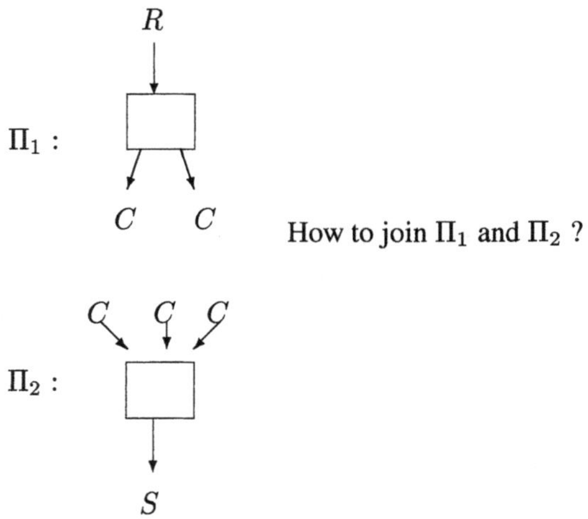

Figure 1.54. The problem of substitution

This kind of problem appears in a multiple-conclusion calculus because we have at the same calculus rules with more than one conclusion as well as rules with more than one premise. There are situations where the junction of two derivations represents a fallacy as in derivation 1 of Table 1.4 whilst in other contexts this operation represents a logically correct reasoning as in derivation 2 of the same table. One alternative to avoid derivations like 1 is to define a condition that no cycles can occur in the proof, as did Kneale. However how can we deal with deductions like 2? The derivation shown in Figure 1.55 is not satisfactory. The example pointed by Shoesmith and Smiley is an instance of

$\dfrac{\dfrac{A \vee B}{A \quad B}}{A \wedge B}$	$\dfrac{\dfrac{A \vee A}{A \quad A}}{A \wedge A}$
1	**2**

Table 1.4. Circuits in a multiple-conclusion calculus

derivation 2. In this case, a reasonable solution must include the allowing of cycles in the derivations. Ungar uses another example to discuss the issue of adequacy in his system. He shows that it is not possible to build a derivation of $*(A, B) \wedge *(C, D)$, where $*(X, Y)$ abbreviates $(X \vee Y) \rightarrow ((X \vee Y \rightarrow X) \vee (X \vee Y \rightarrow Y))$. The derivation of $*(A, B)$ would be as follows.

$$\dfrac{A \vee B}{\dfrac{\dfrac{\dfrac{A}{A \vee B \rightarrow A}}{((A \vee B) \rightarrow A) \vee ((A \vee B) \rightarrow B)}}{*(A, B)} \qquad \dfrac{\dfrac{\dfrac{B}{A \vee B \rightarrow B}}{((A \vee B) \rightarrow A) \vee ((A \vee B) \rightarrow B)}}{*(A, B)}}$$

Although the example of Shoesmith and Smiley is easily derivable in Ungar's system, both examples are instances of the same problem. In Ungar's system the $\rightarrow$*-introduction* rule has only one conclusion while in Kneale developments it has two conclusions.

$$\begin{array}{cccc} \multicolumn{2}{c}{\underline{A \vee A}} & \multicolumn{2}{c}{\underline{A \vee A}} \\ A & \underline{A \quad} & \underline{\quad A} & A \\ & \multicolumn{2}{c}{A \wedge A} & \end{array}$$

Figure 1.55. An attempt to derive $A \vee A \vdash A \wedge A$

5.4.2 Solving the problem of substitution.

To solve the problem of substitution a precise notion of derivations is given. The application of a rule of inference is interpreted as an operation on graphs, called *combination.* Those combinations which generate the class of logically correct derivations are called *substitutions.* The idea is similar to the second solution of Shoesmith and Smiley, where an appropriate number of copies of the derivations to be joined is made as well as the requirement of a circuit-free graph is relaxed. The following example illustrates the technique used to solve this problem.

Example 74 Let Π_1 and Π_2 be derivations of occurrences of A and B respectively, as follows:

$$\Pi_1 : \begin{array}{cc} \Pi'_1 & \\ \vdots & \vdots \\ A & A \end{array} \qquad \Pi_2 : \begin{array}{ccc} & \Pi'_2 & \\ \vdots & \vdots & \vdots \\ B & B & B \end{array}$$

By combining Π_1 and P_2 with the graph $\frac{A \quad B}{A \wedge B}$ as in Figure 1.56 one can reach the conclusion $A \wedge B$. As we can note, similarly to the operation of *multiple junction*, three copies of Π_1 and two copies of Π_2 are glued together.

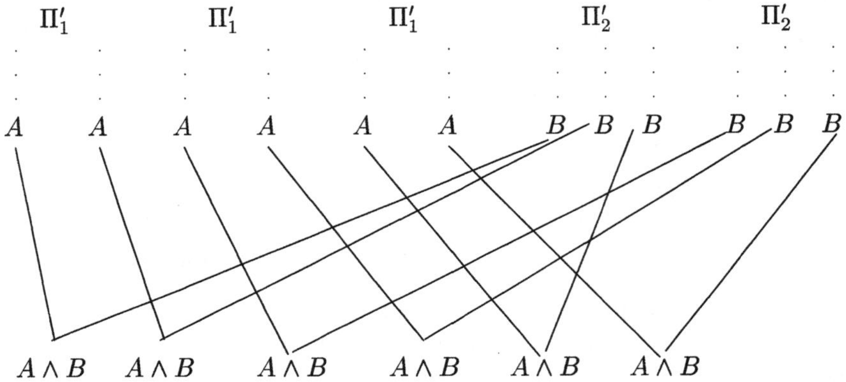

Figure 1.56. Combination on graphs

6. Finale

Girard's intention in proposing the concept of proof-nets ("*the natural deduction of linear logic*") was to study the geometry of deductions. In [40] he presented various arguments in defense of his programme, emphasizing the importance of "finding out the geometrical meaning of the *Hauptsatz, i.e.* what is hidden behind the somewhat boring syntactical manipulations it involves".

What we have seen in Carbone's work is exactly a geometrical study of the cut-elimination process, in which she proposes a combinatorial model to study the expansion of proofs after cut-elimination. Like Girard in his linear logic, Carbone notices the need to impose some restrictions on the use of structural rules and presents an acyclic sequent calculus, called ALK [17], which is not studied here.

We understand the essence of Carbone's work as an attempt to explain the expansion of proofs after cut-elimination by a geometric analysis. What we want to emphasise in the work of Carbone is her presentation of an extensive study of cut elimination and structure of proofs in purely combinatorial terms. In fact she develops a combinatorial model for cut-elimination based on the concept of optical graphs and the duplication operation on graphs. From this model she explains the expansion of proofs after cut-elimination. Moreover, her work evolves to a refined notion of *logical flow graphs* (*core proof and logical core proof*) and some conjectures on the structural properties of proofs of tautologies which are 'hard' to prove. A future survey would include these notions, as well as the concepts of *focal pairs* and *visibility graph*, central to compute lower and upper bounds in the expansion of proofs.

In the context of the theory of proof-nets, we emphasise the proposal of a purely geometrical tool to approach the soundness of a graph of proof.

6.1 Towards a new perspective: N-graphs

Statman's geometrical perspective does not seem to have developed much further than his doctoral thesis, but the fact is that it looks as if the main idea, *i.e.* extracting structural properties of proofs in natural deduction using appropriate geometric intuitions, offers itself as a very promising one. With this in mind, and having at our disposal some interesting and rather novel techniques developed for *proof-nets* and *logical flow graphs*, we have tried to focus our investigation on a research for an alternative proposal for studying the geometry of natural deduction systems. The lack of symmetry in natural deduction systems is one of the aspects which represent a challenge for such a kind of study. Of course, one alternative to deal with the problem of lack of symmetry in ND systems is to look at multiple-conclusion calculi. We already have in the literature different approaches involving such calculi. We mention, for example, Kneale's *tables of development* (studied in depth by Shoesmith & Smiley) and Ungar's multiple-conclusion natural deduction. Our proposal is similar to both Kneale's and Ungar's in various aspects, mainly in the presentation of a multiple conclusion calculus inspired in natural deduction systems. However, we wish to bring our system, which we have named as **N-graphs**, into the current state-of-affairs by incorporating some of the techniques developed for *proof-nets* (*e.g.* splitting theorem, soundness criteria, sequentialisation), and at the same time study its structural properties via a geometrical representation of graphs of proofs. A future development should also include the study of the complexity of ND derivations inspired in the work of Carbone for the sequent calculus.

Drawing on the main approaches reviewed in the paper, we outline a proposal for a formalisation of proofs, called N-graphs, based mainly on the rules of natural deduction formalism, as some structural rules of sequent calculus. Our intention is to present a propositional classical calculus similar to the multiple-conclusion ones studied above. However we want to incorporate the geometrical technique of the theory of proof-nets to define its soundness criterion. Because of the presence of more connectives than in the theory of proof-nets, we face knew difficulties which become even more complicated with the definition of the connective "$\rightarrow$" as in ND and Ungar's system, *i.e.* with a "global" feature.

In our formalism, proofs are represented by graphs which are constructed from a set of *basic links* illustrated in Figures 1.58 and 1.59, which represent the set of schema of rules of the calculus.

6.1.1 Proof-graphs. We adopt the usual propositional language with logical constants: $\wedge$ (conjunction); $\vee$ (disjunction); $\rightarrow$ (implication); $\neg$ (negation); the constant $\perp$ for *absurdity* (or, *the false*); and the constant $\top$ for *truth.* We use the letters $A, B, C, D, E, \ldots$ for arbitrary formulas (or formula-occurrences) in the language. Formula-occurrences play a more important role in our formalism, but sometimes we resort to some *abus de langage* and talk about formulas when in fact formula-occurrences are meant.

The atomic steps in a derivation are represented by the *basic links* shown in Figures 1.58 and 1.59. The class of graphs which includes N-graph derivations are called *proof-graphs.* Before we give the notion of proof-graphs we will introduce some definitions on graphs, inspired on Carbone' s combinatorial model (Section 4, [18, 22]).

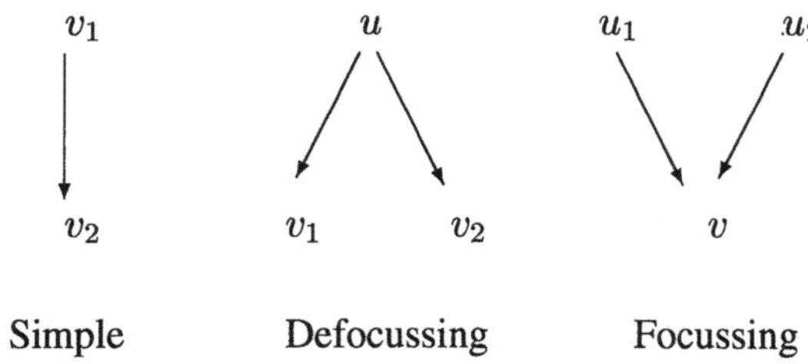

Figure 1.57. Links in a proof-graph

Definition 75 (branch point) A branch point is a vertex in a digraph with two edges oriented towards it or two edges oriented away from it.

Definition 76 (focussing branch point) A focussing branch point is a vertex in a digraph with two edges oriented towards it.

Definition 77 (defocussing branch point) A defocussing branch point is a vertex in a digraph with two edges oriented away from it.

Definition 78 (focussing link) A focussing link is the set $\{(u_1, v), (u_2, v)\}$ in a digraph in which v is a focussing branch point as illustrated in Figure 1.57. The vertices u_1 and u_2 are called premises of the link, while the vertex v conclusion.

Definition 79 (defocussing link) A defocussing link is the set $\{(u, v_1), (u, v_2)\}$ in a digraph in which u is a defocussing branch point as illustrated in Figure 1.57. The vertices v_1 and v_2 are called conclusions of the link, while u premise.

Definition 80 (simple link) A simple link is an edge (u, v) in a digraph that does not belong to a focussing neither a defocussing link. The vertex u is called premise of the link and v conclusion.

Definition 81 (proof-graph) A proof-graph is a connected oriented graph defined as follows:

- each vertex is labelled with a formula-occurrence;
- the edges are of two kinds ("meta" and "solid"), thus they are labelled or not by a letter, say "m" for "meta-edges" ($(u, v)^m$), to identify its kind. The "meta edges" are used to indicate the cancellation of a hypothesis, while the "solid" ones represent the logical relations between formula-occurrences in a deductive step;
- there are three kinds of links (simple, focussing and defocussing) divided into *logical* and *structural* ones. In Figures 1.58 and 1.59 we can see its names and how its vertices are labelled with;
- every vertex in a proof-graph is labelled with a conclusion of a unique link and is the premise of at most one link.

The set of edges in a proof-graph can be empty. In this case the proof-graph represents an axiom.

Definition 82 (conjunctive link) The links $\wedge$-*I*, $\bot$-*link*, $\rightarrow$ -*E*, $\top$-*focussing-weak* and the expansion are called conjunctive.

Definition 83 (disjunctive link) The links $\vee$-*E*, $\top$-*link*, $\bot$- *defocussing-weak* and the contraction are called disjunctive.

Note that all disjunctive links with the exception of the contraction are defocussing ones as well as all conjunctive links are focussing with the exception of the expansion link.

In a proof-graph the direction of the edges defines a logical relation between the vertices rather than a temporal order between then. As we will see below, derivations are built from the components shown in Figures 1.58 and 1.59 called *basic links*. Sometimes we abuse of notation in using the labels of the vertices to denote the vertices themselves, as well as derivations and proof-graphs, and inference rules and basic links.

Definition 84 (solid indegree) The solid indegree of a vertex v in a proof-graph is the number of solid edges oriented towards it.

Definition 85 (solid outdegree) The outdegree of a vertex v in a proof-graph is the number of solid edges oriented away from it.

As expected, the notions of *meta indegree* and *meta outdegree* are defined notions. The edges of the *expansion* and *contraction* links are called expansion and contraction edges, respectively.

The set of vertices with outdegree equal to zero is the set of conclusions of a derivation represented in a proof-graph G and is defined as $CONC(G)$. Similarly, $PREMIS(G)$ is the set of vertices of G with indegree equal to zero which denotes the premises G; as well as $HYPOT(G)$ is the set of vertices in G with solid indegree equal to zero and meta indegree equal to 1, denoting the set of cancelled hypothesis.

6.1.2 Unsound proof-graphs. In a multiple-conclusion calculus we have rules with more than one conclusion ($\vee$-*elimination*) as well as rules with more than one premise ($\wedge$-*introduction* and $\rightarrow$-*elimination*). This allows the existence of cycles in the graphs of proofs, and thus making soundness difficult to prove. For example, the first kind of cycle that one wants to avoid is the one shown in the proof-graph 1 of Figure 1.60. Kneale avoids a corresponding development by defining the condition that formulas which are already connected, either directly by one single horizontal line or indirectly through several horizontal lines, should not be connected again in any way. However, with this restriction, the graphs of proofs should not have cycles, and new difficulties arise. For example, as we have seen, Shoesmith and Smiley provide an example of a tautology, namely $(A \rightarrow A) \wedge (A \vee (A \rightarrow A))$, in which there is not a Kneale development for it.

Shoesmith–Smiley as well as Ungar approach this kind of problem by allowing the existence of cycles in the derivation and by defining operations on derivations with appropriate restrictions.

In the theory of proof-nets the existence of cycles is also allowed, even if in this case we have a less ‘bureaucratic’ approach to deal with the definition of sound proofs. In the treatment given by Shoesmith–Smiley as well as Ungar, the links on the graphs of proofs do not always correspond to a deductive step, sometimes they refer to combination of derivations. In the theory of proof-nets, each link in the proof-structures corresponds to an inference rule, since it represents a logical relation between formulas.

Inspired in the analysis of such a kind of problem in multiple-conclusion calculus given by Shoesmith–Smiley and Ungar, as well as on the simplicity of the technique of the theory of proof-nets, we will discuss how we shall deal

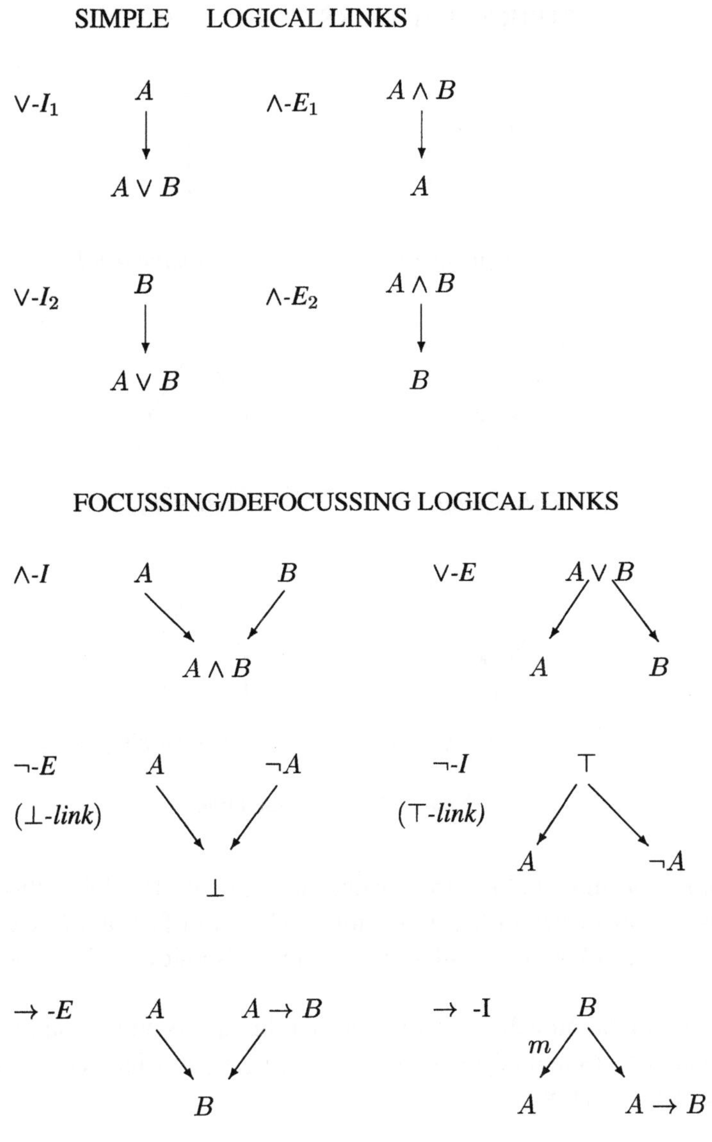

Figure 1.58. Logical links

with the existence of cycles as well as how we will define the class of sound proof-graphs in our system.

As in the theory of proof-nets, every link in an N-graph derivation corresponds to a deductive step.

We use explicit contraction rules (*expansion* and *contraction* links) as well as the relaxation of the condition that the two premises in $\wedge$-*introduction* and

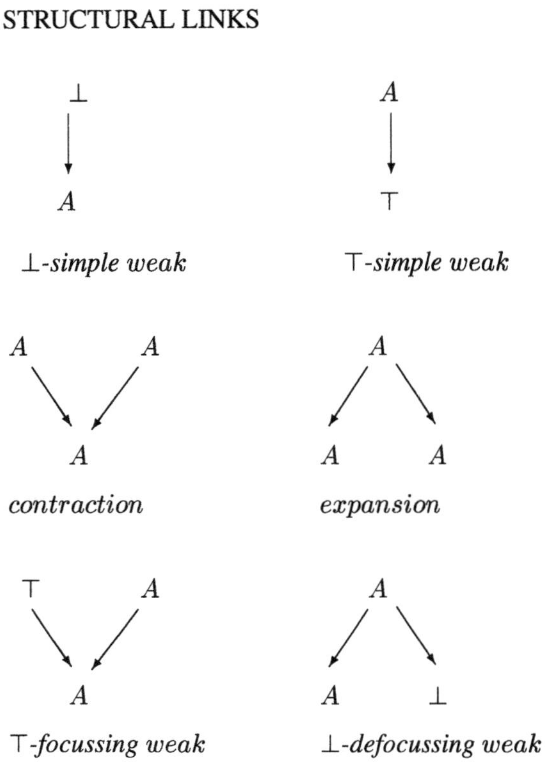

Figure 1.59. Structural links

→-elimination must belong to separate derivations. By doing this we allow the presence of cycles in the derivations. This is in fact an *abus de langage*: sometimes we talk about "cycles" when in fact "semicycles" are meant.

Definition 86 (semipath) A semipath in a digraph is an alternating sequence of distinct vertices and edges $v_0, x_1, v_1, \ldots, x_n, v_n$ in which each edge x_i may be either (v_{i-1}, v_i) or (v_i, v_{i-1}).

Definition 87 (semicycle) An alternating sequence of vertices and edges $v_0, x_1, v_1, \ldots, x_n, v_0$ is a semicycle provided the sequence $v_1, v_2, \ldots, v_n$ is a semipath.

By having an explicit contraction rule (*contraction* link) as illustrated in the proof-graph 2 of Figure 1.60, we allow the grouping of conclusions into equivalence classes. In order to group assumptions we use the *expansion link* which allows us to deal with derivations like the one llustrated in the proof-graph 4 of Figure 1.60.

With the introduction of the focussing contraction and defocussing expansion structural links we break with the pattern of focussing links having a conjunctive flavor as well as defocussing ones having a disjunctive flavor. The contraction link is disjunctive while the expansion is conjunctive. By doing this, we face a problem similar to Girard's with proof-nets. In the case of proof-nets we cannot distinguish a *par link* from a *times link*: although they have dual meaning (one is disjunctive and the other one conjunctive) their graphical representations (*i.e.* the graphs) are isomorphic, *i.e.* both are focussing links. The "key" example used to illustrate this problem in the theory of proof-nets is shown in our formalism as in proof-graph 3 of Figure 1.60. As we can see, the example shown in the proof-graph 2 of the same figure is logically correct while in a similar example, the proof-graph 3, we have a fallacious conclusion.

The expansion link groups assumptions into equivalence classes in the same way as the contraction link groups conclusions into equivalence classes. While the former is conjunctive, the latter is disjunctive. Thus, in Figs. 1.61 and 1.62 the leftmost proof-graph is sound while the other one is not.

The presence of the $\rightarrow$ connective with its introduction rule as in ND systems also complicates the task of identifying sound proof-graphs. For example, the proof-graph 5 in Figure 1.60 is logically sound while the proof-graph 6 is not.

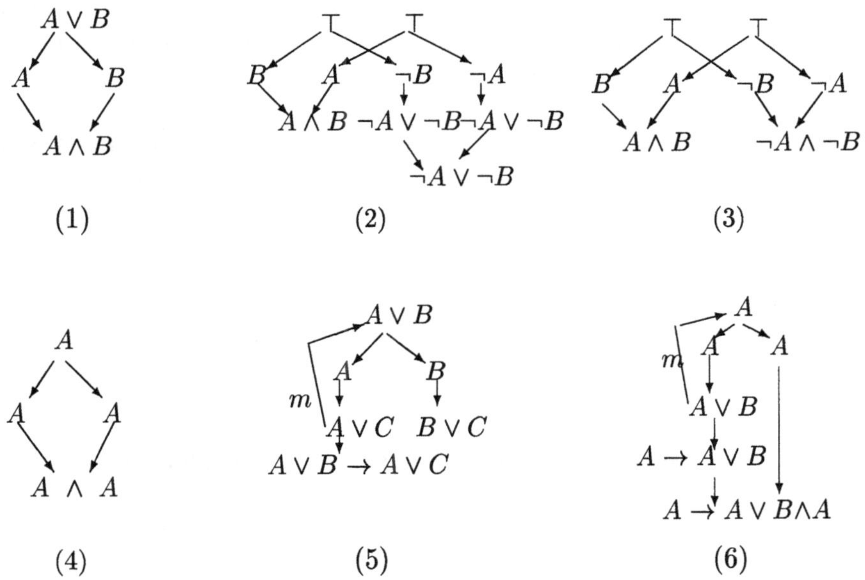

Figure 1.60. Proof-graphs

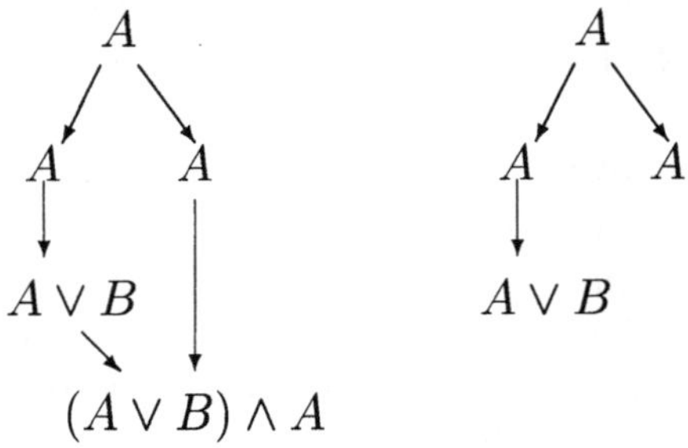

Figure 1.61. Expansion link in proof-graphs

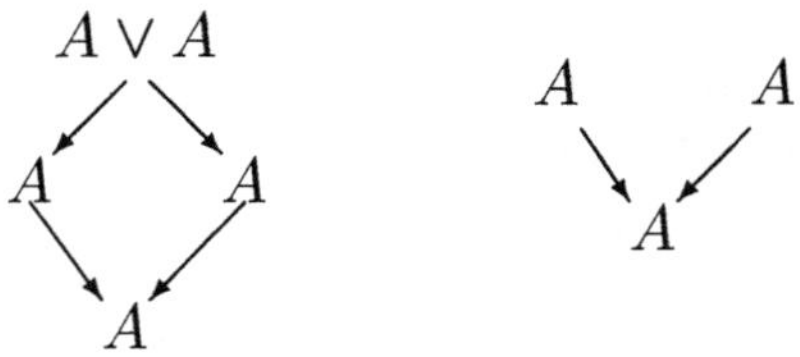

Figure 1.62. Contraction link in proof-graphs

6.1.3 Global soundness criterion. We shall define a geometrical criterion to test whether a proof-graph is logically sound. Like the Danos and Regnier criterion [28] studied in Section 3, we associate a set of graphs to a given proof-graph and then give the notion of N-graph derivations, *i.e.* those proof-graphs which are logically correct.

Definition 88 (switching) Given a proof-graph G, a switching graph S associated with G is a spanning subgraph of G in which the following edges are removed:

- one of the two edges of every expansion link;
- one of the two edges of every contraction link;
- all meta edges.

Definition 89 (switching expansion) Given a proof-graph G, a switching expansion graph S' associated with G is a spanning subgraph of G in which all meta edges are removed as well as one of the two edges of every expansion link is removed.

Definition 90 (meta-condition) Given a proof-graph G, we say that the meta-condition holds for it, iff for every meta-edge $(u, v)^m$ of a defocussing link $\{(u, v)^m, (u, w)\}$ in G, there is a path or a semipath, without passing through

(u, w), from v to u in every switching expansion graph S' associated with G and the solid indegree of v is equal to zero.

Definition 91 (N-graph derivation) A proof-graph G is an N-graph derivations, or N-graph for short, iff the meta-condition holds for G and every switching graph associated with G is acyclic and connected.

6.1.4 The soundness of the criterion. In order to assure that every N-graph derivation represents a proof logically correct we need to prove the following two theorems:

Theorem 92 (map to N-graph [30]) *Given a derivation* Π *of* $A_1, \ldots, A_n \vdash B_1, \ldots, B_m$ *in the classical sequent calculus, it is possible to buil a correspodent N-graph* $NG(\Pi)$ *whose the elements of* $PREMIS(NG(\Pi))$ *and* $CONC(NG(\Pi))$ *are in one-to-one correspondence with the occurrences of formulae* $A_1, \ldots, A_n$ *and* $B_1, \ldots, B_m$ *respectively.*

Theorem 93 (sequentialisation [30]) *Given an N-graph derivation G then there is a sequent calculus derivation* $SC(G)$ *of* $A_1, \ldots, A_n \vdash B_1, \ldots, B_m$ *in the classical sequent calculus whose the occurrences of formulae* $A_1, \ldots, A_n$ *and* $B_1, \ldots, B_m$ *are in one-to-one correspondence with the elements of* $PREMIS(G)$ *and* $CONC(G)$ *respectively.*

Next we formulate our version of the *splitting* theorem which again will be proved in a forthcoming publication.

Theorem 94 (splitting [30]) *Let G be an N-graph whose every initial link is disjunctive defocussing and every final link is conjunctive focussing. Then there must be either (i) some disjunctive defocussing initial or conjunctive focussing final with the split property or (ii) a cut branch point.*

Corollary 95 ([30]) *If G is an N-graph as in theorem 94 then:*

1 If $\{(u, v_1), (u, v_2)\}$ *is a initial disjunctive defocussing link with the split property, then removing it from G we obtain three subgraphs* G_1, G_2 *and* G_3 *which are also N-graphs as follows:*

- G_1 *is the vertex* u;
- G_2 *has* v_1 *among its premises;*
- G_3 *has* v_2 *among its premises;*
- $PREMIS(G) = \{u\} \cup PREMIS(G_2) \cup PREMIS(G_3)$;
- $CONC(G) = CONC(G_2) \cup CONC(G_3)$.

2 if $\{(u_1, v), (u_2, v)\}$ is a final conjunctive focussing link with the split property, then removing it from G we obtain three subgraphs G_1, G_2 and G_3 which are also N-graphs as follows:

- *G_1 is the vertex v;*
- *G_2 has u_1 among its conclusions;*
- *G_3 has u_2 among its conclusions;*
- *$PREMIS(G) = PREMIS(G_2) \cup PREMIS(G_3)$;*
- *$CONC(G) = \{v\} \cup CONC(G_2) \cup CONC(G_3)$.*

3 if s is a cut branch point of a expansion link $\{(s, s_1), (s, s_2)\}$, then removing this link from G we obtain two subgraphs which are also N-graphs as follows:

- *G_1 has s among its conclusions;*
- *G_2 has s_1 and s_2 among its premises;*
- *$PREMIS(G) = PREMIS(G_1) \cup PREMIS(G_2)$;*
- *$CONC(G) = CONC(G_1) \cup CONC(G_2)$;*

6.1.5 Some examples. In order to understand the technique of building proofs using the N-graph formalism we will look at a few examples of deductions in N-graph, ND and sequent calculus.

Example 96 The N-graph deduction of $(A \wedge B) \to (A \wedge B \wedge C) \vee \neg C$ is as in Figure 1.63 while in ND and sequent calculus is respectively as follows:

$$\dfrac{\dfrac{\dfrac{\dfrac{\dfrac{\dfrac{\dfrac{[A\wedge B]^1 \quad [C]^2}{(A\wedge B)\wedge C}}{(A\wedge B)\wedge C\vee\neg C} \quad [\neg((A\wedge B)\wedge C\vee\neg C)]^3}{\bot}}{\neg C}\,2 \qquad \dfrac{\dfrac{[\neg((A\wedge B)\wedge C\vee\neg C)]^3 \quad \dfrac{[\neg C]^4}{(A\wedge B)\wedge C\vee\neg C}}{\bot}}{C}\,4}{\bot}}{(A\wedge B)\wedge C\vee\neg C}\,3}{(A\wedge B)\to(A\wedge B)\wedge C\vee\neg C}\,1$$

$$\dfrac{\dfrac{\dfrac{\dfrac{A\wedge B\vdash A\wedge B \qquad \vdash C,\neg C}{A\wedge B\vdash A\wedge B\wedge C,\neg C}}{A\wedge B\vdash A\wedge B\wedge C, A\wedge B\wedge C\vee\neg C}}{A\wedge B\vdash A\wedge B\wedge C\vee\neg C, A\wedge B\wedge C\vee\neg C}}{A\wedge B\vdash A\wedge B\wedge C\vee\neg C}$$

Example 97 *The N-graph deduction of $(((A \to B) \to A) \to A$ (Peirce law) is as in Figure 1.64 while the ND and sequent calculus deduction is as follows:*

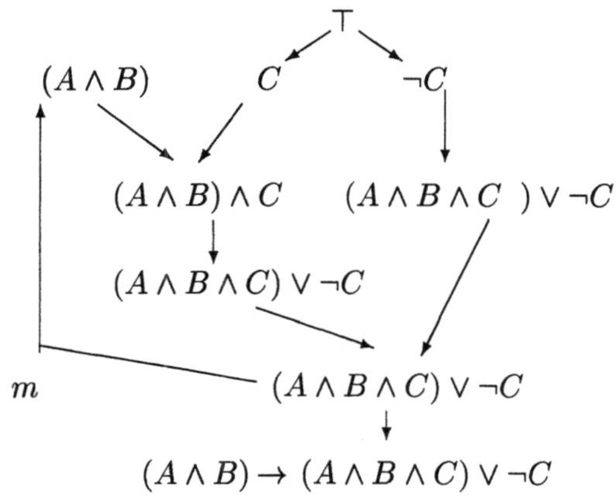

Figure 1.63. N-graph of $(A \wedge B) \to (A \wedge B \wedge C) \vee \neg C$

$$\cfrac{\cfrac{\cfrac{\cfrac{[((A\to B)\to A]^1 \qquad \cfrac{\cfrac{\cfrac{[A]^2 \quad [\neg A]^3}{\bot}}{B}}{A\to B}\,2}{A} \qquad [\neg A]^3}{\bot}}{A}\,3}{(((A\to B)\to A)\to A}\,1$$

$$\cfrac{\cfrac{\cfrac{\cfrac{A\vdash A}{A\vdash B,A}}{\vdash A\to B, A} \qquad A\vdash A}{(A\to B)\to A\vdash A}}{((A\to B)\to A)\to A}$$

Notes

1. Associated with Curry's early discovery of the correspondence between the axioms of intuitionistic implicational logic and the type schemes of the so-called 'combinators' of Combinatory Logic [23], and has been referred to as the *formulae-as-types* interpretation. Howard's [47] extension of the formulae-as-types paradigm to full intuitionistic first order predicate logic meant that the interpretation has since been referred to as the 'Curry–Howard' functional interpretation or Curry–Howard isomorphism.

2. In a recent work [21], Carbone relates the expansion of proofs with quantifiers to the group-theoretical notion of distorsion for finitely presented groups, as wells as the worse-case super-exponential expansion of proofs is related to the super-exponential distorsion of Gersten's finitely presented groups. Exponential expansion relates in similar ways to exponential distorsion in the Baumslag–Solitar's group. Groups are read from proofs out of cyclic structures.

3. The Province of logic. *Contemporary British Philosophy* (ed. H. D. Lewis), third series, pp. 235–261, Aberdeen, 1956.

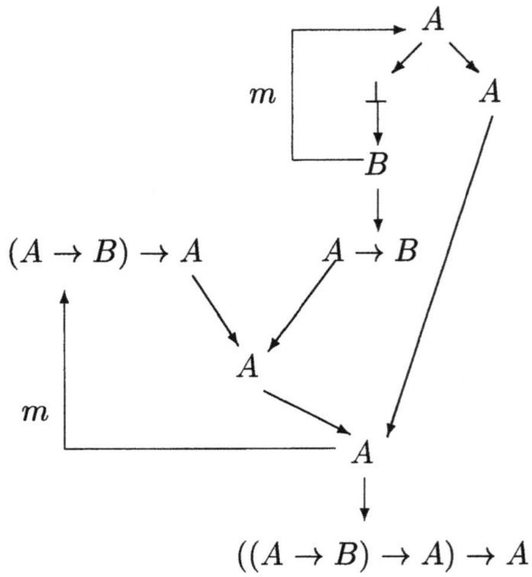

Figure 1.64. N-graph of $\vdash ((A \to B) \to A) \to A$

4. Cf.: 'Linear logic makes us hope that a final answer to the problem of inversion of rules might be found.' [40, p. 100].

5. Cf.: 'In a previous paper [41], we gave a purely geometrical interpretation of Gentzen's *Hauptsatz* for constructive (*i.e.* intuitionistic and linear) logics. This paper is the main piece in a general program of mathematisation of algorithmics, called *geometry of interaction*.' [42, opening lines]

6. Cf. 'Processes as Proofs?', abstract in *Logic Journal of the Interest Group in Pure and Applied Logics* 6(4):660, July 1998, Oxford University Press. Abstract of a paper presented at the *Logic for Concurrency and Synchronisation - First Workshop*, March 4–6 1998, DI–UFPE, Recife.

7. The commutative conversions, also called "permutative conversions" or "permutative reductions" [32, 31] establish the non-interference of the newly open branch of the proof (the local hypothesis open in the elimination of $\exists$ and $\vee$) with the main branch. They are the following basic transformations between proofs:

$$\dfrac{\dfrac{A_1 \vee A_2 \quad \begin{matrix}[A_1]\\ C\end{matrix} \quad \begin{matrix}[A_2]\\ C\end{matrix}}{C}}{W} \quad \rightsquigarrow \quad \dfrac{A_1 \vee A_2 \quad \begin{matrix}[A_1]\\ \dfrac{C}{W}\end{matrix} \quad \begin{matrix}[A_2]\\ \dfrac{C}{W}\end{matrix}}{W}$$

$$\dfrac{\dfrac{\exists x.P(x) \quad \begin{matrix}[P(t)]\\ C\end{matrix}}{C}}{W} \quad \rightsquigarrow \quad \dfrac{\exists x.P(x) \quad \begin{matrix}[P(t)]\\ \dfrac{C}{W}\end{matrix}}{W}$$

8. All along the text we only consider the propositional calculus. However *LFG* are used, both by Buss and Carbone, to find interesting consequences in the context of predicate logic. Buss shows his results on k-provability for first order logic, and Carbone has two papers, [20] and [21] where the dynamics of proofs with quantifiers is analysed together with the super-exponential expansion (occurring, in the worst case).

9. The pigeonhole principle is a largely used example to prove lower bounds in various proof systems, it says that there is no injective function which maps $\{1, 2, \ldots, n\}$ to $\{1, 2, \ldots, n-1\}$. In other words, "*If* $n+1$ *pigeons sit in n holes then some hole contains more than one pigeon.*"

10. NB: Vertical edges in this figure make sense only if you specify that formulas are intended to be specific occurrence in a proof which is "embedded" in the plane.

11. Originally, in order to simplify his proof, Gentzen [36] uses the *mix rule* instead of the cut:

$$\frac{\Gamma \vdash \Theta \quad \Delta \vdash \Lambda}{\Gamma, \Delta^* \vdash \Theta^*, \Lambda}(mix\ rule)$$

where Δ^*, Θ^* are respectively Δ and Θ without all formulas of the form M (mix formula). M may be any arbitrary formula.

One could then ask if the main source of expansion of cut-free proofs is actually the combination of cut and contraction, once in his original proof Gentzen uses the mix rule, instead of the cut one. However we have already made some study of cases and we concluded that no matter if one use the mix or cut rule. In all cases a splitting occurs either by the combination of cut and contraction or by the combination of mix and contraction.

12. There are other cut-elimination procedures which are possible. One of them is presented in [16], where Carbone studies some properties of proofs which is not preserved by the transformations in the usual Gentzen's cut-elimination procedure.

13. The model that Carbone and Semmes propose turns out to be a very general framework to study "expansion", implicit versus explicit descriptions and symmetries in combinatorial structures. It applies to proofs as well as to circuits, automata and graphs in general as studied in [22].

14. Also in [p. 248][59], Prawitz explains why he wanted to keep the Λ_C-rule applied only to atomic formulas and then paying the price of having a "*restricted*" version of first order classical ND system:

> "Note in particular that the principle of indirect proof when not restricted to atomic formulas constitute quite a new principle for inferring compound formulas (which is not at all justified in the terminology of sec. 2.2.2 by the meaning given to the constants by the introduction rules). Our restriction that the conclusion of the application of the rules Λ_I and Λ_C are to be atomic are motivated by these considerations. It is by this restriction that these extra rules do not disturb the pattern of introduction and elimination..."

15. Proposed by Curry in [24] and also in *Foundations of Mathematical Logic*.

16. In [63, p. 26], Shoesmith and Smiley give a notion of adequacy:

> "A set of rules is adequate for a calculus if it is both sound and complete."

In [67, p. 62], Ungar gives the following definition for adequacy:

> "I propose to call a notion of derivation adequate for R if the relation of derivability by R which it determines coincides with the consequence relation for R."

where R is the set of rules of the system.

17. The property of a corner-free graph argument corresponds to the requirements in the definition of proof-structures (in linear logic), that every formula-occurrence in the structure is the conclusion of one and only one link and is the premise of at most one link.

References

[1] S. Abramsky. Computational interpretations of linear logic. *Theoretical Computer Science* 111:3–57, 1993. Revised version of Imperial College Technical Report DoC 90/20, October 1990.

[2] S. Abramsky. Proofs as Processes. *Theoretical Computer Science* 135:5–9, 1994.

[3] S. Abramsky. Interaction categories and communicating sequential processes. In A. W. Roscoe, editor, *A Classical Mind: Essays in honour of*

C.A.R. Hoare pages 1–16. Prentice-Hall International, 1994.

[4] Yuuki Andou. A Normalization-Procedure for the First Order Classical Natural Deduction with Full Logical Symbols. *Tsukuba Journal of Mathematics*, 19(1):153–162, 1995.

[5] A. Asperti. A linguistic approach to deadlock. Rapport de Recherche du Laboratoire d'Informatique de l'École Normale Supérieure de Paris, LIENS-91-15, October 1991.

[6] A. Asperti. Causal Dependencies in Multiplicative Linear Logic with MIX. *Mathematical Structures in Computer Science*, 5:351–380, 1995.

[7] G. Bellin. Mechanizing Proof Theory: Resource-aware Logics and Proof-transformati ons to extract Implicit Information. PhD Thesis, Departament of Philosophy, Stanford University, June 1990.

[8] G. Bellin. Chu's Construction: A Proof-Theoretic Approach. (this volume).

[9] G. Bellin. Two paradigms of logical computation in affine logic?. (this volume).

[10] G. Bellin and P. J. Scott. On the π-calculus and linear logic, (with an introduction by S. Abramsky), *Theoretical Computer Science* 135:11–65, 1994.

[11] G. Bellin and J. van Wiele. Subnets of Proof-nets in MLL^-. In *Advances in Linear Logic*, Girard, Lafont and Regnier, eds., London Math. Soc. Lect. Notes Series **222** Cambridge University Press, pp. 249–270.

[12] S. Buss. Polynomial Size Proofs of the Propositional Pigeonhole Principle *Journal of Symbolic Logic*, 52(4):917–927, 1987.

[13] S. Buss. The undecidability of k-provability. *Annals of Pure and Applied Logic*, 53:72–102, 1991.

[14] A. Carbone. Some Combinatorics Behind Proofs. Report of University of Paris 12, with the website: www.univ-paris12.fr/lacl/ale.

[15] A. Carbone and S. Semmes. Making proofs without Modus Ponens: An introduction to the combinatorics and complexity of cut elimination *Bulletin of the American Mathemathical Society*, 34:131–159, 1997.

[16] A. Carbone. Interpolants, Cut Elimination and Flow Graphs for the Propositional Calculus. *Annals of Pure and Applied Logic*, 83:249–299, 1997.

[17] A. Carbone. Turning cycles into spirals. *Annals of Pure and Applied Logic*, 96:57–73, 1999.

[18] A. Carbone. Duplication of directed graphs and exponential blow-up of proofs. *Annals of Pure and Applied Logic*, 100:1–76, 1999.

[19] A. Carbone. The cost of a cycle is a square. To appear in the Journal of Symbolic Logic.

[20] A. Carbone. Cycling in proofs and feasibility. *Transactions of the American Mathematical Society*, 2000.

[21] A. Carbone. Asymptotic cyclic expansion and bridge groups of formal proofs. To appear in the Journal of Algebra.

[22] A. Carbone and S. Semmes. *A Graphic Apology for Symmetry and Implicitness.* Oxford Mathematical Monographs. Oxford University Press, 2000.

[23] H. B. Curry. Functionality in Combinatory Logic. In *Proceedings of the National Academy of Sciences of USA* 20:584–590, 1934.

[24] H. B. Curry. *A Theory of Formal Deducibility*, Notre Dame University Press, 1950. (third printing (1966) of second (1957)).

[25] S. Cook. The complexity of theorem-proving procedures. In *Proc. 3rd STOC*, 1971, pp. 151–158.

[26] S. Cook and R. Reckhow. On the lengths of proofs in the propositional calculus. *Proceedings of Sixth Annual ACM Symposium on the Theory of Computing*, 1974.

[27] S. Cook and R. Reckhow. The relative efficiency of propositional proof systems. *Journal of Symbolic Logic*, 44:36–50, 1979.

[28] V. Danos and L. Regnier. The structure of multiplicatives. *Archive for Mathematical Logic*, 28:181–203,1989.

[29] Anjolina Grisi de Oliveira. *Proof Transformations for Labelled Natural Deduction via Term Rewriting. (In Portuguese).* Master's thesis, Depto. de Informática, Universidade Federal de Pernambuco, C.P. 7851, Recife, PE 50732-970, Brazil, April 1995.

[30] Anjolina Grisi de Oliveira. *Proofs from a Geometric Perspective.* PhD thesis, Centro de Informática, Universidade Federal de Pernambuco, C.P. 7851, Recife, PE 50732-970, Brazil, February 2001.

[31] Ruy J. G. B. de Queiroz and Dov M. Gabbay. The functional interpretation of the existential quantifier. *Bulletin of the Interest Group in Pure and Applied Logics*, 3(2 and 3):243–290, 1995. Abstract in JSL 58(2):753–754,1993. (Presented at *Logic Colloquium '91*, Uppsala, August 9–16.)

[32] Ruy J. G. B. de Queiroz and Dov M. Gabbay. Labelled natural deduction. In *Logic, Language and Reasoning. Essays in Honor of Dov Gabbay's 50th Anniversary,* H.J. Ohlbach and U. Reyle (eds.), Kluwer Academic Publishers, 1999, pp. 199–278.

[33] L. Franklin. Multiple-conclusion System for Intuitionistic Propositional Proceedings of *XII Encontro Brasileiro de Lógica*, 25 a 28 de maio de 1999, Parque Nacional do Itatiaia, RJ, Brazil.

[34] Dov M. Gabbay. *Labelled Deductive Systems, Volume I - Foundations.* Oxford University Press, 1996.

[35] Jean Gallier. Constructive Logic Part II: Linear Logic and Proof nets. University of Pennsylvania, CIS Dept. tech. Report MS-CIS-91-75, 1991.

[36] Gerhard Gentzen. Untersuchungen über das logische Schliessen. *Mathematische Zeitschirift*, pages 176–210 and 405–431, 1935. English translation:"Investigations into Logical Deduction" in *The Collected Works of Gehard Gentzen*, ed. M.E.Szabo, North-Holland Pub Co.,1969.

[37] R. O. Gandy. Proofs of strong normalization. In *To H. B. Curry: essays in Combinatory Logic, Lambda calculus and Formalism*, ed. J. R. Hindley and J. P. Seldin, Academic Press, 1980.

[38] Jean-Yves Girard. Linear logic. *Theoretical Computer Science*, 50:1–102, 1987.

[39] Jean-Yves Girard. *Proof Theory and Logical Complexity.* Volume 1 of *Studies in Proof Theory*. Bibliopolis, Naples, 1987.

[40] Jean-Yves Girard. Towards a geometry of interaction. In *Categories in Computer science and Logic*, volume 92 of *Contemporary Mathematics*, pages 69–108. AMS Publications, 1989.

[41] Jean-Yves Girard. Geometry of interaction 1: Interpretation of system F. In *Logic Colloquium '88*, R. Ferro et al. (eds), North-Holland, 1989, pp. 221-260.

[42] Jean-Yves Girard. Geometry of interaction 2: Deadlock-free algorithms. In *COLOG '88*, P. Martin-Löf and G. Mints (eds), Lecture Notes in Computer Science 417, Springer, pp. 76–93.

[43] Jean-Yves Girard. A new constructive logic: classical logic. *Mathematical Structures in Computer Science* 1:255–296, 1991.

[44] Jean-Yves Girard. Linear Logic: A Survey. In *Proceedings of the International Summer School of Marktoberdorf.* NATO Advanced Science Institute, series F94, pp. 63–112. L. F. Bauer, M. Brauer and H. Schwichtenberg editors, Springer-Verlag, 1993.

[45] Jean-Yves Girard, Yves Lafont, and Paul Taylor. *Proofs and Types*. Cambridge University Press, 1989.

[46] A. Haken. The intractability of resolution. *Theoretical Computer Science*, 38:297–308, 1985.

[47] W. Howard. The formulae-as-types notion of construction. In *To H. B. Curry: Essays on Combinatory Logic, Lambda Calculus and Formalism*, J. P. Seldin and J. R. Hindley (eds.), Academic Press, London, pp. 479–490.

[48] S. Jaskowski. On the rules of suppositions in Formal Logic. *Studia Logica*, 1:5–32, 1934.

[49] W. Kneale. The Province of Logic. *Contemporary British Philosophy* (ed. H. D. Lewis), third series, pp. 235–261, Aberdeen, 1956.

[50] W. Kneale and M. Kneale. *The development of logic*, Oxford, 1962.

[51] J. Krajicek. *Bounded arithmetic, propositional logic, and complexity theory*. volume 60 of the series *Encyclopedia of Mathematics and Its Applications*, Cambridge University Press, 1995, 343 p. ISBN 0-521-45205-8.

[52] D. Leivant. Assumptions classes in natural deduction. *Zeitschrift für Mathematische Logik und Grundlagen der Mathematik*, 25:1–4, 1979.

[53] P. Martin-Löf. *Intuitionistic Type Theory*, Bibliopolis, Naples, 1984.

[54] Cosme D. B. Massi. *Normalization Proofs for Classical Logic* (in Portuguese: "Provas de Normalização para a Lógica Clássica"). PhD thesis, Departamento de Filosofia do Instituto de Filosofia e Ciências Humanas da Universidade Estadual de Campinas - SP, 1990.

[55] Christos Papadimitriou. NP-Completeness: A Retrospective. In *ICALP'97*, Springer, 1997.

[56] Luiz Carlos Pereira and Cosme Massi. Normalização para a lógica clássica. *O que nos faz pensar*, 2:49–53, 1990.

[57] Jan von Plato. Proof Theory of full classical propositional logic. Technical Report. Departament of Philosophy, University of Helsinki, 1998.

[58] D. Prawitz. *Natural Deduction. A Proof-Theoretical Study*, volume 3 of *Acta Universitatis Stockholmiensis. Stockholm Studies in Philosophy*. Almqvist & Wiksell, Stockholm, 113pp, 1965.

[59] D. Prawitz. Ideas and results in proof theory. In J. E. Fenstad, editor, *Proceedings of the Second Scandinaviam Logic Symposium*. North-Holland, 1971.

[60] D. Prawitz. Validity and normalizability of proofs in first and second order classical and intuitionistic logic. In *Atti del Congresso Nazionale di Logica*, pp. 11–36, Montecatini, 1979.

[61] P. Pudlák. The lengths of proofs. In *Handbook of Proof Theory*. Sam Buss, editor. North-Holland, pp. 547–637, 1998.

[62] Jonathan Seldin. Normalization and Excluded Middle I. *Studia Logic* XLVIII, pp. 193–217, 1989.

[63] D. J. Shoesmith and T. J. Smiley. *Multiple-Conclusion Logic*, Cambridge University Press, 1978.

[64] Gunnar Stalmarck. Normalization Theorems for Full First Order Classical Natural Deduction. *The Journal of Symbolic Logic*, 56(1):129–149, March 1991.

[65] R. Statman. *Structural Complexity of Proofs*. PhD thesis, Stanford University, May 1974.

[66] N. Tennant. *Natural Logic*. University of Edinburgh Press, 1978.

[67] A. M. Ungar. *Normalization, Cut-elimination and the Theory of Proofs*. Number 28 in Lecture Notes. CSLI - Center for the Study of Language and Information, Stanford, 1992.

[68] Alasdair Urquhart. The Complexity of Propositional Proofs. *The Bulletin of Symbolic Logic*, 1(4):425–467, December, 1995.

[69] J. I. Zucker. The Correspondence between Cut-Elimination and Normalization *Annals of Mathematical Logic*, 7:1–156, 1974.

Chapter 2

CHU'S CONSTRUCTION: A PROOF-THEORETIC APPROACH

Gianluigi Bellin*
Facoltà di Scienze
Università di Verona
bellin@sci.univr.it

Abstract The essential interaction between classical and intuitionistic features in the system of linear logic is best described in the language of category theory. Given a symmetric monoidal closed category $\mathcal{C}$ with products, the category $\mathcal{C} \times \mathcal{C}^{op}$ can be given the structure of a $*$-autonomous category by a special case of the Chu construction. The main result of the paper is to show that the intuitionistic translations induced by Girard's trips determine the functor from the free $*$-autonomous category $\mathcal{A}$ on a set of atoms $\{P, P', \ldots\}$ to $\mathcal{C} \times \mathcal{C}^{op}$, where $\mathcal{C}$ is the free monoidal closed category with products and coproducts on the set of atoms $\{P_O, P_I, P'_O, P'_I, \ldots\}$ (a pair P_O, P_I in $\mathcal{C}$ for each atom P of $\mathcal{A}$).

Keywords: Chu spaces, proof-nets, linear logic

1. Preface

An essential aim of linear logic [16] is the study of the dynamics of proofs, essentially normalization (cut elimination), in a system enjoying the good proof-theoretic properties of *intuitionistic* logic, but where the dualities of *classical* logic hold. Indeed *classical linear logic* **CLL** has a denotational semantics and a game-theoretic semantics; proofs are formalized in a sequent calculus, but also in a system of *proof-nets* and in the latter representation cut elimination not only has the strong normalizability property, but is also confluent. Although Girard's main system of linear logic is *classical*, considerable attention in the literature has also been given to the system of *intuitionistic linear logic*

*Research supported by EPSRC senior research fellowship on grant GL/L 33382. We wish to thank Martin Hyland for his essential suggestions and crucial support in the develoment of this work.

R.J.G.B. de Queiroz (ed.), Logic for Concurrency and Synchronisation, 89–109.

ILL, where proofs are also formalized in a sequent calculus and in a natural deduction system. A better understanding of the relations between **CLL** and **ILL** is one of the goals to which the present work is intended as a contribution.

The fact that intuitionistic logic plays an important role in the architecture of linear logic is not surprising: as indicated in the introductory section of Girard's fundamental paper [16], a main source of inspiration for the system was its denotational semantics of coherent spaces, a refinement of Scott's semantics for the λ-calculus. Fundamental decisions about the system **CLL** were made so that **CLL** has a semantics of proofs in coherent spaces in the same way as intuitionistic logic has a semantics of proofs in Scott's domains. But linear logic is not just a refinement of intuitionistic logic, such as **ILL**: there are expectations that **CLL** may tell us something fundamental about classical logic as well, indeed, that through linear logic a deep level of analysis may have been reached from which the "unity of logic" can be appreciated [17]. Therefore the relations between classical and intuitionistic components of linear logic deserve careful investigation.

A natural points of view to look at this issue is *categorical logic*. It has been known for years that monoidal closed categories provide a model for *intuitionistic linear logic*, though a fully adequate formulation of the syntax and of the categorical semantics of **ILL** especially with respect to the exponentials, has required considerable subtlety and effort [4, 5, 6]. It is also well known that $*$-autonomous categories give a model for *classical linear logic* [3]. The appendix to [2] provides a method, due to Barr's student Chu, to construct $*$-autonomous categories starting from monoidal closed ones.

In our proof-theoretic investigation we encounter a special case of Chu's construction, namely $\mathbf{Chu}(\mathcal{C}, \top)$ where $\mathcal{C}$ is a symmetric monoidal closed category with terminal object $\top$. More specifically, given the free $*$-autonomous category $\mathcal{A}$ on a set of objects (propositional variables) $\{P, P', \ldots\}$ and given the symmetric monoidal closed category $\mathcal{C}$ with products, free on the set $\{P_O, P_I, P'_O, P'_I, \ldots\}$ (a pair P_O, P_I in $\mathcal{C}$ for each atom P of $\mathcal{A}$), the category $\mathcal{C}\times\mathcal{C}^{op}$ can be given the structure of a $*$-autonomous category by Chu's construction. Indeed, since the dualizing object is the terminal object, $\mathbf{Chu}(\mathcal{C}, \top)$ is just $\mathcal{C} \times \mathcal{C}^{op}$ and the pullback needed to internalize the homsets is in fact a product. Here the tensor product $(X, X^{op}) \otimes (Y, Y^{op})$ must be an object of the form $(X \otimes Y, (X \multimap Y^{op}) \times (Y \multimap X^{op}))$ and the identity of the tensor must be $(\mathbf{1}, \top)$. Dually, the *par* $(X, X^{op})\wp(Y, Y^{op})$ is defined as $((X^{op} \multimap Y) \times (Y^{op} \multimap X), X^{op} \otimes Y^{op})$ and the identity of the *par* must be $(\top, \mathbf{1})$. Now since $\mathcal{A}$ is free, there is a functor F of $*$-autonomous categories from $\mathcal{A}$ to $(\mathcal{C} \times \mathcal{C}^{op})$ taking P to (P_O, P_I). This is well-known, but so far no familiar construction had been shown to correspond to the functor F given by the abstract theory. The main contribution of this paper is to show that a familiar proof-theoretic construction, namely *Girard's trips* [16] on a proof-net, represent the action of such a functor on the morphisms of $\mathcal{A}$. Of course one could state the same result using Danos–Regnier graphs, as it was done in [8], but as we shall see a simpler definition of orientations is possible in terms of Girard's trips.

The key idea is simple enough and may be illustrated as a logical translation of formulas and proofs in **CMALL** into formulas and proofs in **IMALL**. In the translation a **CMALL** sequent S: $\vdash \Gamma, A$ becomes *polarized*: a selected formula-occurrence A is mapped to a *positive* formula-occurrence A_O in the succedent of an intuitionistic sequent S' (the *output* part of a logical computation); all other formula-occurrences C in Γ are mapped to *negative* C_I in the antecedent of S' (the *input* part). The polarized occurrences of an atom A become A_O, A_I, just two copies of A. Negation changes the polarity. For other complex polarized formulas, the polarization of the immediate subformulas is uniquely determined – for instance, $(A \wp B)_I$ becomes $A_I \otimes B_I$ – except in the cases of $(A \wp B)_O$ and $(A \otimes B)_I$. In these cases we take the product (logically, the *with*) of two possible choices (the "switches" in a proof-net): for instance, $(A \wp B)_O$ is encoded as $(A_I \multimap B_O) \& (B_I \multimap A_O)$. The intuitive motivation is clear: $A \wp B$ has a reading simultaneously as the internalization of the function space $\mathrm{Hom}_{\mathcal{A}}(A^{\perp}, B)$ and of the function space $\mathrm{Hom}_{\mathcal{A}}(B^{\perp}, A)$. The fact that the translation is functorial here means, roughly, that it is defined independently on the formulas (objects) and on the proofs (morphisms) and that it admits the rule of Cut (composition of morphisms); it is also compatible with cut-elimination. In this form the result can be easily proved within the formalisms of the sequent calculi for **CMALL** and **IMALL**. However, when we ask questions about the *faithfulness* and *fullness* of such a functor, thus also asking questions about the identity of proofs in linear logic, we find it convenient to consider the more refined syntax of proof-nets.

On the other hand, proof-nets are also useful to highlight the geometric aspect of certain logical properties; indeed ideas related to the present result have already proved quite useful in the study of what is sometimes called the *géometrie du calcul* (*geometry of computations*). Our own investigation has been motivated by the desire to understand and clarify the notion of a proof-net and the present result appears to reward many collective efforts in this direction. Given a proof-structure, i.e., a directed graph where edges are labeled with formulas, a *correctness criterion* characterizes those proof-structures which correspond to proofs in the sequent calculus. Girard's original condition (*"there are no short trips"*) [16] is *exponential* in time on the size of the proof-structure, but other *quadratic* criteria were found soon after (among others, one was given in [7]). Thus it is natural to ask *what additional information is contained in the construction of Girard's trips other than the correctness of a proof-structure.* The beginning of an answer came in 1992, when Jacques van de Wiele and the author, inspired by Danos' notion of a *pure net*, defined the *trip translation*: every Girard's trip on a cut-free proof-net corresponds to a derivation in the fragment of intuitionistic linear logic with *times* and *linear implication.* But the significance of this result seemed limited by the fact that the treatment of cut was quite cumbersome and the result itself did not seem to extend beyond the multiplicative fragment. A better understanding of its significance – and, as we hope, the possibility of its generalization – has come only from an explicit effort to formulate the trip translation as a *functorial* operation. In this way it became evident that *classical multiplicative* linear logic has to be related to *intuitionistic multiplicative and additive* linear logic and the categorical result followed, for which we gratefully acknowledge the influence and the support of Martin Hyland.

There is a conspicuous literature on Chu's spaces and linear logic. Moreover Chu's construction is related to many other more concrete semantics that yield *full completeness* results for fragments of linear logic, from R. Loader's *Linear logical Predicates* [22] to various game-theoretic semantics (cf. [26]). Clearly this is not the place to survey such a body of literature. It is impossible however not to mention the work of V. Pratt, who has advocated this direction of research for a long time (see, e.g., [25]) and has recently obtained (with Plotkin *et al.* [11]) a full completeness result for multiplicative linear logic (without units) with respect to Chu spaces. Chu's construction $\mathbf{Chu}(\mathcal{H}, \top)$ where $\mathcal{H}$ is a Heyting algebra, is also explicitly used by Anna Patterson in her thesis ([24], Section 6.7.), to show that the *algebra of constructive duality* is a model of **CLL** with Mix (we are grateful to an anonymous referee for this reference).

Among the researchers who have worked on proof-nets and developed ideas related to the trip translation, we should mention F. Lamarche, who introduced the notion of *essential net* for intuitionistic linear logic [20] in the context of his research on the game-theoretic semantics [21]. Arnaud Fleury has considered trips and intuitionistic translations in a non-commutative context with explicit exchange rule [13], giving one of the most interesting and least understood developments in this area. Already in 1992-93 the consideration of trips as translations from classical to intuitionistic linear logic had suggested the possibility of giving a *linear time* correctness condition for proof-nets: after all, just one unsuccessful trip suffices to discard a proof-structure as incorrect, and just one successful trip, if appropriately translated to an intuitionistic derivation, suffices to test the correctness of a proof-net. However only in 1999 Murawski and Ong [23] were able to prove such a conjecture, making essential use of a result of Gabow and Tarjan [15].

When the languages and the aims of different scientific communities meet in a new theory and new structures are identified, the conceptual architecture may look different from the different points of view and it may be hard to say which structures are fundamental. From the point of view of categorical logic Chu's construction seems to suggest a fundamental status to monoidal closed categories with respect to the $*$-autonomous ones. However, such a view must accompanied by the warning that the correspondence established by our interpretation of Chu's construction is not an isomorphism, as the functor F is *not full*, and its faithfulness at the moment is only a conjecture. Research towards a refinement of the present result, in particular with respect to the units and the additives is in progress, as well as towards its extension to the exponentials. Finally, further work is needed to spell out intriguing analogies between our version of Chu's construction and the *game theoretic semantics* of linear logic.

2. The trip translation

In this section, after the basic formal definitions we present our functorial translation from the sequent calculus for **CMALL** to that for **IMALL**, we state the categorical result and sketch the proof.

2.1 Languages, intuitionistic and classical MALL

The syntax of propositional *classical* Linear Logic **CLL** is given in Girard [16]: formulas are in "negation normal form", i.e., they are built from propositional constants $\mathbf{1}$, $\bot$, $\top$ and $\mathbf{0}$, atoms P_i and negations of atoms $(P_i)^\bot$ using the connectives $\otimes$, $\wp$, & and $\oplus$ and the exponential operators ! and ?; linear negation for nonatomic formulas is defined and linear implication is also defined, see in Table 0.

$\bot^\bot =_d \mathbf{1}$	$\mathbf{1}^\bot =_d \bot$	$(A \otimes B)^\bot =_d A^\bot \wp B^\bot$	$(A \wp B)^\bot =_d A^\bot \otimes B^\bot$
$\top^\bot =_d \mathbf{0}$	$\mathbf{0}^\bot =_d \top$	$(A \& B)^\bot =_d A^\bot \oplus B^\bot$	$(A \oplus B)^\bot =_d A^\bot \& B^\bot$
	$(!A)^\bot =_d ?(A^\bot)$	$(?A)^\bot =_d !(A^\bot)$	
	$A \multimap B =_d A^\bot \wp B$		

Table 0: Definition of linear negation and linear implication.

CMALL [**CMLL**] is the fragment of **CLL** without exponentials [without exponentials and additives]; **CMLL**$^-$ is **CMLL** without propositional constants.

Recall that the sequent calculus for propositional **CMALL** is defined by the axioms and rules given in Table 1:

identity rules

logical axiom: $\vdash A^{\perp}, A$

cut:
$$\frac{\vdash \Gamma, A^{\perp} \quad \vdash \Delta, A}{\vdash \Gamma, \Delta}$$

structural rules

exchange:
$$\frac{\vdash \Gamma, A, B, \Delta}{\vdash \Gamma, B, A, \Delta}$$

logical rules

one: $\vdash \mathbf{1}$

nil:
$$\frac{\vdash \Gamma}{\vdash \Gamma, \bot}$$

times:
$$\frac{\vdash \Gamma, A \quad \vdash \Delta, B}{\vdash \Gamma, \Delta, A \otimes B}$$

par:
$$\frac{\vdash \Gamma, A, B}{\vdash \Gamma, A \wp B}$$

true: $\vdash \Gamma, \top$

with:
$$\frac{\vdash \Gamma, A \quad \vdash \Gamma, B}{\vdash \Gamma, A \& B}$$

plus:
$$\frac{\vdash \Gamma, A_i}{\vdash \Gamma, A_0 \oplus A_1} \quad \text{for } i = 0, 1.$$

Table 1: The sequent calculus **CMALL**.

The propositional language of *intuitionistic* Linear Logic **ILL** is built from a set of propositional atoms $\{P_O, P_I, P'_O, P'_I, \ldots\}$, and the propositional constants **1**, $\top$ and **0**, using the connectives $\multimap$ (*linear implication*) and $\otimes$, $\&$ and $\oplus$ and the exponential **!**. (We take the point of view that there is no symbol "$\bot$" for "multiplicative falsity" in **ILL**. Thus P_O and P_I may be regarded as "positive and negative atoms", respectively). Again **IMALL** is the fragment of **ILL** without exponentials, etc.

The sequent calculus for propositional **IMALL** has the axioms and rules given in Table 2.

identity rules

logical axiom:

$$A \vdash A$$

cut:

$$\frac{\Gamma \vdash A \quad A, \Delta \vdash B}{\Gamma, \Delta \vdash B}$$

structural rules

exchange:

$$\frac{\Gamma, A, B, \Delta \vdash C}{\Gamma, B, A, \Delta \vdash C}$$

logical rules

R one:

$$\vdash \mathbf{1}$$

L one:

$$\frac{\Gamma \vdash B}{\Gamma, \mathbf{1} \vdash B}$$

R times:

$$\frac{\Gamma \vdash A \quad \Delta \vdash B}{\Gamma, \Delta \vdash A \otimes B}$$

L times:

$$\frac{\Gamma, A, B \vdash C}{\Gamma, A \otimes B \vdash C}$$

R implication:

$$\frac{\Gamma, A \vdash B}{\Gamma \vdash A \multimap B}$$

L implication:

$$\frac{\Gamma \vdash A \quad B, \Delta \vdash C}{A \multimap B, \Gamma, \Delta \vdash C}$$

R true:

$$\Gamma \vdash \top$$

L zero:

$$\mathbf{0}, \Gamma \vdash A$$

R with:

$$\frac{\Gamma \vdash A \quad \Gamma \vdash B}{\Gamma \vdash A \& B}$$

L with:

$$\frac{\Gamma, A_i \vdash B}{\Gamma, A_0 \& A_1 \vdash B} \quad \text{for } i = 0, 1.$$

plus:

$$\frac{\Gamma \vdash A_i}{\Gamma \vdash A_0 \oplus A_1} \quad \text{for } i = 0, 1.$$

L plus:

$$\frac{A, \Gamma \vdash C \quad B, \Gamma \vdash C}{A \oplus B, \Gamma \vdash C}$$

Table 2: The sequent calculus **IMALL**.

When a derivation π (in **CMALL** or **IMALL**) ends with a *cut* inference and the immediate subderivations are π_1, π_2, we use the notation $(\pi_1|\pi_2)$ for π.

2.2 The functorial trip translation for MALL

Definition 1 (Trip translation) (i) The *trip translation* maps formulas of **CMALL** to pair of formulas of **IMALL**: $A \mapsto (A_O, A_I)$. The formulas of **CMALL** are first *polarized* and then translated into **IMALL** formulas. To say that a formula A is polarized is to say that it is regarded either as an *output* A_O (*positive polarization*) or as an *input* A_I (*negative polarization*).

(ii) The *trip translation* maps polarized atoms P_O, P_I to atomic formulas of **IMALL** (also denoted by P_O, P_I, respectively). For polarized constants and polarized complex formulas the trip translation is defined inductively according to the table in Table 3.

(iii) A *pointed sequent* $\vdash \Gamma, \mathbf{A}$ is a sequent with a selected formula occurrence, i.e., with a *switch* choosing one of its formulas. The chosen formula will be written in **boldface**.

$(P^{\perp})_O$	$= P_I$,	for P atomic;	$(P^{\perp})_I$	$= P_O$;	
$\mathbf{1}_O$	$= \mathbf{1}$	$\mathbf{1}_I = \top$	$\bot_I$	$= \mathbf{1}$	$\bot_O = \top$;
$(A \otimes B)_O$	$= A_O \otimes B_O$,		$(A \wp B)_I$	$= A_I \otimes B_I$;	
$(A \otimes B)_I$	$= (A_O \multimap B_I)\ \&$ $(B_O \multimap A_I)$;		$(A \wp B)_O$	$= (A_I \multimap B_O)\ \&$ $(B_I \multimap A_O)$;	
$\top_O$	$= \top$	$\top_I = \mathbf{0}$;	$\mathbf{0}_O$	$= \mathbf{0}$	$\mathbf{0}_I = \top$;
$(A \& B)_O$	$= A_O \& B_O$,		$(A \& B)_I$	$= A_I \oplus B_I$;	
$(A \oplus B)_O$	$= A_O \oplus B_O$,		$(A \oplus B)_I$	$= A_I \& B_I$	

Table 3: Functorial trip translation, the propositions.

(iv) The polarization of a sequent $\vdash \Gamma, \mathbf{A}$ is defined as follows: the selected formula $\mathbf{A}$ is regarded as an *output* A_O, all other formulas C in Γ are regarded as *inputs* C_I; we write Γ_I to indicate this fact.

(v) The *trip translation* maps a pointed sequent $S = \vdash \Gamma, \mathbf{A}$ of **CMALL** to a sequent $S' = \Gamma_I \vdash A_O$ of **IMALL**, where Γ_I and A_O are translated as in (ii). Notice that since a sequent $\vdash \Delta$ may be regarded as the *par* of all its formulas, by Table 3 the translation $\vdash (\wp(\Delta))_O$ of the pointed sequent $\vdash \wp(\mathbf{\Delta})$ is the product (*with*) of the translations of all the *"pointings"* of Δ.

(vi) The *trip translation* maps sequent derivations of **CMALL** to sequent derivations in **IMALL** according to the definition in Table 4.

Proposition 2 *For any formula A of* **CMALL**, *the translation satisfies $(A^{\perp})_O = A_I$ and $(A^{\perp})_I = A_O$. Therefore the translation of the* cut-*rule is well-defined.*

Proof. By induction on the logical complexity of A. ∎

Theorem 3 *(i) The trip translation maps a* **CMALL** *proposition A to a pair of* **IMALL** *propositions (A_O, A_I) and a* **CMALL** *derivations π of $\vdash \Gamma$ to an* **IMALL** *derivation $\tilde{\pi}$ of $\vdash (\wp(\Gamma))_O$. If $\vdash \Gamma', \mathbf{A}$ is a pointing of $\vdash \Gamma$, then a branch of $\tilde{(\pi)}$ contains a derivation π' of $\Gamma'_I \vdash A_O$.*

(ii) The trip translation is functorial in the sense it preserves the cut rule. Namely, given a derivation $(\pi_1|\pi_2)$ ending with a cut

$$\frac{\begin{array}{cc}\pi_1 & \pi_2 \\ \vdash \Gamma, A & \vdash A^{\perp}, \Delta\end{array}}{\vdash \Gamma, \Delta}$$

and a pointing $\vdash \Xi, \mathbf{C}$ of $\vdash \Gamma, \Delta$ there is a unique pair of pointings of $\vdash \Gamma, A$ and of $\vdash A^{\perp}, \Delta$ such that

$$(\pi_1|\pi_2)' = (\pi_1'|\pi_2')$$

(iii) Moreover, if S is translated to S' and S reduces to S_0 by cut-elimination, then there exists S_0' which is the translation of S_0 and S' reduces to S_0'.

$$\vdash P^{\perp}, \mathbf{P} \quad \Longrightarrow \quad P_O \vdash P_O \qquad\qquad \vdash \mathbf{P}^{\perp}, P \quad \Longrightarrow \quad P_I \vdash P_I$$

$$cut : \frac{\vdash \Gamma, \mathbf{A} \qquad \vdash A^{\perp}, \Delta, \mathbf{C}}{\vdash \Gamma, \Delta, \mathbf{C}} \Longrightarrow cut : \frac{\Gamma_I \vdash A_O \qquad A_I^{\perp}, \Delta_I \vdash C_O}{\Gamma_I, \Delta_I \vdash C_O}$$

$$\otimes_O : \frac{\vdash \Gamma, \mathbf{A} \qquad \vdash \Delta, \mathbf{B}}{\vdash \Gamma, \Delta, \mathbf{A} \otimes \mathbf{B}} \Longrightarrow \otimes - R : \frac{\Gamma_I \vdash A_O \qquad \Delta_I \vdash B_O}{\Gamma_I, \Delta_I \vdash A_O \otimes B_O}$$

$$\otimes_I : \frac{\vdash \Gamma, \mathbf{A} \qquad \vdash B, \Delta, \mathbf{C}}{\vdash \Gamma, A \otimes B, \Delta, \mathbf{C}} \Longrightarrow \multimap - L : \frac{\dfrac{\Gamma_I \vdash A_O \qquad B_I, \Delta_I \vdash C_O}{\Gamma_I, A_O \multimap B_I, \Delta_I \vdash C_O}}{\Gamma_I, (A_O \multimap B_I) \& (B_O \multimap A_I), \Delta_I \vdash C_O}$$

$$\wp_O : \frac{\vdash \Gamma, A, \mathbf{B}}{\vdash \Gamma, \mathbf{A \wp B}} \text{and} \frac{\vdash \Gamma, \mathbf{A}, B}{\vdash \Gamma, \mathbf{A \wp B}} \Longrightarrow \multimap \text{-}R : \frac{\dfrac{\Gamma_I, A_I \vdash B_O}{\Gamma_I \vdash A_I \multimap B_O} \qquad \dfrac{\Gamma_I, B_I \vdash A_O}{\Gamma_I \vdash B_I \multimap A_O}}{\Gamma_I \vdash (A_I \multimap B_O) \& (B_I \multimap A_O)}$$

$$\wp_I : \frac{\vdash A, B, \Gamma, \mathbf{C}}{\vdash A \wp B, \Gamma, \mathbf{C}} \quad \Longrightarrow \quad \otimes - L : \frac{A_I, B_I, \Gamma_I \vdash C_O}{A_I \otimes B_I, \Gamma_I \vdash C_O}$$

$$\vdash \mathbf{1} \quad \Longrightarrow \quad \vdash \mathbf{1}$$

$$\perp_I : \frac{\vdash \Gamma, \mathbf{A}}{\vdash \perp, \Gamma, \mathbf{A}} \quad \Longrightarrow \quad \mathbf{1} - L : \frac{\Gamma_I \vdash A_O}{\mathbf{1}, \Gamma_I \vdash A_O}$$

$$\perp_O : \frac{\vdash \Gamma}{\vdash \Gamma, \boldsymbol{\perp}} \quad \Longrightarrow \quad \Gamma_I \vdash \top$$

$$\vdash \Gamma, \mathbf{T} \quad \Longrightarrow \quad \Gamma_I \vdash \top \qquad\qquad \vdash \top, \Gamma, \mathbf{C} \quad \Longrightarrow \quad \mathbf{0}, \Gamma_I \vdash C_O$$

$$\&_O : \frac{\vdash \Gamma, \mathbf{A} \qquad \vdash \Gamma, \mathbf{B}}{\vdash \Gamma, \mathbf{A \& B}} \Longrightarrow \& - R : \frac{\Gamma_I \vdash A_O \qquad \Gamma_I \vdash B_O}{\Gamma_I, \Delta_I \vdash A_O \& B_O}$$

$$\&_I : \frac{\vdash A, \Gamma, \mathbf{C} \qquad \vdash B, \Gamma, \mathbf{C}}{\vdash A \& B, \Gamma, \mathbf{C}} \Longrightarrow \oplus - L : \frac{A_I, \Gamma_I \vdash C_O \qquad B_I, \Gamma_I \vdash C_O}{A_I \oplus B_I, \Gamma_I \vdash C_O}$$

$$\oplus_O : \frac{\vdash \Gamma, \mathbf{A_i}}{\vdash \Gamma, \mathbf{A_0} \oplus \mathbf{A_1}} \Longrightarrow \oplus - R : \frac{\Gamma_I \vdash (A_i)_O}{\Gamma_I \vdash (A_0)_O \oplus (A_1)_O}$$

$$\oplus_I : \frac{\vdash \Gamma, A_i, \mathbf{C}}{\vdash \Gamma, A_0 \oplus A_1, \mathbf{C}} \Longrightarrow \& - L : \frac{\Gamma_I, (A_i)_I \vdash C_O}{\Gamma_I, (A_0)_I \& (A_1)_I \vdash C_O}$$

Table 4: Functorial trip translation, the proofs.

Remark 4 (i) The concise notation of part *(ii)* of Theorem 3 can be spelt out as follows. Here $(\pi_1|\pi_2)'$ is the trip translation of $(\pi_1|\pi_2)$ restricted to the given pointing $\vdash \Xi, \mathbf{C}$ of $\vdash \Gamma, \Delta$. Depending on whether C is in Γ or in Δ we have either the pair of pointings $\{\vdash A, \Gamma', \mathbf{C}, \quad \vdash \Delta, \mathbf{A}^\perp\}$ or the pair $\{\vdash \Gamma, \mathbf{A}, \quad \vdash A^\perp, \Delta', \mathbf{C}\}$. Moreover, π_1' and π_2' are the trip translations of π_1 and π_2 restricted to the appropriate pointings, i.e., either π_1' is a derivation of $A_I, \Gamma_I' \vdash C_O$ and π_2' a derivation of $\Delta_I \vdash A_I$, or π_1' is a derivation of $\Gamma_I \vdash A_O$ and π_2' a derivation of $A_O, \Delta' \vdash C$. The theorem says that $(\pi_1|\pi_2)'$ is the same as the derivation $(\pi_1'|\pi_2')$ which is obtained by applying *cut* to π_1' and π_2'.

(ii) Notice that there is no anomaly in the $\perp_O$ rule. One pointing of $\vdash \Gamma, \perp$ selects $\perp$ and $\perp_O = \top$, thus one branch in the derivation $\tilde{\pi}$ of $\vdash (\wp(\Gamma, \perp))_O$ is given by the *axiom* $\Gamma_I \vdash \top$. Such a branch adds no information in addition to that contained in the derivation of $\vdash (\wp(\Gamma))_O$ which is also contained in the *other* branches of $\tilde{\pi}$.

Fact 5 *(i) To the trip translation there corresponds an obvious map in the opposite direction, let us write it as* $(\ )^c$*; in proof-theoretic terms it amounts to regarding* **IMALL** *as a fragment of* **CMALL**, *namely:*

- *writing formulas in* "negation normal form", *using De Morgan laws as in Table 0 and then rewriting* $(P_O)^c = (P_I^\perp)^c = P$, $(P_I)^c = (P_O^\perp)^c = P^\perp$ *for propositional letters* P;
- *writing proofs in the sequent calculus with* "right-hand sequents" *only.*

(ii) Unfortunately, in general it is not true that $((A)_O)^c = A$, *e.g., let* A *be* $\perp$.

3. Chu's construction

We follow the categorical semantics for **IMLL** in [5]:

Theorem 6 *Let* A *be the free* $*$*-autonomous category on a set of objects* $\{P, P', \ldots\}$ *and let* C *be the symmetric monoidal closed category with products, free on the set of objects* $\{P_O, P_I, P_O', P_I', \ldots\}$ *(a pair* P_O, P_I *for each atom* P *of* A*).*

(i) We can give $C \times C^{op}$ *the structure of a* $*$*-autonomous category thus:*

$$(X_O, X_I) \otimes (Y_O, Y_I) =_{df} (X_O \otimes Y_O, (X_O \multimap Y_I) \times (Y_O \multimap X_I))$$

$$\text{with unit} \quad (\mathbf{1}, \top) \quad \text{and involution} \quad (X_O, X_I)^\perp =_{df} (X_I, X_O)$$

where $\mathbf{1}$ *is the unit of* $\otimes$ *in* C *and* $\top$ *the terminal object of* C.

Therefore there is a functor F *from* A *to* $(C \times C^{op})$ *sending an object* P *to* (P_O, P_I). *If* $\pi : I \to \wp(\Gamma)$ *is a morphism of* A *represented as a proof-net*

$\mathcal{R}$ with conclusions Γ, then the morphism $(\mathbf{1}, \top) \to (\wp(\Gamma)_O, \wp(\Gamma)_I)$ is given by all the Girard's trips *on $\mathcal{R}$, in the precise sense spelt out in Proposition 15 below.*

(ii) If in addition C has also coproducts, then $(C \times C^{op})$ has also products:

$$(X_O, X_I)\&(Y_O, Y_I) =_{df} (X_O\&Y_O, X_I \oplus Y_I).$$

The functor F preserves also the structure of the products.

PROOF. [Sketch] In $C \times C^{op}$ define $X \wp Y =_{df} (X^\perp \otimes Y^\perp)^\perp$ and $X \multimap Y =_{df} X^\perp \wp Y$ and show that $X \multimap Y$ gives the closed structure for $\otimes$. By the definitions

$$(V,Y) \multimap (W,Z) = ((V \multimap W) \times (Z \multimap Y), V \otimes Z)$$

We exhibit a natural bijection of hom-sets

$$(U,X) \otimes (V,Y) \to (W,Z) \qquad \sim \qquad (U,X) \to (V,Y) \multimap (W,Z)$$

Indeed a map in $(U,X) \otimes (V,Y) \to (W,Z)$ consists of a map $a : U \otimes V \to W$ and of a pair of maps $b : Z \to U \multimap Y$, $c : Z \to V \multimap X$; a map in $(U,X) \to (V,Y) \multimap (W,Z)$ consists of a pair of maps $a' : U \to V \multimap W$, $b' : U \to Z \multimap Y$ and of a map $c' : V \otimes Z \to X$. But $a \mapsto a'$ and $c \mapsto c'$ by the natural bijections given by the symmetric monoidal closed structure of C. Similarly, there is a natural bijection η between $Z \to U \multimap Y$ and $Z \otimes U \to Y$ and also a natural bijection η' between $Z \otimes U \to Y$ and $U \to Z \multimap Y$; composing η' and η we get $b \mapsto b'$. Therefore $\multimap$ is a right adjoint for $\otimes$ in $C \times C^{op}$.

To see how the action of the functor is given by Girard's trips in the case of **CMLL** *without units*, the proof in [8], pp. 37-44, (briefly reviewed in the following section) shows that a Danos–Regnier graph on a cut-free proof-net $\mathcal{R}$ with conclusions Γ, $\mathbf{C}$ (with $\mathbf{C}$ the selected conclusion) determines an orientation (polarization) δ of the formulas in the proof-net and a *reduced translation* of $\mathcal{R}$ into a cut-free derivation π^δ of $\Gamma_I^\delta \vdash C_O^\delta$ *in* **IMLL**$^-$. (The same result holds if we start with a Girard trip, as indicated below; indeed each Girard trip uniquely determines a Danos–Regnier graph.)

Now it is easy to see that a derivation π^δ can also be obtained as follows: consider the **IMALL** derivation $F(\mathcal{R})$ given by the trip translation of (a sequentialization of) $\mathcal{R}$ and remove all *&-left* and *&-right* inferences, modifying the formulas in the derivation accordingly. The result is a family $\mathcal{F}$ of **IMLL**$^-$ derivations, a pair of derivations for each *&-right* application in $F(\mathcal{R})$. Every derivation π^δ determined by some Danos–Regnier switching is equivalent *modulo permutations of inferences* with one derivation in $\mathcal{F}$ and, moreover, every derivation in $\mathcal{F}$ is equivalent to a derivation π^δ for some δ induced by a Danos–Regnier switching. (In the terminology of Proposition 15 below we have: $\mathcal{F} = \{\sigma_{s_1\mathcal{R}}[F(\mathcal{R})], \ldots, \sigma_{s_k\mathcal{R}}[F(\mathcal{R})]\}$ where $s_1, \ldots, s_k$ are all the Danos–Regnier switchings of $\mathcal{R}$.)

A simpler formulation of the above result could be given by mapping proof-nets for *classical* $\mathbf{MLL}^-$ to F. Lamarche and A. Tan's proof-nets for *intuitionistic* $\mathbf{MLL}^-$ (see [20, 26]). For an extension to **CMLL** *with units* and to **CMALL**, see Remarks 7 and 17 below. ∎

Remark 7 (i) The functor F is *not full.* For instance, $F(\bot) = (\top, \mathbf{1})$, and there is a morphism $F(\mathbf{1}) \to F(\bot)$ in $C \times C^{op}$, namely, the map $t \times t'$: $(\mathbf{1}, \top) \to (\top, \mathbf{1})$, where $t' : \top \to \mathbf{1} =_{df} t : \mathbf{1} \to \top$; however, the free category A does not have a morphism $\mathbf{1} \to \bot$, i.e., $\bot$ is not provable in **MLL**. The system **CLL** + $\bot = \mathbf{1}$ has been studied, e.g., in [14], and perhaps this extension of linear logic deserves further consideration, but it is not the answer to our concern here. The task is rather to characterize a subcategory of $C \times C^{op}$ for which the functor F is full.

(ii) Is the functor F is faithful? This question raises the issue of the identity of proofs in linear logic. Proof-nets provide a solution to this problem for the multiplicative fragment without units, because in this fragment proof-nets represent sequent derivations up to permutations of inferences and the Church-Rosser property holds strictly (see the next section). Notice that in [5] cut reductions in **IMALL** correspond to $\beta\eta$ equality of the terms which express the maps in the free symmetric monoidal closed category and commutative conversions correspond to natural isomorphisms between them. If in $C \times C^{op}$ we consider terms up to $\beta\eta$ equivalence, then faithfulness is obtained by making the corresponding assumption about A, i.e., by stipulating that the morphisms of A are represented by proof-nets *up to cut-elimination and η-equivalence.* For an extension to this result to **MLL** with units, see the note in Section 4.4. However, in the case of proof-nets for **MALL** the notion of identity of proofs is not well-understood, and thus the issue of faithfulness must be left to further research.

4. Proof-nets, trips and translations

In this section we summarize some results about proof-nets that illustrate the geometric connections between Girard's trips and translations into fragments of intuitionistic linear logic.

4.1 Proof-nets: basic definitions

Definition 8 (i) Proof-structures for **CMALL** are directed graphs with at least one external point and where each vertex is typed and has the form indicated in Figure 2.1. The dashed line in a $\bot$-vertex is called an *attachment.* When a

$\perp$-link is attached to an edge A, we may think of the attachment as resulting from an axiom $\perp$, $\mathbf{1}$ linked to a *times* link $\mathbf{1} \otimes A$, i.e., as an application of the isomorphism $A \sim A \otimes \mathbf{1}$.

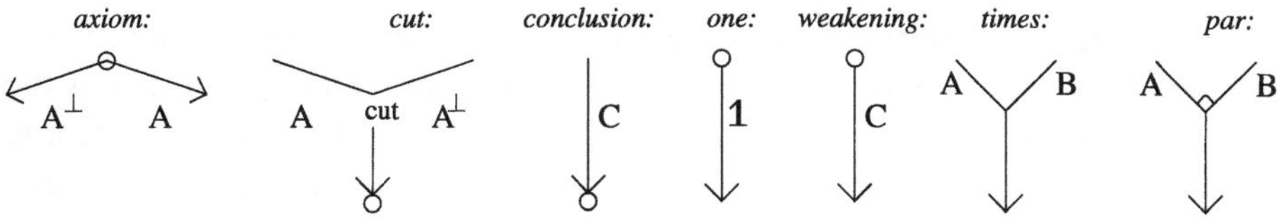

Figure 2.1. Links

(ii) On proof-structures for **CMLL** *Girard's trips* are defined according to the drawings in Figure 2.2; the choice of the form of a trip at a *par* link is called a *left* or *right switch*. *Proof-nets* for **MLL** are proof-structures satisfying Girard's *no-short-trip* condition, namely, that for every switching of the *par* links and every conclusion C, the trip starting at C returns to C after visiting each edge precisely twice (cf. [16])

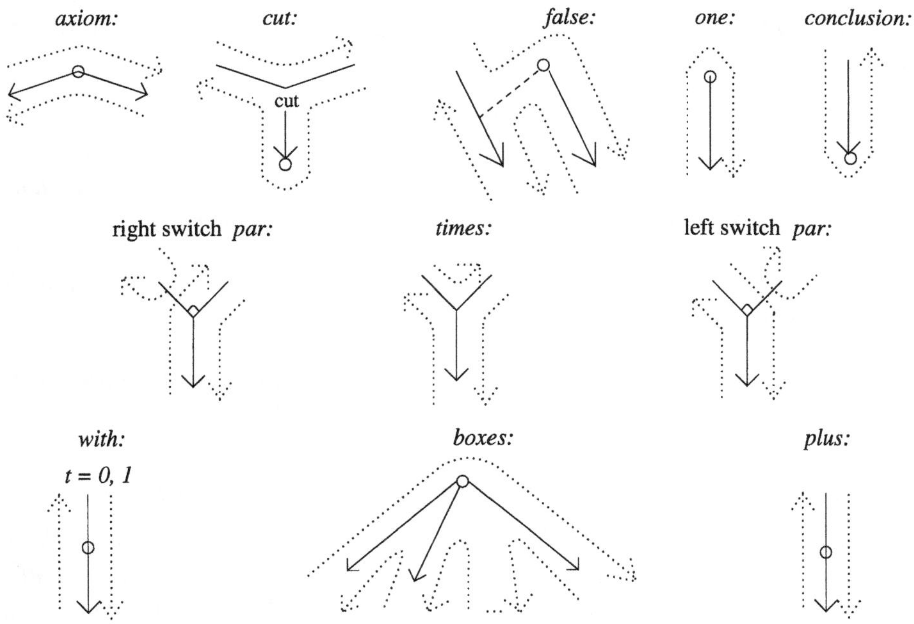

Figure 2.2. Girard's trips

(iii) Perhaps more familiar is the equivalent characterization of proof-nets for $\mathbf{MLL}^-$ in terms of *Danos–Regnier graphs* [10] (which is readily extended to **MLL** *with units* using attachments as above). Given a proof-net $\mathcal{R}$ for $\mathbf{MLL}^-$, *a switching s of $\mathcal{R}$ in the sense of Danos–Regnier* is an assigment to each *par* link in $\mathcal{R}$ of a choice of the *left* or *right* premise and, moreover, a "pointing" of the conclusions Γ of $\mathcal{R}$. Given a proof-net $\mathcal{R}$ and a switching s of it, the *Danos–Regnier* graph $s\mathcal{R}$ (determined by s) is the graph resulting from $\mathcal{R}$ by removing the edge which enters a *par* link from the premise which is *not* selected by s. The standard correctness condition for **MLL**$^-$ proof-nets is the following: a proof-structure $\mathcal{R}$ for **MLL**$^-$ is a proof-net if for every switching s the Danos–Regnier graph $s\mathcal{R}$ is acyclic and connected (an undirected tree).

In the case of **CMALL** we have *true boxes* which behave like $n+1$-ary axioms. Moreover, a *boolean valued polynomial* is associated with each edge, a distinct *boolean variable* x is associated with each *with* link, x and $1-x$ being added to the polynomials associated with the left and right premise of the *with* link in question. The polynomial of the conclusion of a link is the sum of the polynomials of the premises. All the conclusions of the proof-net must have 1 as associated polynomial. A proof structure is *sliced* by substituting arbitrary boolean values for the variables and erasing the edges whose polynomial evaluates to 0; in a sliced proof-structure additive links are all unary. (For a more precise definition, of additive proof-nets, see Girard [18].) All said, a **CMALL** proof-structure is a proof-net if for every evaluation of the polynomials, the resulting slice has no short trip.

What matters here is that the following theorem can be proved:

Girard's Theorem. *There exists a 'context-forgetting' map $(\ )^-$ from sequent derivations in* **MALL** *to proof-nets with the following properties:*

(a) *Let* d *be a sequent derivation of* $\vdash \Gamma$*; then* $(\mathrm{d})^-$ *is a proof-net with conclusion* Γ;

(b) (Sequentialization) *If $\mathcal{R}$ is a proof-net with conclusion Γ, then there is a sequent calculus derivation* d *of* $\vdash \Gamma$ *such that* $\mathcal{R} = (\mathrm{d})^-$.

About proof-nets for **MLL**$^-$ more can be proved (see [9], Theorem 2):

Permutability of Inferences Theorem. *Let* d *and* d$'$ *be a pair of derivations of the same sequent* $\vdash \Gamma$ *in* **MLL**$^-$*. Then* $(d)^- = (d')^-$ *if and only if there exists a sequence of derivations* $d = d_1, d_2, \ldots, d_n = d'$ *such that* d_i *and* d_{i+1} *differ only for a permutation of consecutive inferences.*

Remark 9 A corollary of the latter theorem is that the syntax of proof-nets for **MLL**$^-$ solves the problem of identity of proofs in this fragment; for this reason proof-nets have found applications to coherence problems in category theory.

4.2 Trips and linear λ terms

A trip in the sense of Girard induces the structure of a λ-term in any suitable graph with a selected external point (*pointed graph*):

Theorem 10 (J. van de Wiele) *Every* connected pointed graph *with vertices of incidence* 1 *and* 3 *corresponds to a linear λ-term (and vice versa). The correspondence is established in linear time by a* trip *starting with the selected external point.*

We perform a trip in the style of Girard according to the figure below; during the trip we determine whether a vertex is to be regarded as (1) a variable, (2) an application or (3) a λ-abstraction. Case (2) occurs when during the *second visit* to a vertex of incidence 3 the trip enters the vertex through *the same edge* from which it had exited after the first visit; case (3) occurs when the second visit is through the *other* edge; case (1) is that of an external vertex different from the selected one.

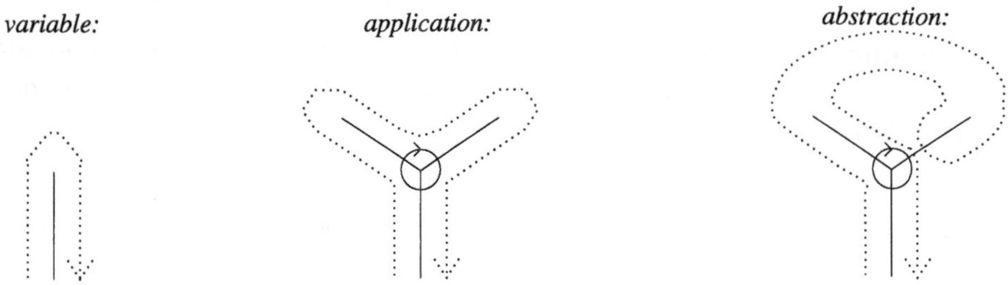

Figure 2.3. Linear λ-terms

The proof is by induction on the number of vertices. Notice that if we remove the first vertex of incidence 3 encountered during the trip then the resulting graph is disconnected in case (2) but remains connected in case (3). Different variables are assigned to different external points in case (1). Since linear λ-terms are always typable, van de Wiele's result also shows that every such connected pointed graph corresponds to a proof in the implicative fragment of intuitionistic linear logic.

4.3 Reduced translations from $\mathbf{CMLL}^-$ to $\mathbf{IMLL}^-$

Essentially the same technique applied to a *proof-net* for $\mathbf{CMLL}^-$ yields a translation of the proof-net into a derivation in the *implication and tensor* fragment of intuitionistic multiplicative linear logic. These translations shall be called *reduced trip translations* or simply *reduced translations*. Trips always have a *starting point* in a conclusion and the *order* of the passages across a link matters. To state our result we need a definition.

Definition 11 Given a proof-net $\mathcal{R}$ for **CMLL**$^-$ and a selected conclusion A, we say that a Girard trip starting from A is *covariant* on an edge if the *second passage* of the trip is in the same direction as the edge; otherwise, the trip is *contravariant* on the edge. Now a trip starting from A induces an *input-output orientation* $\delta : \mathcal{R} \to \{I, O\}$ thus: an edge X is an *output* X_O or an *input* X_I depending on whether the trip is covariant or contravariant on it.

Theorem 12 (Bellin and van de Wiele) *(i) Every* Girard's trip *on a* cut-free *proof-net for* **CMLL**$^-$ *starting from a selected conclusion corresponds to a sequent calculus derivation in* **IMLL**$^-$.
(ii) Conversely, every sequent derivation in **IMLL**$^-$ *corresponds to a trip on a proof-net.*

The following proposition follows almost immediately from the definition of orientation and the basic properties of trips.

Proposition 13 *Every orientation makes the selected conclusion an output, all other conclusions are inputs. Every link is oriented in one of the admissible ways indicated in Figure 2.4.*

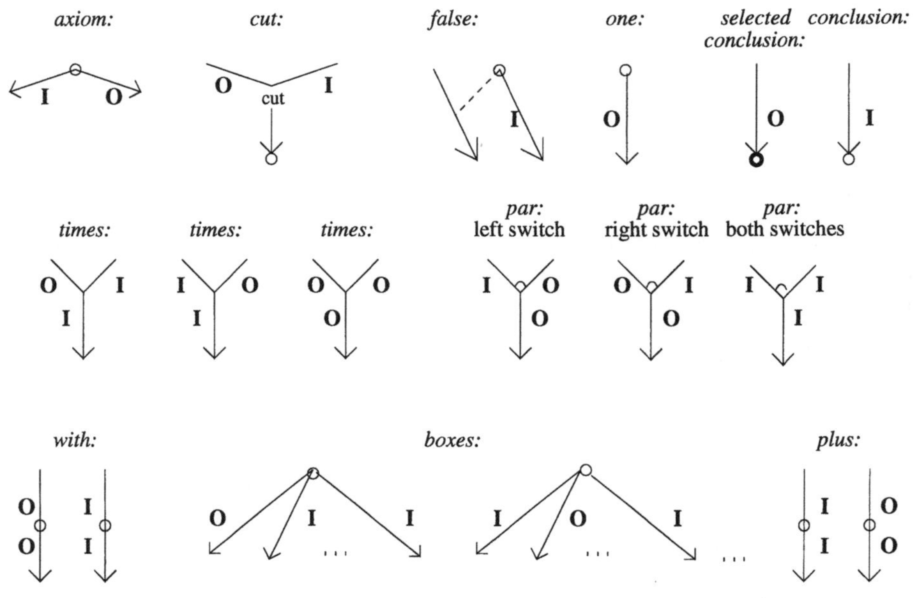

Figure 2.4. Admissible orientations.

Given an orientation $\delta : S \to \{I, O\}$, the *formulas* in the proof-net are translated as follows:

$$\begin{array}{rcll@{\qquad}rcl}
(P^{\perp})_O & = & P_I, & \text{for } P \text{ atomic} & (P^{\perp})_I & = & P_O; \\
(A \otimes B)_O & = & A_O \otimes B_O, & & (A \wp B)_I & = & A_I \otimes B_I; \\
(A \otimes B)_I & = & A_O \multimap B_I, & \text{if } \delta(A) = O, & & & \\
 & = & B_O \multimap A_I, & \text{if } \delta(B) = O. & & & \\
(A \wp B)_O & = & A_I \multimap B_O, & \text{if } \delta(B) = O, & & & (\textit{right switch}) \\
 & = & B_I \multimap A_O, & \text{if } \delta(A) = O. & & & (\textit{left switch})
\end{array}$$

For further details, see [8], pp. 37–44. ∎

Remark 14 (i) Theorem 12 could be stated in terms of *Danos–Regnier graphs*, as it was done in [8]; notice that every Girard's trip determines a unique Danos–Regnier graph [10]. Girard's trips allow us to give a more concise definition of orientation, but Danos and Regnier's characterization yields the refinements in Proposition 15 below.

(ii) Reduced translations of formulas are not *functorial*, in the sense that they depend not only on the given **CMLL** formula, but also on a trip on a given proof; i.e., the map on objects depends also on morphisms. Reduced translations of proofs are not functorial, in the sense that they may not be compatible with *cut*. Indeed, the orientation induced by a switching may be *computationally inconsistent*: e.g., consider the orientation on the cut formulas $A_O \otimes B_I$ and $A_O^{\perp} \wp B_I^{\perp}$ induced by a *left switch* on the *par* link.

(iii) The above result does not extend to full **CMLL**: let $\perp$ be the selected conclusion in the cut-free proof-net with conclusion P, $P^{\perp}$, $\perp$.

(iv) The above result does not extend to **CMALL**: consider the cut-free proof-net with conclusions $A \& B$, $A^{\perp} \otimes B^{\perp}$, $A \oplus B$.

However, reduced translations suffice to characterize the action of the functor F of theorem 6 on a derivation in the fragment $\mathbf{MLL}^-$ in the following sense. Let s be a switching *in the sense of Danos–Regnier* on a proof-net R for $\mathbf{MLL}^-$. Let $\mathbf{IMALL}^-$ be **IMALL** without *plus*. Consider the set of maps $\sigma : \mathbf{IMALL}^- \to \mathbf{IMLL}^-$ with the following properties:
(a) σ acts on the propositions as follows:

$$(i) \qquad \sigma[(A_I \multimap B_O) \& (B_I - o A_O)] = \sigma[A_I - o B_O] \quad \text{or} \quad \sigma[B_I \multimap A_O]$$

$$(ii) \qquad \sigma[(A_O \multimap B_I) \& (B_O \multimap A_I)] = \sigma[A_O \multimap B_I] \quad \text{or} \quad \sigma[B_O \multimap A_I]$$

(b) σ acts on $\mathbf{IMALL}^-$ derivations π by removing all &-right and &-left inferences

Clearly, given such a σ defined arbitrarily on propositions we do not know whether there is a proof π of $\vdash (\wp(\Gamma))_O$ such that $\sigma(\pi)$ is a proof of $\vdash \sigma[(\wp(\Gamma))_O]$. Moreover, given any derivation π in $\mathbf{IMALL}^-$, $\sigma[\pi]$ needs not be a derivation in **IMLL**$^-$ (for instance, if the map on derivation removes a &-left inference with active formula $\sigma[A_O \multimap B_I]$ and the map on propositions yields $\sigma[B_O \multimap A_I]$). However, Danos–Regnier switchings allow us to define well-behaved maps σ as functions of a proof-net and of a switchings.

Proposition 15 *Let $\mathcal{R}$ be a cut-free proof-net for* **MLL**$^-$ *with conclusions* Γ *and let s be a Danos–Regnier switching of $\mathcal{R}$. Let π be the derivation of* $\vdash (\wp(\Gamma))_O$ *in* **IMALL**$^-$ *given by the Chu functor. Then there exists a map $\sigma_{s\mathcal{R}}$ such that $\sigma_{s\mathcal{R}}[\pi]$ is a derivation of $\sigma_{s\mathcal{R}}[(\wp(\Gamma))_O]$ in* **IMLL**$^-$. *Moreover,*

$$\sigma_{s\mathcal{R}}[(A_I \multimap B_O)\&(B_I \multimap A_O)] = \sigma_{s\mathcal{R}}[A_I \multimap B_O] \quad \textit{iff} \quad s(A\wp B) = \text{right}.$$

Remark 16 (i) It can be shown that and the map $\sigma_{s\mathcal{R}}$ depends only on the values of s on the *par* links which in a Girard trip are reached from below, i.e., the *par* links whose conclusion is oriented as an "output" and which correspond to formula-occurrences of type (i) in π.

(ii) Let π be a reduced translation in **IMLL**$^-$ of a derivation d in **MLL**$^-$ and let $(\pi)^c$ be its translation back to **MLL**$^-$ according to Fact 5. Then d and $(\pi)^c$ are equal (possibly modulo permutations of inferences).

4.4 Chu's construction in MLL with units

As indicated in the Preface, one of the original motivations for this paper was to find a *functorial* definition of the trip translation, in view of a possible extension to the whole system **CLL** and given the fact that the *reduced trip translation* does not extend beyond **MLL**$^-$. We have now a functorial translation and a satisfactory explanation of its meaning in terms of Chu's construction. But what about extensions to **MLL** with units and **CMALL**?

As indicated in Remark 7, the problem with *fullness* may require a basic reformulation of the construction, e.g., the definition of a subcategory of $C \times C^{op}$ for which the functor is full. Moreover, *faithfulness* for **CMALL** requires a reconsideration of additive proof-nets. On the other hand, the proof of *faithfulness* for **MLL** *with units* seems at hand, thanks to A. Tan's thesis [26], although we cannot spell out the details here.

Remark 17 (MLL *with units*) (i) We do not know how to define proof-nets for **MLL** *with units* so as to extend the theorem on Permutability of Inferences to **MLL** with units, thus it is no longer true that the the proof-net representation solves the problem of identity of proofs in **MLL** with units (cf. Remark 9). Any permutation of the *nil* rule with other inferences in a derivation d results in a *rewiring* of $(d)^-$, i.e., in a modification of the 'attachment' of the corresponding $\perp$-link. (Of course, this problem would not occur in the system **MLL**

with the axiom $\bot = \mathbf{1}$.) Therefore the obvious way to characterize the identity of proofs for **MLL** *with units* is to give explicit equations between proof-nets.

(ii) A similar problem occurs for the representation of proofs in **IMLL**: in fact the systems of Natural Deduction or Sequent Calculi with term-assignments for **ILL** in [4, 5, 6] are given together with an axiomatic characterization of the identity of proofs in the form of an *equational theory* of terms. Similarly, Lamarche's proof-nets for **ILL** [20] require a theory of *rewiring* already in the case of **MLL** *with units*. This work has been done in Chapter 6 of A. Tan's thesis [26]: after a careful definition of the correspondence between sequent calculus with term assignments and proof-nets for **IMLL**, the process of *rewiring* is defined so that it does preserves the correctness criterion, it does not affect the (equivalence classes of) terms which the proof-net interprets, it is strongly normalizing and confluent and, moreover, the process of cut-elimination, *incorporating unit rewirings*, remains strongly normalizing and confluent.

(iii) Rewiring in **CMLL** proof-nets is also defined in such a way that it preserves the correctness criterion. Since classical proof-nets may have several conclusions, it is not obvious how to define a canonical element in each equivalence class of proof-nets.

(iv) Let us consider the again action of the functor F : **CMLL** $\to$ **IMALL** *on the units*. If π is a proof of $\vdash \Gamma, \bot$, then $F(\pi) = \Gamma_I \vdash \top$, an *axiom*, i.e., the proof π is *erased*. It follows that a single *reduced translation*, regarded as a **CMLL** proof, no longer contains the same information as the original proof (cf. Remark 16.(ii)).

(v) Considering the definition of reduced translations from Danos–Regnier graphs, we may follow the hint of Definition 8.(i) and define the orientation as if an attachment resulted from an *axiom* $\bot, \mathbf{1}$ where the edge $\mathbf{1}$ enters a *times* link with conclusion $\mathbf{1} \otimes A \sim A$. The definition extends without problems when the orientation of $\bot$ is $\bot_I$: indeed the corresponding **IMLL** proof-net has a $\mathbf{1}$-link with a suitable attachment. If the orientation is $\bot_O$ we may no longer have a coherent orientation for the edge A (in the case where we would give the different orientations A_O, $(\mathbf{1} \otimes A)_I$): but this is still fine, because $F(\bot_O) = \top$ and we may certainly take an axiom $\top, A_I, A_O$ in the reduced translation.

(vi) Finally, let us consider the effect of rewiring of **CMLL** proof-nets $\mathcal{R} \mapsto \mathcal{R}'$ on a reduced trip translation of $\mathcal{R}$. Given a $\bot$ link in $\mathcal{R}$, the rewiring in question may

(1) preserve the orientation $\bot_I$ or
(2) preserve the orientation $\bot_O$ or
(3) change an orientation $\bot_I$ into $\bot_O$ or
(4) vice versa.

In case (1) the effect of the rewiring is either null or a rewiring in **IMLL** as described in [26]. In case (2) the effect is either null or a commutation of a $\top$-axiom, in accordance with standard equations between **IMALL**-proofs. Only cases (3) and (4) do reserve some surprises; e.g., in case (iii) the **IMLL** proof-net resulting from a switching $s\mathcal{R}'$ may be obtained from the **IMLL** proof-net corresponding to $s\mathcal{R}$ only through some complicated *"surgery"*.

References

[1] Samson Abramsky and Radha Jagadeesan. Games and Full Completeness for Multiplicative Linear Logic. *J. Symb. Logic* 59(2):543–574, 1994.

[2] M. Barr. $*$-Autonomous categories, *Lecture Notes in Mathematics* 752, Springer-Verlag, 1979, Berlin, Heidelberg, New York.

[3] M. Barr. $*$-Autonomous categories and linear logic, *Mathematical Structures in Computer Science* 1:159–178, 1991.

[4] N. Benton, G. Bierman, V. de Paiva, M. Hyland. A term calculus for intuitionistic linear logic. Springer LNCS 664, pp. 75–90, 1993.

[5] N. Benton, G. Bierman, V. de Paiva, M. Hyland. Linear λ-Calculus and Categorical Models Revisited, Preprint, Comp. Lab., Univ. of Cambridge.

[6] G. M. Bierman. What is a Categorical Model of Intuitionistic Linear Logic? In *Proceedings of the International Conference on Typed Lambda Calculi and Applications.* April 10-12, 1995. Edinburgh, Scotland. Springer LNCS.

[7] G. Bellin. *Mechanizing Proof Theory: Resource-Aware Logics and Proof-Transformations to Extract Implicit Information*, Phd Thesis, Stanford University. Available as: Report CST-80-91, June 1990, Dept. of Computer Science, Univ. of Edinburgh.

[8] G. Bellin and P. J. Scott. *Theor. Comp. Sci.* 135(1):11–65, 1994.

[9] G. Bellin and J. van de Wiele. Subnets of Proof-nets in $\mathbf{MLL}^-$, in *Advances in Linear Logic*, Girard, Lafont and Regnier eds., London Math. Soc. Lect. Note Series 222, Cambridge University Press, 1995, pp. 249–270.

[10] V. Danos and L. Regnier. The Structure of Multiplicatives, *Arch. Math. Logic* 28:181–203, 1989.

[11] H. Devarajan, D. Hughes, G. Plotkin and V. Pratt. Full Completeness of the multiplicative linear logic of Chu spaces. Porceedings of *LICS 1999.*

[12] Valeria C.V. de Paiva. *The Dialectica Categories.* PhD thesis. DPMMS, University of Cambridge, 1988. Available as Comp. Lab. Tech. Rep. 213, 1990.

[13] A. Fleury. *La règle d'échange.* Thèse de doctorat, 1996, Équipe de Logique, Université de Paris 7, Paris, France.

[14] A. Fleury and C. Retoré. The Mix Rule, *Mathematical Structures in Computer Science* 4:273–85, 1994.

[15] H. N. Gabow and R. E. Tarjan. A Linear-Time Algorithm for a Special Case of Disjoint Set Union. *Journal of Computer and System Science* 30:209–221, 1985.

[16] J-Y. Girard. Linear Logic, *Theoretical Computer Science* 50:1–102, 1987.

[17] J-Y. Girard. On the unity of logic. *Ann. Pure App. Log.* 59:201–217, 1993.

[18] J-Y. Girard. Proof-nets: the parallel syntax for proof-theory. In *Logic and Algebra*, New York, 1995. Marcel Dekker.

[19] M. Hyland and L. Ong. Fair games and full completeness for Multiplicative Linear Logic without the Mix rule. ftp-able at `theory.doc.ic.ac.uk` in `papers/Ong`, 1993.

[20] F. Lamarche. Proof Nets for Intuitionistic Linear Logic 1: Essential Nets. Preprint ftp-able from *Hypatia* 1994.

[21] F. Lamarche. Games Semantics for Full Propositional Logic. *Proceedings of LICS 1995.*

[22] R. Loader. *Models of Lambda Calculi and Linear Logic: Structural, Equational and Proof-Theoretic Characterisations*, PhD Thesis, St. Hugh's College, Oxford, UK, 1994.

[23] A. S. Murawski and C.-H. L. Ong. A Linear-time Algorithm for Verifying MLL Proof Nets via Lamarche's Essential Nets. Preprint, OUCL, Wolfson Building, Parks Road, Oxford OX1 3QD, UK.
`andrzej@comlab.ox.ac.uk`,
http://www.comlab.ox.ac.uk/oucl/people/luke.ong.html, 1999.

[24] A. Patterson. *Implicit Programming and the Logic of Constructible Duality*, PhD Thesis, University of Illinois at Urbana-Champaign, 1998. http://www.formal.stanford.edu/annap/www/abstracts.html#9

[25] V. Pratt. Chu spaces as a semantic bridge between linear logic and mathematics. Preprint ftp-able from
http://boole.stanford.edu/chuguide.html,
1998.

[26] A. Tan. *Full completeness for models of linear logic.* PhD Thesis, King's College, University of Cambridge, UK, October 1997.

Chapter 3

TWO PARADIGMS OF LOGICAL COMPUTATION IN AFFINE LOGIC?

Gianluigi Bellin*
Facoltà di Scienze
Università di Verona
bellin@sci.univr.it

Abstract We propose a notion of *symmetric reduction* for a system of proof-nets for *Multiplicative Affine Logic with Mix* (**MAL** + Mix) (namely, multiplicative linear logic with the mix-rule the unrestricted weakening-rule). We prove that such a reduction has the strong normalization and Church–Rosser properties. A notion of irrelevance in a proof-net is defined and the *possibility* of cancelling the irrelevant parts of a proof-net without erasing the entire net is taken as one of the *correctness conditions*; therefore purely *local* cut-reductions are given, minimizing cancellation and suggesting a paradigm of *"computation without garbage collection"*. Reconsidering Ketonen and Weyhrauch's decision procedure for affine logic [15, 4], the use of the mix-rule is related to the non-determinism of classical proof-theory. The question arises, whether these features of classical cut-elimination are really irreducible to the familiar paradigm of cut-elimination for intuitionistic and linear logic.

Keywords: affine logic, proof-nets

A Silvia Baraldini

*Paper submitted in 1999, revised in 2001. Research supported by EPSRC senior research fellowship on grant GL/L 33382. This research started during a visit to the University of Leeds in 1997: thanks to Stan Wainer, John Derrick and Michael Rathjen and Diane McMagnus for their hospitality. Thanks to Martin Hyland, Edmund Robinson and especially Arnaud Fleury for extremely useful discussions during the final revision.

R.J.G.B. de Queiroz (ed.), Logic for Concurrency and Synchronisation, 111–144.

1. Introduction

1. *Classical Multiplicative Affine Logic* is classical multiplicative linear logic with the unrestricted rule of *weakening*, but without the rule of *contraction.* Classical affine logic is a much simpler system than classical logic, but it provides similar challenges for *logical computation*, both in the sense of *proof-search* and of *proof normalization* (or *cut-elimination*). For instance, the problem of *confluence* of cut-elimination (the *Church–Rosser property*) is already present in affine logic, but here we do not have the problem of *non-termination.* Affine logic is also simpler than linear logic from the point of view of proof-search: e.g., propositional linear logic is undecidable, yet becomes decidable when the unrestricted rule of weakening is added. Provability in *constant-only multiplicative linear logic* is NP-complete, yet it is decidable in linear time for *constant-only multiplicative affine logic*, as it is shown below.

The tool we will use here, proof-nets for affine logic, is older than the notion of a proof-net for linear logic. In a 1984 paper [15], J. Ketonen and R. Weyhrauch presented a decision procedure for first-order affine logic (called then *direct logic*) which essentially consists in building cut-free proof-nets, using the unification algorithm to determine the axioms. The 1984 paper is sketchy and it has been corrected (see [3, 4], where the relation between the decision procedure and proof-nets for $\mathbf{MLL}^-$ are discussed), but it contains the main ideas exploited in the present paper, namely, the construction of proof-nets *free from irrelevance* through *basic chains.* Yet neither the 1984 paper nor its 1992 revisitation contained a treatment of cut-elimination.[1]

2. The problem of non-confluence for classical affine logic is non-trivial: the following well-known example (given in Lafont's Appendix to [14]) reminds us that the Church–Rosser property is non-deterministic under the familiar *asymmetric* cut-reductions.

Example 1

$$\dfrac{\dfrac{\begin{array}{c} d_1 \\ \vdots \\ \vdash \Gamma \end{array}}{\vdash \Gamma, A}\,w_1 \qquad \dfrac{\begin{array}{c} d_2 \\ \vdots \\ \vdash \Delta \end{array}}{\vdash \Delta, \neg A}\,w_2}{\vdash \Gamma, \Delta} \quad \text{reduces to} \quad \dfrac{\begin{array}{c} d_1 \\ \vdots \\ \vdash \Gamma \end{array}}{\dfrac{\textit{weakenings}}{\vdash \Gamma, \Delta}} \quad \text{or to} \quad \dfrac{\begin{array}{c} d_2 \\ \vdots \\ \vdash \Delta \end{array}}{\dfrac{\textit{weakenings}}{\vdash \Gamma, \Delta}}$$

Asymmetric reductions.

Indeed classical logic gives no justification for choosing between the two indicated reductions, the first commuting the cut-rule with the *left* application of the weakening-rule ("pushing d_2 up into d_1", thus erasing d_2), the second commuting the cut-rule with the *right* application of the weakening-rule ("pushing d_1 up into d_2", thus erasing d_1). Therefore the cut-elimination process in **MAL**, *a fortiori* in **LK**, is non-deterministic and non-confluent.

Compare this with normalization in intuitionistic logic. In the typed λ calculus a *cut / left weakening* pair corresponds to substitution of $t : A$ for a variable $x : A$ which does not occur in $u : B$; such a substitution is unambiguously defined as $u[t/x] = u$. Moreover in Prawitz's natural deduction **NJ** [19] the rule corresponding to *weakening-right* is the rule *"ex falso quodlibet"* and the normalization step for such a rule involves a form of η-expansion:

$$\dfrac{\begin{array}{c} d \\ \vdots \\ \bot \end{array}}{A \wedge B} \qquad \textit{reduces to} \qquad \dfrac{\dfrac{\begin{array}{c} d \\ \vdots \\ \bot \end{array}}{A} \quad \dfrac{\begin{array}{c} d \\ \vdots \\ \bot \end{array}}{B}}{A \wedge B}$$

Such a reduction does *not* yield cancellation. Thus the cut-elimination procedure for the intuitionistic sequent calculus **LJ** inherits one sensible reduction strategy from natural deduction: *"push the left derivation up into the right one"*. In the case of a *weakening / cut* pair it is always the *left* dedution to be erased.

3. Here we are interested in exploring an obvious remark: for classical logic in addition to the *asymmetric* reductions of Example 1, there is a *symmetric* possibility, the "*Mix*" of d_1 and d_2.

Example 1 cont.

$$\dfrac{\dfrac{\begin{array}{c} d_1 \\ \vdots \\ \vdash \Gamma \end{array}}{\vdash \Gamma, A}\, w_1 \qquad \dfrac{\begin{array}{c} d_2 \\ \vdots \\ \vdash \Delta \end{array}}{\vdash \Delta, \neg A}\, w_2}{\vdash \Gamma, \Delta} \qquad \text{reduces to} \qquad \dfrac{\begin{array}{c} d_1 \\ \vdots \\ \vdash \Gamma \end{array} \quad \begin{array}{c} d_2 \\ \vdots \\ \vdash \Delta \end{array}}{\vdash \Gamma, \Delta}\, Mix$$

Symmetric reduction.

Instead of choosing a direction where to "push up" the cut-rule, we do both asymmetric reductions, using the mix-rule.

The idea is loosely related to a procedure well-known in the literature for the case when both cut-formulas result from a contraction-rule, with the name *cross-cut reduction.* Let d_1 and d_2 be derivations of the left and right premises of the cut-rule:

Example 2

$$
d_1 = \quad \frac{\dfrac{\begin{array}{c} d'_1 \\ \vdots \\ \vdash \Gamma, A, A \end{array}}{\vdash \Gamma, A} \qquad \dfrac{\begin{array}{c} d'_2 \\ \vdots \\ \vdash \neg A, \neg A, \Delta \end{array}}{\vdash \neg A, \Delta}}{\vdash \Gamma, \Delta} \quad = d_2
$$

Let d_ℓ be obtained by commuting the cut-rule with the *left* application of the contraction-rule and symmetrically, let d_r be obtained by commuting the cut-rule with the *right* application of the contraction-rule. The *cross-cut* reduction is defined as follows:

$$
d_\ell = \quad \dfrac{\dfrac{\dfrac{\dfrac{\begin{array}{c} d'_1 \\ \vdots \\ \vdash \Gamma, A, A \end{array} \qquad \dfrac{\begin{array}{c} d'_2 \\ \vdots \\ \vdash \neg A, \neg A, \Delta \end{array}}{\vdash \neg A, \Delta}}{\vdash \Gamma, \Delta, A}\, cut_1 \qquad \dfrac{\dfrac{\begin{array}{c} d'_1 \\ \vdots \\ \vdash \Gamma, A, A \end{array}}{\vdash \Gamma, A} \qquad \begin{array}{c} d'_2 \\ \vdots \\ \vdash \neg A, \neg A, \Delta \end{array}}{\vdash \neg A, \Gamma, \Delta}\, cut_2}{\vdash \Gamma, \Gamma, \Delta, \Delta}\, cut}{}}{\vdash \Gamma, \Delta}\, contractions \quad = d_r
$$

Cross-cut reduction.

4. As it stands the *symmetric* reduction of Example 1 could not be taken very seriously as a confluent notion of cut-reduction. One issue is the fact that a *weakening* inference may be *permuted* with many other inferences in a sequent derivation, and such permutations may considerably modify the structure of the proof; it would therefore be useful to have some notion of a *normal form for weakening*. Two standard notions can be found in the literature: these amount to applying the weakening-rule either (i) as *high* as possible, i.e., at the level of axioms, or (ii) as *low* as possible. The first solution is not available for *multiplicative* connectives without the contraction-rule; the second solution still leaves room for many ambiguities.

But there is one case where no ambiguity is possible, that of a *weakening / multiplicative disjunction* pair, where the weakening-rule introduces a formula which is active in the disjunction-rule and the other active formula has ancestors in axioms (or, similarly, the case of a weakening-rule introducing a conclusion of the derivation). It turns out that every derivation can be transformed into one where all applications of the weakening-rule are of this form, through *weakening-reductions*. This property may adopted as a notion of *weakening normal form*, but there are two problems: first, if we apply the cut-rule to two derivations in weakening normal form, the resulting derivation may not be in weakening normal form and, second, if we define weakening-reductions like permutations of inferences in the sequent calculus, then they do not yield a unique normal form (Section 2).

A second issue is the nature of the mix-rule: this rule does not simply represent distinct possibilities of proof-transformation. On the contrary, it contributes to create proofs with very rich and complicated structure. However using the unrestricted weakening-rule if a sequent is derivable *with Mix* then it is derivable *without Mix*, i.e., the structural rule Mix is eliminable:

$$\frac{\begin{array}{cc} d_1 & d_2 \\ \vdots & \vdots \\ \vdash \Gamma & \Delta \end{array}}{\vdash \Gamma, \Delta}\,Mix \quad \text{may be transformed into} \quad \dfrac{\dfrac{\begin{array}{c} d_1 \\ \vdots \\ \vdash \Gamma \end{array}}{\textit{weakenings}}}{\vdash \Gamma, \Delta} \quad \text{but also into} \quad \dfrac{\dfrac{\begin{array}{c} d_2 \\ \vdots \\ \vdash \Delta \end{array}}{\textit{weakenings}}}{\vdash \Gamma, \Delta}$$

The ambiguity may be resolved by taking *both* reducts. More generally, for all n we may introduce an n-ary rule *Additive Mix* building a derivation d of $\vdash \Gamma$ out of n derivations $d_1, \ldots, d_n$ of $\vdash \Gamma$:

$$\frac{d_1, \ldots, d_n}{\vdash \Gamma}\quad AM(n).$$

Then every application of Additive Mix may be permuted below other inferences. Thus the replacement of the *multiplicative* mix-rule with the *additive* mix-rule seems to capture the practice of disentangling simpler and more basic arguments from a more complicated one; conversely, the use of the multiplicative mix-rule may be explained as a compact notation unifying different ways of proving the same conclusions (Section 3). But the procedure **Sep** which eliminates the multiplicative mix-rule and permutes occurrences of Additive Mix below other inferences is computationally very expensive: therefore it would be desirable to find a more effective procedure to eliminate the multiplicative mix-rule.

5. If we look again at the *direct logic* decision procedure [15, 4], we see that two of our problems had already been solved there by the notion of a *chain*. Given a sequent Γ and a formula C in Γ, the procedure selects one (positive or negative) atomic subformula P in C and tries to find another subformula P of opposite polarity in some $D \in \Gamma$; in the terminology of proof-nets, it builds an axiom $\overline{P, P^{\perp}}$. Then the paths of subformulas from C to P and from $P^{\perp}$ to D are included in the chain; moreover, if a conjunct is in the chain, say A in $A \otimes B$, but the other conjunct B is not, then the procedure selects an atomic subformula Q of B and tries to find another axiom $\overline{Q, Q^{\perp}}$, and so on. The procedure will stop if all conjuncts have been matched by exactly one axiom. Now we apply this procedure *within a proof-net* $\mathcal{R}$ for *multiplicative affine logic with Mix* (**MAL** + Mix): it yields a path through the proof-net; if only one premise of a *par* link in in the path, then we introduce the other premise by a weakening-link; similarly for conclusions which are not reached by the path. The substructure $\mathcal{S}$ obtained in this way is a *proof-net* for **MAL**, *multiplicative affine logic without Mix*. If we repeat this procedure for all possible choices of axioms, we have a more efficient proof-net counterpart of the **Sep** procedure above. Moreover, the proof-net $\mathcal{S}$ corresponds to a sequent derivation in *weakening normal form*: we call it a proof-net *free from irrelevance*.

6. Given a proof-net for multiplicative affine logic with Mix, we may define a linear-time *pruning* algorithm, which yields a proof-net free from irrelevance, as follows ([4], Section 7.5):

(0) a weakening-formula is irrelevant;

(1) if the conclusion of a link is irrelevant, all its premises are irrelevant;

(2) if a formula in a logical axiom is irrelevant, so is the other;

(3) if one premise of a times-link or of a cut-link is irrelevant, so is the conclusion and the other premise;

(4) if both premises of a par-link are irrelevant, then the conclusion is irrelevant.

Clearly this resembles Girard's definition of the *empire* of a formula in $\mathbf{MLL}^-$ without Mix (cf. [11], Facts 2.9.4). What is important for us is that from the linear-time pruning algorithm we can draw two consequences, one relevant to problems of computational complexity, the other to the definition of confluent normalization procedures.

Given a proof-structure $\mathcal{S}$ without attachments for the weakening-links and satisfying the acyclicity property (of every Danos–Regnier graph), consider the problem of deciding whether attachments may be added to the weakening-links of $\mathcal{S}$ so that the resulting proof-structure $\mathcal{S}'$ is a proof-net. By a result of P. Lincoln and T. Winkler [16] this problem is NP-complete in the strong sense, thus it is as hard as the problem of finding a proof of Γ, given any $\mathbf{MLL}^-$ sequent Γ. But if we consider only proof-nets without irrelevance the problem of attaching the weakening-links is trivial, as they can be attached to a premise of a *par-link* or to a conclusion: the elimination of irrelevance algorithm yields a proof-structure free from irrelevance in linear time. It follows that provability in *constant-only affine logic* is decided in linear time by the pruning algorithm: since axioms are precisely the subformulas **1**, a sequent is provable if and only if the result of pruning the subtree of its formuals is non-empty.

Another remarkable feature of the pruning algorithm is that, regarded as a notion of reduction of proof-nets for **MAL** + Mix, it is *confluent*. It should be noticed that this algorithm *cannot be defined by permutations of inferences in* *the sequent calculus*: this follows from the study of the subnets of proof-nets [3], see examples in Section 4.3. It is this remark that allowed the research to take off, as it showed that proof-nets could be used to identify some invariants of proofs with respect to the behaviour of the *weakening-rule*.

7. The use of the proof-net representation of proofs and the notion of the *irrelevant part* $I(\mathcal{R})$ in a proof-net $\mathcal{R}$ are essential also to define a confluent notion of *cut-reduction*. Irrelevance is not stable under cut-elimination: if $\mathcal{R}_1$ reduces to $\mathcal{R}_2$ and $\mathcal{R}_2$ reduces to $\mathcal{R}_3$ then we may have $I(\mathcal{R}_1) \subset I(\mathcal{R}_2)$ but also $I(\mathcal{R}_3) \subset I(\mathcal{R}_2)$. Therefore we cannot actually eliminate irrelevance during cut-elimination: we may metaphorically say that we need *cut-elimination without garbage collection*. The goal is to eliminate cut through stricly *local* operations, e.g., by reducing the logical complexity of the cut-formula in a *weakening / cut* pair. This could be achieved if we could *expand weakening-links* following the model of natural deduction. Unfortunately such an expansion is *incorrect* in the sequent calculus: let P be atomic, then $\vdash \mathbf{1}, P \otimes P$ is derivable in **MAL** using one weakening-rule, but there is no way to expand the application of the weakening-rule into two weakening-rules with atomic conclusion P.

The solution proposed here is to define *proof-nets modulo irrelevance*. Given a proof-structure $\mathcal{S}$ without attachments for the weakening-links and satisfying the acyclicity property of every Danos–Regnier graph, we require $\mathcal{R} \setminus I\mathcal{R} \neq \emptyset$ as an additional *correctness condition*; namely, we require that *it should be possible to eliminate irrelevance without annihilating the proof-structure*, but we do not actually eliminate irrelevance as in [15, 4]. Then the "incorrect" expansions can be introduced, since they belong to the irrelevant part; the additional weakening-links introduced in this way will eventually be annihilated by the cut-elimination procedure. As a consequence, in the intermediate steps of the cut-elimination procedure only the *pruning* of $\mathcal{R}_i$, i.e., $\mathcal{R}_i \setminus I\mathcal{R}_i$ will be sequentializable. Moreover, during the cut-elimination procedure some weakening-links may be annihilated whose attachments guaranteed the connectedness condition for the original proof-net; therefore we may start with a proof-net with cut links in *multiplicative affine logic without Mix* and obtain a cut-free one in *multiplicative affine logic with Mix*.

8. Many researchers have studied the cut-elimination process for classical logic and defined well-behaved notions of reduction and reduction strategies, enjoying the confluence and strong normalization properties. A common feature of many works is that they recognize the non-determinism of classical cut-elimination as the root of both non-confluence and non-termination and then try to to eliminate non-deter- minism by *disambiguating* classical logic. A way to do so is through translations into intuitionistic and linear logic. Such a use of linear logic is exemplified in Girard [12] and it has been developed extensively by V. Danos, J-B. Joinet and H. Schellinx, see, e.g., [8, 9]. Through the modalities of linear logic the area of a proof-net which may be erased in a *weakening / cut* reduction is always determined *modulo* permutation of !-boxes. Interesting properties of classical cut-elimination are identified in this way, which can be related to known reduction strategies for the λ-calculus. However, it should be clear that the translations of classical logic into linear logic have the same function as the translations into intuitionistic logic, namely to extend the computational paradigm of intuitionistic logic to classical logic. In the case of affine logic, such translations yield *"computations with garbage collection"*.

Another approach is to look for dynamical properties of classical logic that are alternative and irreducible to those of intuitionistic and linear logic. It would be natural to expect that the classical cut-elimination should reflect the *symmetries* that arise from the fact that classical negation is an involution. In Girard [12] a mathematically precise notion of symmetry is given for the notion of cut-reduction in the proof-net representation. The set of substitutions $[p_i/p_i^{\perp}]$ where p_i is an atom in a proof-net may be regarded as the set of generators of a group acting on the graph. A notion of cut-reduction is *symmetric* if every proof-net which is invariant under the action of the group of substitutions is transformed by such a reduction into a proof-net which is also invariant. An example in [12] shows that a symmetric cut-reduction for classical logic cannot be defined without the use of the mix-rule. Our definition of cut-reduction for **MAL** with Mix would appear to be symmetric this technical sense. The cross-cut reduction of Example 2 is not symmetric, but a symmetric variant of it may perhaps be defined using the mix-rule (see Section 6 below).

Most recently, Bierman and Urban [7] have developed a term calculus for classical logic which represents a very large variety of strongly normalizing, but non-confluent reduction strategies for classical sequent calculus. Bierman and Urban also regard *non-determinism* as the distinguishing feature of classical dynamics. The treatment of Mix in the present paper, in particular the elimination of the multiplicative mix-rule through Additive Mix, suggests that our proof-net representation may provide a concise notation for classical non-determinism. Indeed the notion of *cut-elimination without garbage collegion* may also have a common-sense explanation in terms of non-determinism: if we cannot predict the future development of a process, we should be very conservative about what may be discarded or not.

Can we conclude with a positive answer to the question in the title? Do we really have a notion of cut-elimination for classical affine logic whose properties are alternative and irreducible to the familiar reduction strategies for intuitionistic and linear logic? It may be premature to give an answer. But the time may be ripe for a reconsideration of proof-nets for affine logic. Indeed an increasing number of researchers (e.g., [21]) seem to agree today that although the dynamics of classical logic may not be as elegant and pleasing as that of intuitionistic logic, a task of research is to develop tools to study it as it is, allowing the possibility that it may be very different from what we already know.

2. Sequent calculus of MAL + Mix

Definition 3 (i) The propositional language of Multiplicative Affine Logic **MAL** is built from the propositional constants $\mathbf{1}$ and $\bot$ and propositional variables, using the connectives $\otimes$ (*times*) and $\wp$ (*par*) of Multiplicative Linear Logic.

(ii) The language of *Constant-only* **MAL** is the fragment of **MAL** consisting of the formulas which do not contain propositional variables.

(iii) The sequent calculus for propositional **MAL** + Mix is given by the following axioms and rules:

identity rules

logical axiom: $\vdash A^{\perp}, A$ *1 axiom:* $\vdash \mathbf{1}$ *cut:*
$$\frac{\vdash \Gamma, A^{\perp} \quad \vdash \Delta, A}{\vdash \Gamma, \Delta}$$

structural rules

exchange: *weakening:* *mix:*
$$\frac{\vdash \Gamma, A, B, \Delta}{\vdash \Gamma, B, A, \Delta} \qquad \frac{\vdash \Gamma}{\vdash \Gamma, A} \qquad \frac{\vdash \Gamma \quad \vdash \Delta}{\vdash \Gamma, \Delta}$$

logical rules

times: *par:*
$$\frac{\vdash \Gamma, A \quad \vdash \Delta, B}{\vdash \Gamma, \Delta, A \otimes B,} \qquad \frac{\vdash \Gamma, A, B}{\vdash \Gamma, A \wp B}$$

The sequent calculus for classical **MAL** + Mix.

Given a derivation d in **MAL** + Mix, let us write $\#ax$, $\#\mathbf{1}$ for the number of occurrences of logical axioms and of $\mathbf{1}$ axioms, and $\#\otimes$, $\#cut$, $\#mix$, $\#w$ and $\#\wp$ for the number of applications of the *times-*, *cut-*, *mix-*, *weakening-* and *par-rules* and finally $\#conc$ for the number of conclusions in d.

Lemma 4 *Every sequent derivation in* **MAL** + *Mix satisfies the following equations:*

$$\#\text{ax} = \# \otimes + \#\text{cut} + \#\text{mix} + 1$$

$$\#\wp + \#\text{conc} + \#\mathbf{1} = \# \otimes + \#\text{w} + \#2\text{mix} + 2$$

Proof. By induction on the length of the derivation. For instance, let the last inference of d be a *times-rule* with immediate subderivations d_1 and d_2; let

$\#ax_1$, $\#ax_2$ and $\#ax$ be the number of logical axioms in d_1, d_2 and in d, respectively and similarly for the other axioms and rules. Then we have

$$\begin{aligned}
\#ax &= \#ax_1 + \#ax_2 \\
&= (\#\otimes_1 + \#\otimes_2) + (\#cut_1 + \#cut_2) + (\#mix_1 + \#mix_2) + 2 \\
&= (\#\otimes_1 + \#\otimes_2 + 1) + \#cut + \#mix + 1 \\
&= \#\otimes + \#cut + \#mix + 1 \\
\wp + \#conc + \#\mathbf{1} &= (\#\wp_1 + \#\wp_2) + (\#conc_1 + \#conc_2 - 1) + (\#\mathbf{1}_1 + \#\mathbf{1}_2) \\
&= (\#\wp_1 + \#\wp_2) + (\#conc_1 + \#conc_2) + (\#\mathbf{1}_1 + \#\mathbf{1}_2) - 1 \\
&= (\#\otimes_1 + \#\otimes_2) + (\#w_1 + \#w_2) + 2(\#mix_1 + \#mix_2) + 4 - 1 \\
&= (\#\otimes_1 + \#\otimes_2 + 1) + \#w + 2\#mix + 2 \\
&= \#\otimes + \#w + 2\#mix + 2
\end{aligned}$$

■

2.1 Mix and weakening permutations

Definition 5 (i) Let m/r be a pair of consecutive inferences in a derivation d, where m is an instance of Mix and either r is not a *par-rule* or r is a par-rule but all the ancestors of its active formulas occur in the same branch above the mix-rule. Then the inferences m/r are permutable, i.e., we may obtain a derivation d' which is like d, except for having a consecutive pair of inferences r/w in place of w/r. Thus the only exceptions to the permutability of the mix-rule *below* other inferences are of the following form:

$$\dfrac{\dfrac{\begin{matrix} d_1 \\ \vdots \\ \vdash \Gamma, A \end{matrix} \quad \begin{matrix} d_2 \\ \vdots \\ \vdash B, \Delta \end{matrix}}{\vdash \Gamma, A, B, \Delta}\, Mix}{\vdash \Gamma, A \wp B, \Delta}\, par$$

For every occurrence m of the mix-rule in d there is a set of applications of the par-rule $p_1, \ldots, p_n$ such that m cannot be permuted below $p_1, \ldots, p_n$.

(ii) Similarly, if r/m is a consecutive pair of inferences, where m is an instance of the mix-rule, then the inferences are permutable, but if r is a *times-rule, cut-rule* or *mix-rule* rule, then there are *two* ways of permuting, i.e., pushing the mix-rule up to the left premise of r or to the right one.

(iii) Let w/r be a pair of consecutive inferences in a derivation d, where w is an instance of the weakening-rule introducing a formula X and X is not active in r; then the inferences w/r are permutable, i.e., we may obtain a derivation d' which is like d, except for having a consecutive pair of inferences r/w in place of w/r.

(iv) Similarly, if r/w is a consecutive pair of inferences, where w is an instance of the weakening-rule, then the inferences are permutable, but if r is a *times-rule, cut-rule* or *mix-rule*, then there are *two* ways of permuting, i.e., pushing the weakening-rule up to the left premise of r or to the right one.

Remark 6 Weakening and Mix permutations are reversible.

2.2 Weakening reductions and expansions

Definition 7 (i) In the sequent calculus for **MAL** we have the following *weakening-reductions:*

weakening / par *reduction:*

$$\dfrac{\dfrac{\dfrac{\vdash \Gamma}{\vdash \Gamma, A}\,weak}{\vdash \Gamma, A, B}\,weak}{\vdash \Gamma, A \wp B}\,par \qquad \text{reduces to} \qquad \dfrac{\vdash \Gamma}{\vdash \Gamma, A \wp B}\,weak$$

weakening / times *reduction to the right:*

$$\dfrac{\vdash \Gamma, A \quad \dfrac{\vdash \Delta}{\vdash \Delta, B}\,weak}{\vdash \Gamma, \Delta, A \otimes B}\,times \qquad \text{reduces to} \qquad \dfrac{\dfrac{\vdash \Delta}{\text{weakenings}}}{\vdash \Gamma, \Delta, A \otimes B}$$

weakening / times *reduction to the left:*

$$\dfrac{\dfrac{\vdash \Gamma}{\vdash \Gamma, A}\,weak \quad \vdash \Delta, B}{\vdash \Gamma, \Delta, A \otimes B}\,times \qquad \text{reduces to} \qquad \dfrac{\dfrac{\vdash \Gamma}{\text{weakenings}}}{\vdash \Gamma, \Delta, A \otimes B}$$

weakening / cut *reduction to the right:*

$$\dfrac{\vdash \Gamma, A \quad \dfrac{\vdash \Delta}{\vdash \Delta, A^{\perp}}\,weak}{\vdash \Gamma, \Delta}\,cut \qquad \text{reduces to} \qquad \dfrac{\dfrac{\vdash \Delta}{\text{weakenings}}}{\vdash \Gamma, \Delta}$$

weakening / cut *reduction to the left:*

$$\dfrac{\dfrac{\vdash \Gamma}{\vdash \Gamma, A}\,weak \quad \vdash \Delta, A^{\perp}}{\vdash \Gamma, \Delta}\,cut \qquad \text{reduces to} \qquad \dfrac{\dfrac{\vdash \Gamma}{\text{weakenings}}}{\vdash \Gamma, \Delta}$$

(ii) The inference eliminated by any one of the above reduction will be called *irrelevant.* A sequent derivation which contains no irrelevant inference is said to be in *weakening normal form.*

Lemma 8 *In the sequent calculus for* **MAL** *every sequent derivation d can be transformed into a derivation d′ in weakening normal form. The derivation d′ is not unique.*

(ii) A derivation in weakening normal form has the following properties:

1 every formula introduced by an application of the weakening-rule becomes active only in the premise of an application of the par-rule *where the other active formula has an ancestor in an axiom;*

2 *both premises of an application of the* times-rule *have ancestors in (different) axioms.*

PROOF. (i) For every instance w of the weakening-rule, permute w below all inferences until either (a) the formula X introduced by w is active in the inference r immediately below, or (b) there is no logical inference below w. In the first case a weakening-reduction applies, unless r is a *par-rule* and the other active formula has not been introduced by a *weakening-rule*. The process is *non-deterministic*: in cases when *weakening / times* reductions to the right and to the left are both applicable, a random choice is required. (ii) By induction on the length of a derivation in weakening normal form. ∎

Remark 9 (i) *Weakening / par reductions* are always reversible, i.e., we have the following expansion rule:

weakening / par *expansion:*

$$\dfrac{\vdash \Gamma}{\vdash \Gamma, A \wp B}\ weak \qquad \text{expands to} \qquad \dfrac{\dfrac{\dfrac{\vdash \Gamma}{\vdash \Gamma, A}\ weak}{\vdash \Gamma, A, B}\ weak}{\vdash \Gamma, A \wp B}\ par$$

(ii) All other reductions are *irreversible* and produce a genuine loss of information.

(iii) In the sequent calculi for classical or intuitionistic logic the use of the weakening-rule can be standardized in two ways, by prescribing that all applications of the weakening-rule must occur either *as low as possible* or *as high as possible* (i.e., at the level of the *axioms*) in a derivation. In **MAL**, in absence of the *contraction-rule* only the first option is available for *multiplicative* connectives. In particular, in **MAL** + Mix we cannot assume that every formula introduced by Weakening is atomic: consider a cut-free derivation of $\vdash \mathbf{1}, P \otimes P$, where P is an atom different from $\mathbf{1}$.

(iv) On the other hand, for *additive* connectives in absence of the *contraction-rule* only the second option is available, i.e., applying the weakening-rule as high as possible.

More precisely, we could define the propositional language of Additive Affine Logic **AAL** using the additive connectives & (*with*) and ⊕ (*plus*):

$$\textit{with:}\ \dfrac{\vdash \Gamma, A \quad \vdash \Gamma, B}{\vdash \Gamma, A \& B,} \qquad \textit{plus}_i:\ \dfrac{\vdash \Gamma, A_i}{\vdash \Gamma, A_0 \oplus A_1} \qquad i = 0, 1.$$

Proposition 10 *In the sequent calculus for* **AAL** *every sequent derivation d can be transformed into a unique derivation d′ with the property that every formula introduced by the weakening-rule is atomic.*

PROOF. Repeatedly apply permissible permutations of inferences and the following *expansions:*

weakening / with *expansion:*

$$\dfrac{\begin{array}{c} d \\ \vdots \\ \vdash \Gamma \end{array}}{\vdash \Gamma, A\&B}\ weak \qquad \text{expands to} \qquad \dfrac{\dfrac{\begin{array}{c} d \\ \vdots \\ \vdash \Gamma \end{array}}{\vdash \Gamma, A}\ weak \qquad \dfrac{\begin{array}{c} d \\ \vdots \\ \vdash \Gamma \end{array}}{\vdash \Gamma, B}\ weak}{\vdash \Gamma, A\&B}\ times$$

weakening / plus *expansion:*

$$\dfrac{\vdash \Gamma}{\vdash \Gamma, A_0 \oplus A_1}\ weak \qquad \text{expands to} \qquad \dfrac{\dfrac{\vdash \Gamma}{\vdash \Gamma, A_i}\ weak}{\vdash \Gamma, A_0 \oplus A_1}\ plus$$

■

3. Additive mix

For all n we have an n-ary rule *Additive Mix* which builds a derivation d of $\vdash \Gamma$ out of n derivations $d_1, \ldots, d_n$ of $\vdash \Gamma$. We use the following notation:

$$\frac{d_1, \ldots, d_n}{\vdash \Gamma}\quad AM(n).$$

We have the following reduction:

Mix-elimination

$$\dfrac{\begin{array}{c} d_1 \\ \vdots \\ \vdash \Gamma \end{array} \quad \begin{array}{c} d_2 \\ \vdots \\ \Delta \end{array}}{\vdash \Gamma, \Delta}\ Mix \qquad \text{reduces to} \qquad \dfrac{\dfrac{\dfrac{\begin{array}{c} d_1 \\ \vdots \\ \vdash \Gamma \end{array}}{weakenings}}{\vdash \Gamma, \Delta} \qquad \dfrac{\dfrac{\begin{array}{c} d_2 \\ \vdots \\ \vdash \Delta \end{array}}{weakenings}}{\vdash \Gamma, \Delta}}{\vdash \Gamma, \Delta}\ AM(2)$$

3.1 Properties of additive mix

The rule *Additive Mix* has the following properties:

1. *The order of the premises is immaterial*, i.e., we assume an *AM-exchange* rule of the form

$$\frac{d_1, \ldots, d_i, d_{i+1}, \ldots d_n}{\vdash \Gamma} \quad \sim \quad \frac{d_1, \ldots, d_{i+1}, d_i, \ldots d_n}{\vdash \Gamma}$$

2. *The derivations of the premises are paiwise distinct*, i.e., we assume an *AM-contraction* rule of the form

$$\frac{d_1, \ldots, d_{i-1}, d_i, d_i, d_{i+1}, \ldots d_n}{\vdash \Gamma} \quad \text{reduces to} \quad \frac{d_1, \ldots, d_{i-1}, d_i, d_{i+1}, \ldots d_n}{\vdash \Gamma}$$

3. *Consecutive applications of AM can be unified*, i.e., we assume an *AM-merging* rule of the form

$$\frac{\dfrac{d_1, \ldots, d_m}{\vdash \Gamma}\, AM(m) \qquad \dfrac{d_{n+1}, \ldots, d_{n+m}}{\vdash \Gamma}\, AM(n)}{\vdash \Gamma}\, AM(2) \quad \text{reduces to} \quad \frac{d_1, \ldots, d_{n+m}}{\vdash \Gamma}\, AM(n+m)$$

4. *The rule AM-contraction can be strengthened according to different notions of identity of proofs.* For instance, let us put $d \sim d'$ if and only if $(d)^-$ and $(d')^-$ are the same proof-net with attachments. Then the *AM-contraction* rule becomes:

$$\frac{d_1, \ldots, d_i, d_i', \ldots d_n}{\vdash \Gamma} \quad \text{where } d_i \sim d_i' \text{ reduces to} \quad \frac{d_1, \ldots, d_i, \ldots d_n}{\vdash \Gamma}$$

5. A sequent derivation in **MAL** + AM is in *AM normal form* if it has at most one application of Additive Mix as the last inference.

Convention. We assume that the *AM-contraction* in the strong form (4) is always applied without mention, and similarly for the *AM-merging* rules. We work with derivations *modulo* the AM-exchange rule.

3.2 Additive mix: reductions and permutations

AM / exchange *permutation:*

$$\frac{\dfrac{d_1, \ldots, d_n}{\vdash \Gamma, A, B}\, AM(n)}{\vdash \Gamma, B, A}\, exch \quad \text{reduces to} \quad \frac{\dfrac{\overset{d_1}{\vdots}}{\dfrac{\vdash \Gamma, A, B}{\vdash \Gamma, B, A}}\, exch \quad \ldots \quad \dfrac{\overset{d_n}{\vdots}}{\dfrac{\vdash \Gamma, A, B}{\vdash \Gamma, B, A}}\, exch}{\vdash \Gamma, B, A}\, AM(n)$$

AM / weakening *permutation:*

$$\frac{\dfrac{d_1, \ldots, d_n}{\vdash \Gamma}\, AM(n)}{\vdash \Gamma, A}\, weak \quad \text{reduces to} \quad \frac{\dfrac{\overset{d_1}{\vdots}}{\dfrac{\vdash \Gamma}{\vdash \Gamma, A}}\, weak \quad \ldots \quad \dfrac{\overset{d_n}{\vdots}}{\dfrac{\vdash \Gamma}{\vdash \Gamma, A}}\, weak}{\vdash \Gamma, A}\, AM(n)$$

AM / par *permutation:*

$$\dfrac{\dfrac{d_1,\ldots,d_n}{\vdash \Gamma, A, B}\,AM(n)}{\vdash \Gamma, A\wp B}\,par \qquad \text{reduces to} \qquad \dfrac{\dfrac{\begin{array}{c} d_1 \\ \vdots \\ \vdash \Gamma, A, B\end{array}}{\vdash \Gamma, A\wp B}\,par \quad \ldots \quad \dfrac{\begin{array}{c} d_n \\ \vdots \\ \vdash \Gamma, A, B\end{array}}{\vdash \Gamma, A\wp B}\,par}{\vdash \Gamma, A\wp B}\,AM(n)$$

AM / times *permutation:*

$$\dfrac{\dfrac{d_1,\ldots,d_m}{\vdash \Gamma, A}\,AM(m) \quad \dfrac{d'_1,\ldots,d'_n}{\vdash B, \Delta}\,AM(n)}{\vdash \Gamma, A\otimes B, \Delta}\,times \qquad \text{reduces to} \qquad \dfrac{\ldots\dfrac{\begin{array}{c} d_i \\ \vdots \\ \vdash \Gamma, A\end{array} \quad \begin{array}{c} d_j \\ \vdots \\ \vdash B, \Delta\end{array}}{\vdash \Gamma, A\otimes B, \Delta}\,cut\ldots}{\vdash \Gamma, A\otimes B, \Delta}\,AM(m\cdot n)$$

for $i \leq m, j \leq n$.

AM / cut *permutation:*

$$\dfrac{\dfrac{d_1,\ldots,d_m}{\vdash \Gamma, A}\,AM(m) \quad \dfrac{d'_1,\ldots,d'_n}{\vdash A^{\perp}, \Delta}\,AM(n)}{\vdash \Gamma, \Delta}\,cut \qquad \text{reduces to} \qquad \dfrac{\ldots\dfrac{\begin{array}{c} d_i \\ \vdots \\ \vdash \Gamma, A\end{array} \quad \begin{array}{c} d_j \\ \vdots \\ \vdash A^{\perp}, \Delta\end{array}}{\vdash \Gamma, \Delta}\,cut\ldots}{\vdash \Gamma, \Delta}\,AM(m\cdot n)$$

for $i \leq m, j \leq n$.

We define a procedure **Sep** of elimination of Mix, which takes a derivation d in **MAL** + Mix and returns a derivation **Sep**(d) in **MAL** + AM without Mix, in AM normal form:

1 Given a derivation d and a mix-rule m in d, permute m as low as possible, obtaining a derivation d_m where either m is the last inference or m is immediately followed by a par-rule p such that m cannot be permuted below p.

2 Apply the mix-elimination rule to the mix-rule m in d_m, yielding a derivation d'_m, which contains an application AM(2) of Additive Mix.

3 Permute the inference AM(2) below all inferences of d'_m, thus obtaining a derivation d^*_m ending with an application of Additive Mix.

4 Let **Sep**$(d) = d^*_m$ if d^*_m contains no mix-rule; otherwise, select a new application of the mix-rule in d^*_m and start again with (1).

The procedure **Sep** has the desirable property that it does not create irrelevant inferences.

Theorem 11 *(i) Let d be a derivation of* Γ *in* **MAL** + *Mix. The result of applying the procedure* **Sep** *to d*

$$\mathbf{Sep}(d) = \frac{d_1, \ldots, d_n}{\Gamma}$$

is a derivation in **MAL** + *AM* without *Mix in AM normal form.*

(ii) If d is in weakening normal form, then so is **Sep**(d).

PROOF. (ii) If at the end of step 1 the mix-rule m is the last inference, then no irrelevant inference is created at steps (2) or (3), so d^*_m is still in weakening normal form. Otherwise at the end of step (1) the Mix m is followed by some *par-rule* $p_1, \ldots, p_k$ which are not permutable above m, therefore none of their active formulas is introduced by a weakening-rule; since d is in weakening normal form, it follows from Lemma 8.(ii) that all the formulas active in all p_i have ancestors in axioms. Therefore at the end of step (2) the Additive Mix AM(2) is immediately followed by an application of the par-rule which is not irrelevant and, moreover, after one application of the *AM / par* permutation both conclusions of the new par-rules have ancestors in some axiom. Finally, if a times-rule t occurs below m in d_m at the end of step (1), then a formula active in t is descendant of the conclusion of one of the p_i: otherwise, in step 1 t would be permutable above m. The same holds for cut-rules occurring in d_m below m. It follows that the Additive Mix permutations at step (3) do not introduce irrelevant inferences, i.e., d^*_m is still in weakening normal form. ■

4. Proof-nets for MAL + Mix

4.1 Proof-structures and proof-nets

Definition 12 (i) A *proof-structure* for **MAL** + Mix is a directed graph whose edges are labelled with formulas and whose vertices (*links*) are of one of the forms in Figure 3.1.

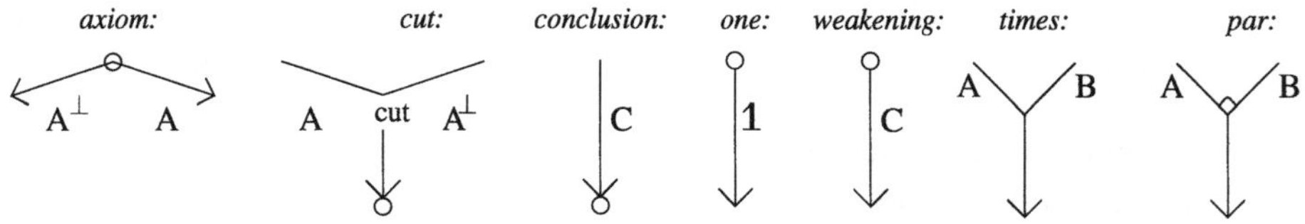

Figure 3.1. Links

In link the incoming edges are called the *premises* and the outgoing edges are the *conclusions* of the link. The transitive closure $\prec$ of the relation between links (equivalently, between edges) determined by the direction of the edges is a partial ordering (since the edges are typed with propositional formulas) and it is called the *structural orientation.* $A \prec B$ reads "A is a hereditary premise of B" or simply "A is above B". A *bottom* link is just a special case of a *weakening-link.*

(ii) Given a proof-structure $\mathcal{S}$ for **MAL** + Mix an *attachment* of a *weakening-link* w is an edge from the weakening-link to a new vertex which lies in another edge ℓ of $\mathcal{S}$. We write $\mathcal{S}^a$ for $\mathcal{S}$ toghether with a set of attachments for all the weakening-links of $\mathcal{S}$. The geometric properties of attachments are fully determined as follows:

- a *weakening-link* w with its attachment $a(w)$ has the same properties as a *logical axiom*;
- the new vertex where the attachment $a(w)$ and the edge ℓ meet has the same properties as a *times-link.*

(iii) A *switching* for a proof-structure $\mathcal{S}$ is a choice for every *par-link* in $\mathcal{S}$ of one of its premises.

(iv) Let $\mathcal{S}^a$ be a proof-structure *with attachments* for the weakening-links. Given a switching s for $\mathcal{S}^a$, the *Danos–Regnier graph* $s\mathcal{S}^a$ is the graph resulting from $\mathcal{S}^a$ by deleting from each *par-link* the edge which is *not* the premise chosen by s.

(v) A proof-structure $\mathcal{S}^a$ with attachments for the weakening-links is a *proof-net* if for every switching s the Danos–Regnier graph $s\mathcal{R}^a$ is acyclic (*correctness criterion* for **MAL** + Mix).

Theorem 13 *(i) There exists a map* $(\ )^-$ *from sequent derivations to proof-structures together with choices* $a_1, \ldots, a_r$ *of attachments for the weakening-links such that* $(d)^- = \mathcal{R}^{a_i}$ *is a proof-net, the attachments being given by any* a_i*, for* $i \leq r$.

(ii) (Sequentialization Theorem) Conversely, if $\mathcal{R}^a$ *is a proof-net, then there exists a sequent derivation* d *and a choice* $a_1, \ldots, a_n$ *of attachments for the weakening-links such that* $(d)^- = \mathcal{R}^a$*, and* $a = a_i$ *for some* $i \leq n$.

PROOF. (ii) For every *weakening-link* w in $\mathcal{R}^a$, let A be its conclusion and let ℓ be the edge, labelled with B, which $a(w)$ is attached to: replace w, $a(w)$ and ℓ with a logical axiom *ax* followed by a *times-link* t. The proof-structure $\mathcal{R}'$ thus obtained is labelled with different formulas than $\mathcal{R}^a$, but it has exactly the same geometric properties, so it is a proof-net in $\mathbf{MLL}^-$ + Mix and can be sequentialized as usual. In a sequent derivation d' such that $(d')^- = \mathcal{R}'$, let t' be the times-rule whose principal formula $A^\perp \otimes B$ corresponds to the new times-link t. Since the only active ancestor of $A^\perp$ is in the axiom *ax*, we may permute t' upwards to the left so that ax and t' are consecutive inferences and then replace them both with a *weakening-rule*:

$$\frac{\vdash A, A^{\perp} \quad \begin{array}{c}\vdots\\ \vdash B, \Gamma\end{array}}{\vdash A, A^{\perp} \otimes B, \Gamma} \quad \text{is replaced by} \quad \frac{\begin{array}{c}\vdots\\ \vdash B, \Gamma\end{array}}{\vdash A, B, \Gamma}$$

By repeating this procedure for all weakening-links of $\mathcal{R}^a$ we clearly obtain a derivation d such that $(d)^- = \mathcal{R}^a$. ■

4.2 Computation and elimination of irrelevance

One of the uses of proof-nets for multiplicative logic *without Weakening* **MLL**$^-$, is to *classify sequent derivations:* we prove that given sequent derivations d and d' in **MLL**$^-$ we have $(d)^- = (d')^-$ if and only if there exists a sequence of derivations $d = d_1, d_2, \ldots, d_n = d'$ such that for all $i < n$, d_i and d_{i+1} differ only for a permutation of two consecutive inferences (cf. [6]). A similar result holds for **MLL**$^-$ + Mix (cf. [3]).

In presence of the weakening-rule, thus already in **MLL** with Weakening restricted to a $\perp$-rule, we do not know how to obtain such a theorem. In [11] proof-nets are defined using *weakening-boxes*, thus the position of each weakening-rule in a given sequential proof d is fixed in the proof-net $(d)^-$.

The standard solution, which we have followed above, is the use of *attachments* for the weakening-links. But from the point of view of the classification of sequent derivations the notion of proof-nets with attachments is perhaps worse that that of proof-nets with weakening-boxes: given a box, there are several ways of making the attachment. In any event, this notion fails to identify sequential proofs *modulo permutations of weakening-rule.*

We turn now to the notion of a proof-net *modulo irrelevance*, which at least succeeds in minimizing the disturbances caused by weakening-links. As recalled in the Introduction, given a proof-structure S without attachments for the weakening-links and satisfying the acyclicity property (of every Danos–Regnier graph), the problem of deciding whether attachments may be added to the weakening-links of S so that the resulting proof-structure S' is a proof-net is NP-complete in the strong sense [16]. On the other hand in a proof-net free from irrelevance the weakening-links can only be attached to a premise of a par-link or to a conclusion. Moreover, the following algorithm identifies and possibly eliminates irrelevance in linear time.

Definition 14 (*computation of irrelevance algorithm*, cf. [11, 4]) Let S be a proof-structure *without attachments.*

(i) The *irrelevant part* $i(w)$ of $\mathcal{S}$ *determined by* a weakening-link w is the smallest subgraph closed under the following rules:

- the weakening-link w and its conclusion are in $i(w)$;
- if the premises A and B of a *par-link* v are in $i(w)$, then v and its conclusion $A \wp B$ are in $i(w)$;
- if either premise A or B of a *times-link* v is in $i(w)$, then v and its conclusion $A \otimes B$ are in $i(w)$; similarly, if A and B are premises of a *cut-link*;
- if $u \prec v$ and v is in $i(w)$, then u and its conclusions are in $i(w)$; in particular, if u is a *logical axiom* in $i(w)$, then both its conclusions are in $i(w)$.

(ii) The *doors* of $i(w)$ in $\mathcal{S}$ are the edges which are in $i(w)$ and are premises of links whose conclusions are not in $i(w)$. Clearly all the doors of $i(w)$ are either conclusions of $\mathcal{S}$ or premises of *par-links*.

(iii) Let w be a weakening-link in $\mathcal{S}$. Write $\mathcal{S}^w$ for $\mathcal{S} \setminus i(w)$, with the addition of a weakening-link w_D with conclusion D for every door D of $i(w)$. If $\mathcal{S}$ has attachments and w is attached to an edge ℓ through $a(w)$, then we may assume that $\ell \notin i(w)$, otherwise such an attachment would generate a cyclic D–R-graph and $\mathcal{S}$ could not be a *proof-net*. Therefore we may define a set of attachments a' for $\mathcal{S}^w$ as follows:

- $a'(w') = a(w')$, if w' occurs both in $\mathcal{S}$ and in $\mathcal{S}^w$;
- $a'(w_D)$ is attached to ℓ, for all doors D of $i(w)$, otherwise.

(iv) The *pruning map* given by $\mathcal{S} \mapsto \mathcal{S}^w$ for $w \in \mathcal{S}$ may be regarded as a notion of reduction, which has the Church–Rosser property.

Lemma 15 *Let w_1 and w_2 be weakening-links in $\mathcal{S}$. Then $\mathcal{S}^{w_1 w_2} = \mathcal{S}^{w_2 w_1}$.*

PROOF. Let D_1 be a door of $i(w_1)$, thus a premise of a par-link p. Suppose that the other premise D_2 of the same par-link is in $i(w_2)$ and apply the irrelevance computation algorithm and eliminate $i(w_2)$ in $\mathcal{S}^{w_1}$ and $i(w_1)$ in $\mathcal{S}^{w_2}$. In both cases the conclusion D of p is included in the irrelevant part $i(w_2)$ or $i(w_1)$ and the par-link p is removed both in $\mathcal{S}^{w_1 w_2}$ and in $\mathcal{S}^{w_2 w_1}$. ∎

Definition 16 (Definition 14 cont.) (v) Let $w_1, \ldots, w_n$ be all the weakening-links in $\mathcal{S}$. By Lemma 15 $\mathcal{S}^{w_1,\ldots,w_n} = \mathcal{S}^{w_{\sigma(1)},\ldots,w_{\sigma(n)}}$ for any permutation σ of $1, \ldots, n$. The *irrelevant part* $I(\mathcal{S})$ of the proof-structure $\mathcal{S}$ is defined as

$\mathcal{S} \setminus \mathcal{S}^{w_1,\ldots,w_n}$, where $w_1, \ldots, w_n$ are all the weakening-links of $\mathcal{S}$. A proof-structure $\mathcal{R}$ is *irrelevance free* if $\mathcal{R} = \mathcal{R} \setminus I(\mathcal{R})$. We write $\mathbf{P}(\mathcal{R}) = \mathcal{R} \setminus I(\mathcal{R})$ for the *pruning operation* given by the computation of irrelevance algorithm.

(vi) A proof-structure $\mathcal{R}$ for **MAL** + Mix is a *proof-net modulo irrelevance* if $\mathcal{R} \setminus I(\mathcal{R})$ is non-empty and $\mathcal{R}$ (without attachments) satisfies the acyclicity condition for all D–R graphs.

Remark 17 Notice that if $\mathcal{S} = \mathbf{P}(\mathcal{R})$, then a *canonical attachment* a for $\mathcal{S}$ is given simply by a choice of a conclusion C to which we may attach all conclusions of $\mathcal{S}$ which are introduced by a weakening-link. Indeed all conclusions of a weakening-link which are premises of a par-link may be canonically attached to the other premise ℓ (which is not introduced by a weakening-link). Therefore if $\mathcal{S}$ has just one conclusion, there is only one canonical attachment a and we may omit mentioning it.

Theorem 18 *There exists a 'context-forgetting' map* $(\ )^-$ *from sequent derivations in* **MAL** + *Mix to proof-nets with the following properties:*

(i) Let d be a sequent derivation, and let $\mathcal{R}^a$ be the proof-net such that $(d)^- = \mathcal{R}^a$. Then $\mathbf{P}(\mathcal{R}^a)$ is non-empty (in fact, it contains at least one axiom and at least one conclusion); if d is in weakening normal form, then $\mathbf{P}(\mathcal{R}) = \mathcal{R}$

(ii) (Sequentialization) If $\mathcal{R}$ is a proof-net modulo irrelevance, then there is a sequent calculus derivation d in weakening normal form such that $\mathbf{P}(\mathcal{R}) = (d)^-$.

PROOF. (see [4] for the special case **MAL** without Mix.) (i) By induction on the length of d. (ii) In $\mathbf{P}(\mathcal{R})$ weakening-links can be given canonical attachments: as a consequence, the usual proof of sequentialization for $\mathbf{MLL}^-$ + Mix goes through without the *detour* used in the proof of Theorem 13 (ii). ∎

Corollary 19 *Provability in Constant-only* **MAL** + *Mix is in* **P**.

PROOF. In Constant-only **MAL** we may consider only proofs whose axioms are **1**-axioms (a logical axiom can be replaced by a **1** axiom together with a $\bot$-*weakening*). Thus to test whether $\vdash \Gamma$ is provable, we need only to check whether the proof-structure $\mathcal{R}$ which consists of the tree of subformulas of Γ is a proof-net *modulo irrelevance*. To decide this in linear time, apply the computation of irrelevance algorithm to $\mathcal{R}$ and check whether $\mathcal{R} \neq I(\mathcal{R})$. ∎

4.3 Examples and properties of irrelevance elimination

We need to justify our algorithm for the computation of irrelevance. For this purpose we recall some facts about the structure of subnets of proof-nets in $\mathbf{MLL}^-$ with Mix (cf. [3]).

Definition 20 (i) A *non-logical axiom* in a proof-structure is a link with no premise and n conclusions, for some n. Danos–Regnier graphs for proof-nets with non-logical axioms are defined as before and so is the notion of a *proof-net with non-logical axioms* for **MAL** + Mix (with attachment of *weakening-link*).

(ii) Let $\mathcal{R}^a$ be a proof-structure for **MAL** + Mix and let $\mathcal{S}$ be a substructure of $\mathcal{R}^a$ with conclusions $C_1, \ldots, C_n$. The *complementary substructure* $\overline{\mathcal{S}}$ of $\mathcal{S}$ in $\mathcal{R}^a$ consists of all edges and links in $\mathcal{R}^a \setminus \mathcal{S}$ with the addition of a non-logical axiom with conclusions $C_1, \ldots C_n$.

(iii) Let $\mathcal{R}^a$ be a proof-net for **MAL** + Mix. A subnet $\mathcal{S}$ of $\mathcal{R}^a$ is a *normal subnet* if the complementary substructure $\overline{\mathcal{S}}$ of $\mathcal{S}$ in $\mathcal{R}^a$ is a proof-net with a non-logical axiom. A normal subnet $\mathcal{S}$ of $\mathcal{R}^a$ has the property that there exists a sequent calculus derivation d and a subderivation d_0 of d such that $(d)^- = \mathcal{R}^a$ and $(d_o)^- = \mathcal{S}$.

(iv) Let $\mathcal{R}^a$ be a proof-net for **MAL** + Mix; let A be an edge and let v_0 be a link in $\mathcal{R}^a$. The *kingdom* kA [or the *empire* eA] of A in $\mathcal{R}^a$ is the smallest [the largest] *normal* subnet of $\mathcal{R}^a$ which has A as a conclusion. The *kingdom* kv_0 [or the *empire* ev_0] of v_0 in $\mathcal{R}^a$ is the smallest [the largest] *normal* subnet of $\mathcal{R}^a$ which has v_0 as a lowermost link (so that the conclusions of v_0 are all conclusions of $\mathcal{R}^a$).

(v) The kingdom of A in $\mathcal{R}^a$ is the smallest set closed under the inductive conditions:

1 $A \in kA$;

2 If v is an axiom with conclusions $X_1, \ldots, X_n$, then

$kv = kX_1 = \ldots = kX_n$;

3 If v is a *times-link* with premises X, Y, then

$kv = k(X \otimes Y) = kX \cup kY \cup \{X \otimes Y\}$;

4 If v is a *par-link* with premises X, Y, then

$kv = k(X \wp Y) = \bigcup_s \bigcup_{Z \in \mathrm{path}_s(X,Y)} kZ \cup \{X \wp Y\}$,

where s ranges over all switchings of $\mathcal{R}$.

(v) The empire of a link v in $\mathcal{R}^a$ is the smallest subnet of $\mathcal{R}^a$ containing the set of all links v' such that for no switching s there is a path $path_s(v, v')$ reaching v "from below" (in the structural orientation).

Remark 21 (i) The computation of irrelevance algorithm was presented first in Section 7.5 of [4]. It is similar to Girard's characterization of the *empire* of a formula in a proof-net for $\mathbf{MLL}^-$, cf. Facts 2.9.4 in [11]. More precisely, for any edge A in a proof-structure $\mathcal{R}$ write e^-A for the part of $\mathcal{R}$ satisfying Girard's characterization. Now let w be any weakening-link in a proof-structure $\mathcal{R}$ for **MAL** without attachments and let us introduce an attachment $a(w)$ connecting w to some other edge. Then $i(w) = e^-(a(w))$, i.e., $i(w)$ may be regarded as the result of applying Girard's algorithm for the empire of the edge $a(w)$. Notice that $e^-(A)$ has different properties in $\mathbf{MLL}^-$ and in $\mathbf{MLL}^-$ + Mix. In $\mathbf{MLL}^-$ + Mix e^-A is a subnet, but not necessarily a *normal* subnet (cf. [3] Section 2.3), i.e., it may not be possible to find a sequentialization d of $\mathcal{R}$ such that e^-A corresponds to a subderivation of d; this remains true in the case of **MAL** + Mix as it is shown by the following examples.

(ii) Consider the proof-net in Figure 3.2 with conclusions $(\mathbf{1}\wp B) \otimes (\mathbf{1}\wp C)$, $(C^\perp \wp B^\perp) \otimes A$, $\mathbf{1}$, where the irrelevant part (marked with a solid broken line) is a subnet which *is not normal*:

Example 22

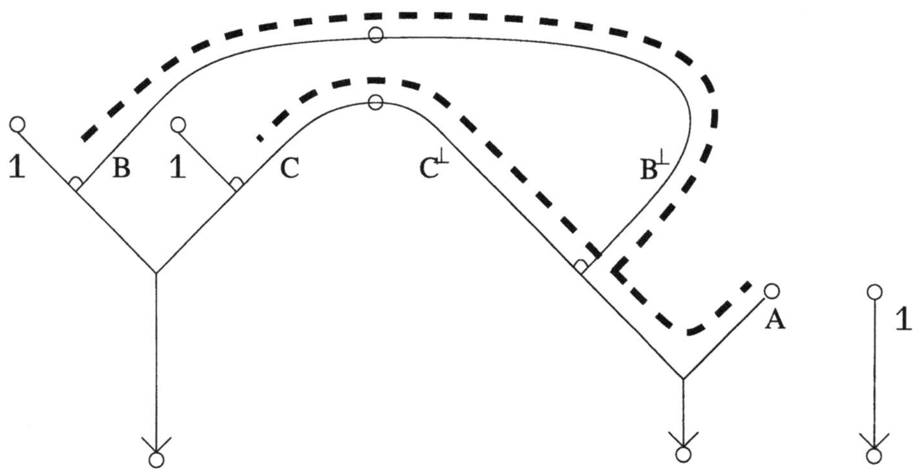

Figure 3.2. Prune

A sequentialization of such a proof-net is the following derivation d:

$$\dfrac{\dfrac{\dfrac{\dfrac{\dfrac{\vdash \mathbf{1} \quad \vdash B, B^{\perp}}{\vdash \mathbf{1}, B, B^{\perp}}\,Mix}{\vdash \mathbf{1}\wp B, B^{\perp}} \quad \dfrac{\dfrac{\vdash \mathbf{1} \quad \vdash C, C^{\perp}}{\vdash \mathbf{1}, C, C^{\perp}}\,Mix}{\vdash \mathbf{1}\wp C, C^{\perp}}}{\vdash (\mathbf{1}\wp B)\otimes(\mathbf{1}\wp C), C^{\perp}, B^{\perp}}}{\vdash (\mathbf{1}\wp B)\otimes(\mathbf{1}\wp C), C^{\perp}\wp B^{\perp}} \quad \dfrac{\vdash \mathbf{1}}{\vdash \mathbf{1}, A}}{\vdash (\mathbf{1}\wp B)\otimes(\mathbf{1}\wp C), (C^{\perp}\wp B^{\perp})\otimes A, \mathbf{1}}$$

The proof-net obtained by pruning the irrelevant part is sequentialized in the following derivation, which is not obtainable from d by permutation of inferences:

$$\dfrac{\dfrac{\dfrac{\dfrac{\vdash \mathbf{1}}{\vdash \mathbf{1}, B}}{\vdash \mathbf{1}\wp B} \quad \dfrac{\dfrac{\vdash \mathbf{1}}{\vdash \mathbf{1}, C}}{\vdash \mathbf{1}\wp C}}{\vdash (\mathbf{1}\wp B)\otimes(\mathbf{1}\wp C)} \quad \dfrac{\vdash \mathbf{1}}{\vdash (C^{\perp}\wp B^{\perp})\otimes A, \mathbf{1}}}{\vdash (\mathbf{1}\wp B)\otimes(\mathbf{1}\wp C), (C^{\perp}\wp B^{\perp})\otimes A, \mathbf{1}}\,Mix$$

(iii) Suppose *irrelevance computation algorithm* had been defined using the notion of *kingdom*: if the conclusion $A\wp B$ of a *par-link* is in $i(w)$, then we should let $k(A\wp B) \subset i(w)$. We show that this procedure, regarded as a notion of reduction, does not have the Church–Rosser property. Consider the following proof-net with conclusions $\mathbf{1}$, $X \otimes (A^{\perp}\wp B^{\perp})$, $(B\wp C)\otimes(A\wp D^{\perp})$, $(D\wp(G^{\perp}\wp G))\otimes((H\wp H^{\perp})\wp E)$, $E^{\perp}\otimes F$, $(F\wp C^{\perp})\otimes Y, \mathbf{1}$.

Example 23

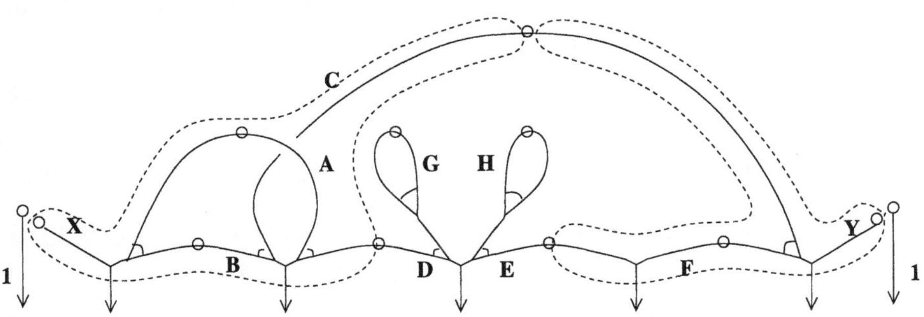

Figure 3.3. Big prune

(a) If we eliminate irrelevance starting from the weakening-link with conclusion X, then $k(A^{\perp}\wp B^{\perp})$ is included in $i(X)$ and pruned first, so the link $(B\wp C)\otimes(A\wp D^{\perp})$ is removed. Next we proceed to the weakening-link with conclusion Y, so $k(F^{\perp}\wp C^{\perp})$ is pruned: but now the only links in it are the par-link itself, a weakening-link with conclusion $C^\perp$ and an axiom $\overline{F, F^\perp}$. Finally we prune $i(F)$ and we are left with a non-empty subnet with conclusions $(D\wp(G^\perp\wp G)) \otimes ((H\wp H^\perp)\wp E)$, in addition to two **1** links (and weakening-links).

(b) If we eliminate irrelevance starting with $i(Y)$ then $k(F^\perp\wp C^\perp)$ is pruned, and the times-links $(B\wp C)\otimes(A\wp D^\perp)$, $(D\wp(G^\perp\wp G))\otimes((H\wp H^\perp)\wp E)$ and $E^\perp \otimes F$ are removed as they belong to $k(F^\perp\wp C^\perp)$. Therefore after pruning $i(X)$ we are left with nothing else than two **1**-links (and weakening-links).

Notice that our official computation of irrelevance algorithm yields in the more conservative pruning which is also given by (a).

4.4 Mix-elimination

Definition 24 Let $\mathcal{R}^{a_0}$, $\mathcal{S}^{a_1}$ be proof-nets with attachments for *weakening-links*.

(i) A map $\varphi : \mathcal{R} \to \mathcal{S}$ *preserves links* (different from weakening-links) if it is a morphism of labelled directed graphs with respect to the links other than weakening-links. In other words, for every vertex v other than a weakening-link in $\mathcal{R}$, if v has incoming arrows labelled $A_1, \ldots, A_m$ and outgoing arrows labelled $B_1, \ldots, B_n$, then $\varphi(v)$ is a vertex of $\mathcal{R}'$ with incoming arrows also labelled $A_1, \ldots, A_m$ and outgoing arrows also labelled $B_1, \ldots, B_n$. However, φ may map a *weakening-link* to a link of another kind.

(ii) A one-to-one map of labelled graphs $\varphi : \mathcal{R}^a \to \mathcal{S}^b$ is an *embedding* if it preserves links and whenever a *weakening link* w of $\mathcal{R}^a$ has a non-canonical attachment $a(w) = \ell$, then $b(\varphi(w)) = \varphi(\ell)$.

(iii) A *covering* of $\mathcal{R}^a$ is a set of embeddings $\varphi_i : \mathcal{S}_i^{a_i} \to \mathcal{R}^a$, for $i \leq n$, such that every edge in $\mathcal{R}$ is in the image of some φ_i.

(iv) The same definitions obviously apply to proof-nets without attachments.

We are going to define a procedure **S** that given a proof-net $\mathcal{R}^a$ for **MAL** + Mix generates a covering $\{\varphi_i : \mathcal{S}_i^{a_i} \to \mathcal{R}^a\}_{i\leq n}$ (cf. [15, 4]).

(I) Let C be a conclusion of $\mathcal{R}^a$. We generate a data structure *chain*(C) as follows:

1 Select a link v with no premises such that $v \prec C$ and let $path_{C,v} \subset$ *chain*(C);

   i. if v is a logical axiom or a **1** axiom, then go to step (2);

   ii. if v is a weakening-link w, then consider the edge ℓ which $a(w)$ is attached to: if ℓ' is the lowermost edge such that $\ell \preceq \ell'$ and

$\ell' \notin chain(C)$ then let $path_{\ell',\ell} \subset chain(C)$; finally, repeat step (1), choosing a link v' with no premises such that $v' \prec \ell$.

2 For every *times-link* or *cut-link* t in $chain(C)$, select a link v with no premises such that $v \prec t$ and let $path_{t,v} \subset chain(C)$; if v is a weakening-link, then go to step (1.ii.);

3 For every logical axiom *ax* in $chain(C)$, if B is a conclusion of *ax* which is *not* in $chain(C)$, then let ℓ be the lowermost edge such that $B \preceq \ell$ and $\ell \notin chain(C)$ and let $path_{\ell,ax} \subset chain(C)$. Then return to step (2).

(II) We repeat this procedure making different choices at steps 1, 2, 3 and with different conclusions C', until no new chain is generated. Let $\{chain_1^{a_1}, \ldots, chain_n^{a_n}\}$ be the chains eventually obtained, where a_i is the restriction of a to $chain_i$.

(III) For $i \leq n$, we transform $chain_i$ into a proof-structure $\mathcal{S}_i^{a_i}$ by adding weakening-links with canonical attachments:

- let p be a vertex in $chain_i$ corresponding to a par-link of $\mathcal{R}$ such that one premise ℓ is in $chain_i$, but the other premises C is not: introduce a weakening-link w labelled C in $chain_i$, together with an attachment $a_i(w)$ to the premise ℓ:

- let C' be a conclusion of $\mathcal{R}$ which does not belong to $chain_i$: introduce a weakening-link w labelled C' in $chain_i$, with an attachment $a_i(w)$ to a selected conclusion C.

(IV) Let $\{\mathcal{S}_1^{a_1}, \ldots, \mathcal{S}_n^{a_n}\}$ be the data thus obtained. We denote by $\mathbf{S}$ the operation such that $\mathbf{S}(\mathcal{R}) = \{\mathcal{S}_1^{a_1}, \ldots, \mathcal{S}_n^{a_n}\}$. Let $\mathcal{F} = \{\mathcal{S}_1^{a_1}, \ldots, \mathcal{S}_n^{a_n}\}$: we write $\mathbf{P}(\mathcal{F}) = \{\mathbf{P}(\mathcal{S}_1^{a_1}), \ldots, \mathbf{P}(\mathcal{S}_n^{a_n})\}$.

Theorem 25 *(i) Let $\mathcal{R}^a$ be a proof-net with conclusions Γ in* **MAL** + *Mix. The operation* $\mathbf{S}(\mathcal{R}) = \{\mathcal{S}_1^{a_1}, \ldots, \mathcal{S}_n^{a_n}\}$ *yields a covering* $\{\varphi_i : \mathcal{S}_i^{a_i} \to \mathcal{R}^a\}_{i \leq n}$, *where the $\mathcal{S}_i^{a_i}$ are proof-nets in* **MAL** without Mix.

(ii) $\mathbf{PS}(\mathcal{R}^a) = \mathbf{SP}(\mathcal{R}^a)$.

PROOF. (i) Let $\mathcal{R}^a$ be a proof-net with conclusions Γ. Each $\mathcal{S}_i$ is a proof-structure: the inclusion map $\iota : \mathcal{S}_i \to \mathcal{R}$ preserves *times-links* and *cut-links* by step (I.2.), *axiom* links by step (I.3.) and each $\mathcal{S}_i^{a_i}$ has conclusions Γ by step (III). Moreover $\iota : \mathcal{S}_i^{a_i} \to \mathcal{R}^a$ is an embedding because the attachments a_i contain the restriction of a to $chain_i$ and all other weakening-links have canonical attachments. Also by construction it is clear that every edge v in $\mathcal{R}^a$ is in the image of some φ_i. Therefore $\mathbf{S}(\mathcal{R})$ yields a covering.

Each $\mathcal{S}_i^{a_i}$ is a proof-net: a cyclic Danos–Regnier graph in $\mathcal{S}_i^{a_i}$ would be one in $\mathcal{R}$. We need to show that $\mathcal{S}_i^{a_i}$ is a proof-net in **MAL** without Mix. The number of axioms, *times-links* and *cut-links* in $\mathcal{S}_i^{a_i}$ is determined by the procedure **S**, which after the first axiom selects an axiom for each *times-link* or *cut-link* encountered. Therefore by Lemma 4 any sequentialization of $\mathcal{S}_i^{a_i}$ is a derivation with no mix-rule.

(ii) Notice that $\mathbf{SP}(\mathcal{R}^a) \subset \mathbf{S}(\mathcal{R}^a)$. Indeed the chains obtained by the procedure **S** applied to $\mathbf{P}(\mathcal{R}^a)$ are also obtained by the procedure **S** applied to $\mathcal{R}^a$ starting with a conclusion C in $\mathbf{P}(\mathcal{R}^a)$ and remaining in $\mathbf{P}(\mathcal{R}^a)$ by selecting a logical axiom or a **1** axiom at steps (1) and (2) of the procedure; this is always possible since $C \in \mathbf{P}(\mathcal{R}^a)$. The chains $chain_1, \ldots, chain_k$ obtained in this way will be called *basic chains*. Let $\mathbf{SP}(\mathcal{R}) = \{\mathcal{S}_1^{a_1}, \ldots, \mathcal{S}_k^{a_k}\}$.

Furthermore notice that the procedure **P** applied to $\mathbf{S}(\mathcal{R}^a)$ transforms chains into basic chains. Indeed computing and eliminating irrelevance from a link w considered at a step (I.ii.) has the effect of removing the part of the chain visited up to that step: but this would also have been achieved by starting with the conclusion C' below ℓ' and by selecting $path_{C',\ell'}$ and so on. We conclude that **P** transforms $\mathbf{S}(\mathcal{R}^a)$ into $\mathbf{SP}(\mathcal{R}^a)$, and therefore $\mathbf{PS}(\mathcal{R}^a) = \mathbf{SP}(\mathcal{R}^a)$. ∎

Definition 26 Let $\mathcal{R}^a$ be a proof-structure for **MAL** + Mix such that $\mathcal{R} \neq I(\mathcal{R})$. Let *ax* be an axiom of $\mathcal{R}^a$ and let $\mathcal{R}^a_{ax}$ be the result of replacing *ax* in $\mathcal{R}^a$ with two weakening-links if *ax* is a logical axiom or with one weakening-link if *ax* is a **1** axiom. The set of axioms in $\mathcal{R}^a$ is *minimal* if for every *ax* in $\mathcal{R}^a$, $\mathcal{R}_{ax} = I(\mathcal{R}_{ax})$.

We give a direct proof that the connectedness condition of all D–R graph is equivalent to the minimality condition on axioms. We need the following fact:

Fact 27 *Given a proof-net with attachments $\mathcal{R}^a$ for* **MAL** + *Mix and any D–R switching s, any connected component of $s\mathcal{R}^a$ containing a weakening-link w contains also an axiom of $\mathcal{R}^a$, which is reached from w by crossing attachments and then always proceeding upwards in the structural orientation.*

PROOF. Let the attachment $a(w)$ be connected to an edge ℓ_1; let $\mathcal{S}_{\ell_1}$ be the smallest substructure ending with ℓ_1 and consider the uppermost links in $\mathcal{S}_{\ell_1}$ which are connected to ℓ_1 in $s\mathcal{R}^a$. If any such vertex is an axiom link, then we are done. Otherwise given a *weakening-link* w_1, let ℓ_2 be the edge which $a(w_1)$ is connected to, and consider the substructure ending with ℓ_2, and so on. In every case we proceed upwards in the structural orientation. Eventually we must reach an axiom. ∎

Proposition 28 ([4], Section 6) *Let $\mathcal{R}^a$ be a proof-net for* **MAL** + *Mix.* $\mathcal{R}^a$ *be a proof-net for* **MAL** *without Mix if and only if its set of axioms is minimal. Thus the minimality condition on axioms is equivalent to the condition of connectedness of all D–R graphs.*

PROOF. Since $\mathcal{R}^a$ is a proof-net, by Theorem 18 $I(\mathcal{R}) \neq \mathcal{R}$. Suppose $s\mathcal{R}^a$ is disconnected, for some switching s. If $\mathcal{R}^a$ consists of disconnected proof-structures, then the removal of an axiom in one substructure does not affect the computation of irrelevance in another disconnected substructure so $\mathcal{R}^a$ does not satisfy the minimality condition on axioms.
If $\mathcal{R}^a$ as a proof-structure is connected, then we may assume that there are par-links $p_1, \ldots p_k$ such that their premises belong to different connected components of $s\mathcal{R}^a$. Select a switching s_1 so that if the D–R graph determined by s_1 reaches a premise ℓ_1 of p_i then s_1 choses the other premise ℓ_2 and a switching s_2 which makes the opposite choices. Therefore the switchings s_1 and s_2 determine two connected components α and β of $s_1\mathcal{S}_i^{a_i}$ and $s_2\mathcal{S}_i^{a_i}$ respectively, where α and β contain different premises of the links p_i. It is easy to see that the removal of one axiom in α may at most make α irrelevant, but not β.

Conversely, if every D–R graph $s\mathcal{R}^a$ is connected, then $\mathcal{R}^a$, as a proof-net with attachments, has properties similar to those of proof-nets for **MLL**$^-$, and it is easy to show by induction on the ordering of the kingdoms (cf. Lemma 3 in [6]) that the removal of one axiom in $\mathcal{R}^a$ induces $\mathcal{R}^a = I(\mathcal{R}^a)$. ∎

Example 29 (See Figure 3.4.)

Example 29 shows a proof-net $\mathcal{R}$ in **MLL**$^-$ + Mix (top figure), representing a proof of

$$\vdash D \wp C^\perp, C \otimes (B^\perp \wp (B \wp A)), (A^\perp \wp D^\perp) \otimes E, E^\perp,$$

together with $\mathbf{S}(\mathcal{R})$, namely, $\{\mathcal{S}_1, \mathcal{S}_2, \mathcal{S}_3\}$ (below, from left to right). Here:

- $\mathcal{S}_1$ is determined by the chain which starts either with $D \wp C^\perp$ or with $(A^\perp \wp D^\perp) \otimes E$ and reaches the axioms $\overline{D, D^\perp}$, and $\overline{E, E^\perp}$;

- $\mathcal{S}_2$ is determined by the chain which starts either with $D \wp C^\perp$ or with $C \otimes (B^\perp \wp (B \wp A))$, reaches the axioms $\overline{C^\perp, C}$ and chooses $\overline{B^\perp, B}$;

- $\mathcal{S}_2$ is determined either (i) by the chain which starts with $C \otimes (B^\perp \wp (B \wp A))$, reaches the axioms $\overline{C^\perp, C}$ and $\overline{E, E^\perp}$ after choosing $\overline{A, A^\perp}$ or (ii) by the chain which starts with $(A^\perp \wp D^\perp) \otimes E$ and reaches the axioms $\overline{E, E^\perp}$, $\overline{A, A^\perp}$ and $\overline{C^\perp, C}$.

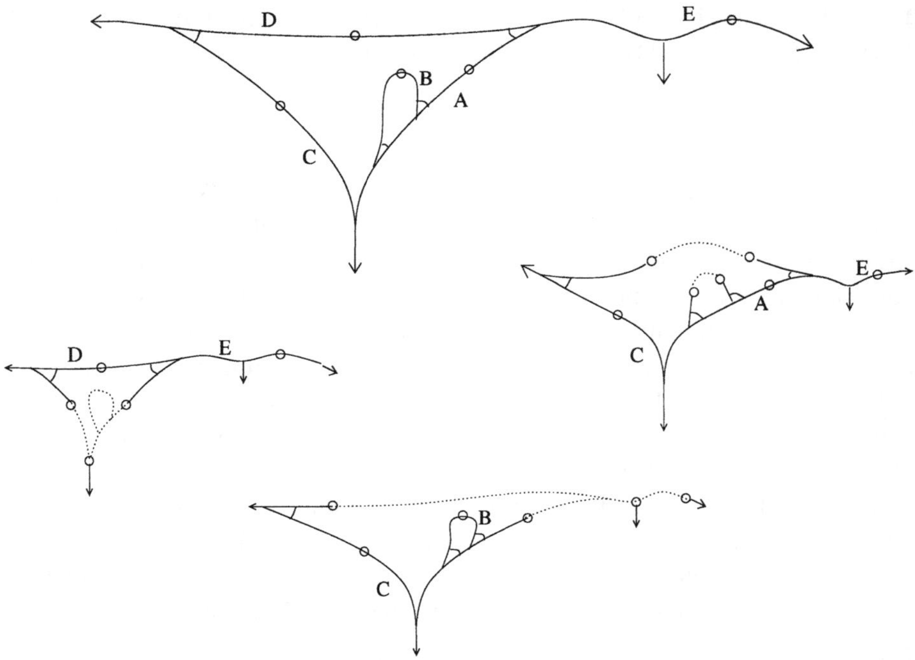

Figure 3.4. Cover

5. Cut-elimination modulo irrelevance

In this section we define a procedure of cut-elimination for proof-nets *modulo irrelevance*: this notion of reduction is based on a notion of *weakening-expansion* which applies to the irrelevant part of a proof-net, but is not admissible in general. Therefore this notion of reduction does not extend to proof-nets with attachments of the *weakening-links*.

It is essential to notice that the notion of the *irrelevant part* of a proof-net is highly unstable under our notion of reduction: if $\mathcal{R}_1$ reduces to $\mathcal{R}_2$ and $\mathcal{R}_2$ reduces to $\mathcal{R}_3$, it may very well be the case that $I(\mathcal{R}_1) \subset I(\mathcal{R}_2)$ but also $I(\mathcal{R}_2) \supset I(\mathcal{R}_3)$. Therefore in the cut-elimination process we will *mark* the irrelevant part of a proof-net, not *remove* it.

The cut-reduction for proof-nets *modulo irrelevance* for **MAL** + Mix (Figure 3.5) are the same as those of **MLL** without units. In addition, there is the *two weakenings / cut* reduction which annihilates a cut-link whose premises are both conclusions of weakening-links. Moreover there are the *weakening-expansions*:

- if a premise $A \otimes B$ $[A^{\perp} \wp B^{\perp}]$ of a *cut-link* is conclusion of a *times-link* [*par-link*] and the other premise of the cut-link is conclusion of a *weakening-link* w, then w is replaced by a *par-link* [*times-link*] whose premises are conclusions of *weakening-links* w_1 and w_2.

Theorem 30 *(i) If $\mathcal{R}$ is a proof-net modulo irrelevance and $\mathcal{R}$ reduces to $\mathcal{R}'$, then $\mathcal{R}'$ is a proof-net modulo irrelevance.*

(ii) The cut-elimination process for proof-nets modulo irrelevance for **MAL** + *Mix has the strong normalization and Church–Rosser property.*

PROOF. (i) If $s\mathcal{R}$ is acyclic for every switching s, where there is no attachment for weakening links in $\mathcal{R}$, then the usual argument for $\mathbf{MLL}^-$ + Mix shows that $s\mathcal{R}'$ is acyclic for every switching s. If $\mathbf{P}(\mathcal{R})$ is non-empty, then $\mathbf{P}(\mathcal{R}')$ is non-empty by Lemma 31. (ii) Notice that if a cut-link occurs in $\mathcal{R}$, then exactly one among the five reduction or expansions applies to it. Moreover, such operations are strictly *local.* ∎

Lemma 31 *Let $\mathcal{R}$ be a proof-net modulo irrelevance for* **MAL** + *Mix.* If $\mathcal{R}$ *reduces to $\mathcal{R}'$, then $\mathbf{P}(\mathcal{R}') = \mathcal{R}' \setminus I(\mathcal{R}')$ is non-empty.*

PROOF. [of Lemma 31] First notice that *weakening-expansions* cannot make any relevant part of the proof-net irrelevant. Also in a *times /par* reduction if one premise of the times-link is in the irrelevant part, then the *cut-link* itself is irrelevant, and similarly if both premises of the par-link are irrelevant. Consider a *times /par* reduction where the par-link with conclusion $A^\perp \wp B^\perp$ has one and only one irrelevant premise, e.g., $B^\perp$ is conclusion of a weakening link w. Suppose both cut-links resulting from the reduction were irrelevant in $\mathcal{R}'$. Since $\{A, A^\perp\} \subset I(\mathcal{R}')$ but $A, A^\perp \notin I(\mathcal{R})$ it must be the case that by computing $I(w)$ in $\mathcal{R}'$ we reach either A or $A^\perp$. Now the algorithm for the computation of $I(w)$ starts with $B^\perp$, the conclusion of w, and continues as in the computation of $e^-(B)$: but since $\mathcal{R}'$ is like $\mathcal{R}$ except for the local rewriting of the *cut-link* in question, it follows that not only in $\mathcal{R}'$ but already in $\mathcal{R}$ we must have $A \in e^-B$ or $A^\perp \in e^-B$. It is easy to show that for some switching s there exists a path $path_s(A, B)$ or $path_s(A^\perp, B)$ in $s\mathcal{R}$. But this contradicts the acyclicity condition of the Danos–Regnier graphs for $\mathcal{R}$. ∎

Remark 32 Let $\mathcal{R}^a$ be a proof-net with attachments and suppose a *weakening-expansion* is applied to $\mathcal{R}$, yielding $\mathcal{S}$ with a new *times-link*: then there may be no system of attachments b such that $\mathcal{S}^b$ is a proof-net. This can be seen by Lemma 4: if $\mathcal{R}$ satisfies $\#ax = \#\otimes + \#cut + 1$ then in $\mathcal{S}$ the number of times-links is increased by one *coeteris paribus.* A stronger result would be to show that if $\mathcal{S}$ is the *cut-free* proof-structure resulting from cut-elimination, then there exists a system of attachments b such that $\mathcal{S}^b$ is a proof-net. We will not pursue the matter here.

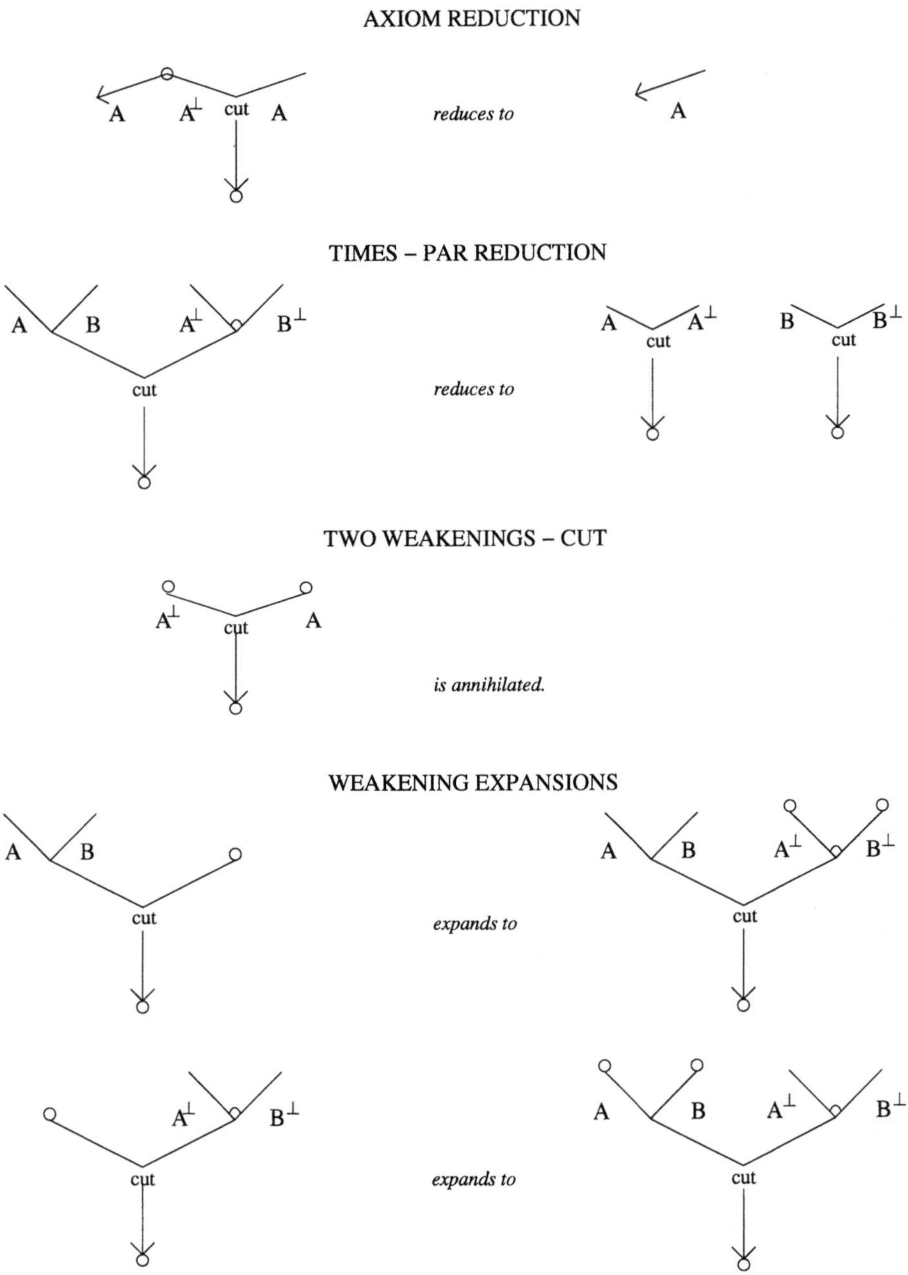

Figure 3.5. Cut

Example 33

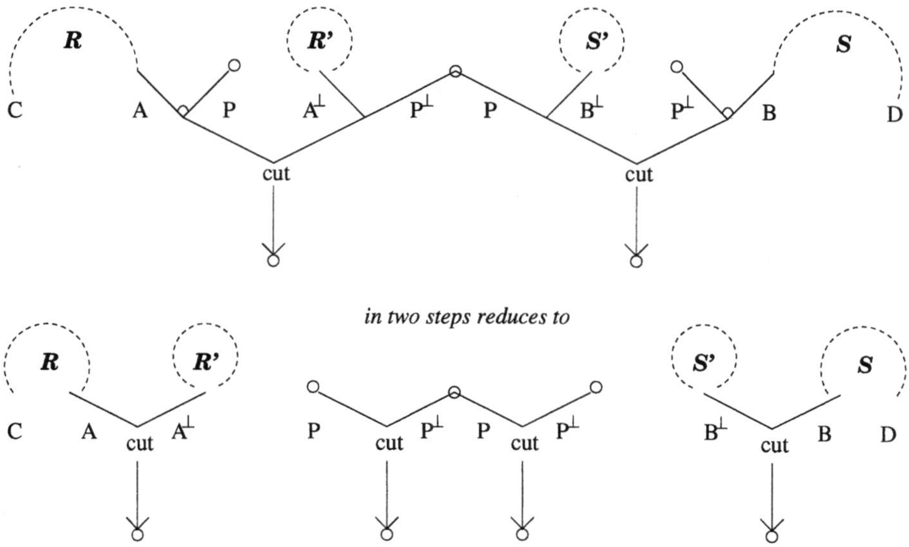

Figure 3.6. Example 33

Remark 34 Example 33 shows that the Church–Rosser property is lost if we *erase* the irrelevant part, instead of just *marking* it, during the cut-elimination process: e.g., after one reduction step, we would erase either $e(P)$ or $e^{-}(p^{\perp})$. Notice that after two more steps, (an *axiom reduction* and a *two weakenings / cut reduction*) we obtain two disconnected proof-nets: therefore our cut-elimination process essentially requires the use of the *mix-rule*.

6. Symmetric reductions require Mix

Can we define a symmetric reduction for *proof-nets* with contraction-links? An example in the Appendix B2 of Girard [12] shows that in order to do so we need the rule Mix. Let a and b atomic formulas and consider the proof-net π associated with a proof of $\vdash a \wedge b, a \wedge \neg b, \neg b \wedge a, \neg a \wedge \neg b$, and the proof-net π' to which π reduces as indicated in Figure 3.7.

Example 35

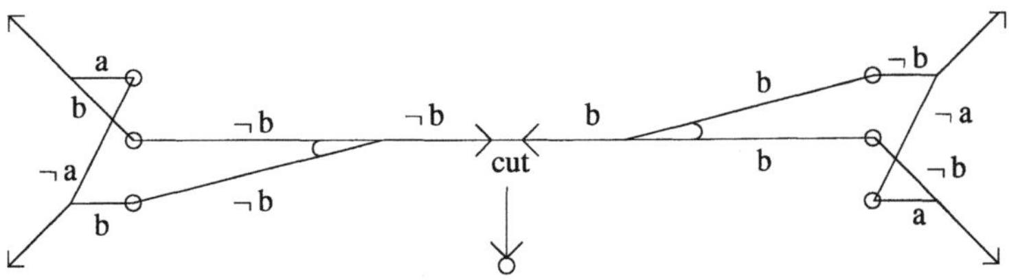

reduces to

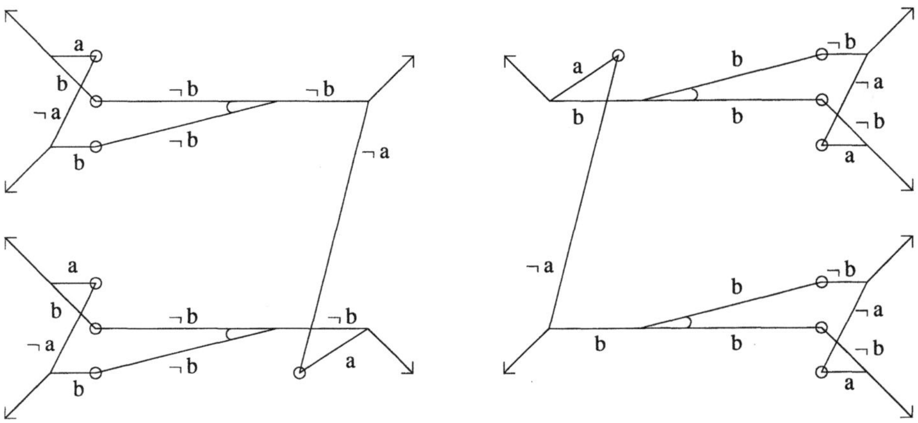

Figure 3.7. Mix

The reduction in Example 35 is the only one which has the following property of symmetry. Let $a\backslash\neg a$ the substitution operation which replaces every edge a with an edge $\neg a$ throughout the proof-net. Consider the group of substitutions **S** consisting of

$$\{\text{identity},\ a\backslash\neg a,\ b\backslash\neg b,\ a\backslash\neg a \text{ and } b\backslash\neg b\}$$

on the graphs π and π'. This group acts transitively on both π and π'; both π and π' are invariant under the substitutions of **S** – under the assumption that axioms, cut-links and contraction-links are symmetric, i.e., not ordered. It is easy to see that every *asymmetric* reduction, sending π to one of the connected components of π', does *not* preserve the property of invariance under substitution. Therefore Girard's example shows that there cannot be any symmetric cut-elimination for classical logic which does not use the rule Mix. It is also an easy exercise to see that (the proof-net translation of) the cross-cut reduction (as described in the intoduction) applied to the proof-net π yields a proof-net π'' which is *not* invariant under some substitution. The problem of finding a proof-net representation for classical logic is therefore entirely open.

Notes

1. Ketonen's notion of a chain was independently rediscovered by A. Asperti [1] in his characterization of proof-nets for **MLL**$^-$ + Mix through distributed processes. Therefore all the results in this paper can be interpreted in terms of Asperti's distributed processes.

References

[1] A. Asperti. Causal Dependencies in Multiplicative Linear Logic with MIX. *Mathematical Structures in Computer Science* 5:351–380, 1995.

[2] G. Bellin. *Mechanizing Proof Theory: Resource-Aware Logics and Proof-Transformations to Extract Implicit Information*, Phd Thesis, Stanford University. Available as: Report CST-80-91, June 1990, Dept. of Computer Science, Univ. of Edinburgh.

[3] G. Bellin. Subnets of proof-nets in multiplicative linear logic with Mix, *Mathematical Structures in Computer Science* 7:663–699, 1997.

[4] G. Bellin and J. Ketonen. A Decision Procedure Revisited: Notes on Direct Logic, Linear Logic and its Implementation, *Theoretical Computer Science* 95(1):115–142, 1992.

[5] G. Bellin and P. J. Scott. Selected papers of the meeting on Mathematical Foundations of Programming Semantics (MFPS 92), Part 1 (Oxford 1992), *Theoretical Computer Science* 135(1):11–65, 1994.

[6] G. Bellin and J. van de Wiele. Subnets of Proof-nets in **MLL**$^-$, in *Advances in Linear Logic*, J-Y. Girard, Y Lafont and L. Regnier editors, London Mathematical Society Lecture Note Series 222, Cambridge University Press, 1995, pp. 249–270.

[7] C. Urban and G. M. Bierman. Strong Normalization of Cut-Elimination in Classical Logic, *Fundamenta Informaticae* XX:1–32, 2000.

[8] V. Danos, J-B. Joinet and H. Schellinx. LKQ and LKT: Sequent calculi for second order logic based upon linear decomposition of classical implication, in *Advances in Linear Logic*, J.-Y. Girard, Y. Lafont and L. Regnier editors, London Mathematical Society Lecture Note Series 222, Cambridge University Press, 1995, pp. 211–224.

[9] V. Danos, J-B. Joinet and H. Schellinx. A new deconstructive logic: linear logic. Preprint n.52, Équipe de Logique Mathématique, Université Paris VII, Octobre 1994.

[10] A. Fleury and C. Retoré. The Mix Rule. *Mathematical Structures in Computer Science* 4:273–85, 1994.

[11] J-Y. Girard. Linear Logic. *Theoretical Computer Science* 50:1–102, 1987.

[12] J-Y. Girard. A new constructive logic: classical logic. *Mathematical Structures in Computer Science* 1:255–296, 1991.

[13] J-Y. Girard. On the unity of logic. *Annals of Pure and Applied Logic* 59:201–217, 1993.

[14] J-Y. Girard, Y.Lafont, and P. Taylor. *Proofs and Types*. Cambridge Tracts in Theoretical Computer Science 7, Cambridge University Press, 1989.

[15] J.Ketonen and R.Weyhrauch. A Decidable Fragment of Predicate Calculus, *Theoretical Computer Science* 32:297–307, 1984.

[16] P. Lincoln and T. Winkler. Constant-only multiplicative linear logic is NP-complete. Selected papers of the meeting on Mathematical Foundations of Programming Semantics (MFPS 92), Part 1 (Oxford 1992), *Theoretical Computer Science* 135(1):155–169, 1994.

[17] M. Parigot. Free Deduction: an analysis of computation in classical logic. In Voronkov, A., editor, *Russian Conference on Logic Programming*, pp. 361–380. Sprinver Verlag. LNAI 592, 1991.

[18] M. Parigot. Strong normalization for second order classical natural deduction. in *Logic in Computer Science*, pp. 39–46. IEEE Computer Society Press. Proceedings of the Eight Annual Symposium LICS, Montréal, June 19–23, 1992.

[19] D. Prawitz. *Natural deduction. A proof-theoretic study.* Almquist and Wiksell, Stockholm, 1965.

[20] D. Prawitz. Ideas and Results in Proof Theory, in *Proceedings of the Second Scandinavian Logic Symposium*, ed. Fenstad, North-Holland, 1971.

[21] Edmund Robinson. Proof Nets for Classical Logic, preprint, Dagstuhl workshop 1141/1 on Semantic Foundations of Proof-search, April 2001. `http://www.dcs.qmw.ac.uk/~edmundr`.

Chapter 4

PROOF SYSTEMS FOR π-CALCULUS LOGICS

Mads Dam
Dept. of Teleinformatics
Royal Institute of Technology, Sweden
mfd@sics.se

Abstract We study the problem of verifying general temporal and functional properties of mobile and dynamic process networks, cast in terms of the π-calculus. Much of the expressive power of this calculus derives from the combination of name generation and communication (to handle mobility) with dynamic process creation. In the paper we introduce the π-μ-calculus, an extension of the modal mu-calculus with name equality, inequality, first-order universal and existential quantification, and primitives for name input and output as an appropriate temporal logic for the π-calculus. A compositional proof system is given with the scope of verifying dynamic networks of π-calculus agents against properties specified in this logic. The proof system consists of a local part based, roughly, on the classical sequent calculus extended with data structures for private names, and rules to support process structure dependent reasoning. In addition the proof system contains a rule of discharge to close well-founded cycles in the proof graph. The proof system is shown to be sound in general and weakly complete for the non-recursive fragment of the specification logic. We also obtain a weak completeness result for recursive formulas against finite-control calculus processes. Two examples are considered. The first example is based on Milner's encoding of data types into the π-calculus, specifically the natural numbers. This encoding is interesting from the point of view of verification, since it makes essential use of all the distinguishing features of the π-calculus, including dynamic process creation. Corresponding to the encoding of natural numbers into the π-calculus we propose an encoding of the type of natural numbers into the π-μ-calculus and establish some type correctness properties. As the second example we consider a garbage-collecting unbounded buffer (which dynamically create and destroy buffer cells) and show how to establish absence of spurious output of such a system.

Keywords: π-calculus

R.J.G.B. de Queiroz (ed.), Logic for Concurrency and Synchronisation, 145–212.

1. Introduction

In this paper we study the problem of verifying general temporal and functional properties of mobile and dynamic process networks. In such a network processes can be created, and process interconnection topology can be modified during execution. Mobility is often achieved by a mechanism for generating inter-process links, and passing them between processes. For instance, in the π-calculus [12] links are primitive names of communication channels, and in the programming language Erlang [3] links serve as both communication channels and process identifiers. The combination of mobility with dynamic creation of processes is very powerful. For the case of π-calculus this is witnessed by the encodings into the π-calculus of, e.g., data types, functions, objects, and higher-order processes [10, 11, 19, 16]. But it is also a power which is used extensively in everyday programming practice, to dynamically set up data and process structures, to adapt applications and systems to change in their environments, to support fault tolerance and code replacement in running systems, to name just a few scenarios (cf. [3]).

The cost of this power is the unbounded and essential growth of state spaces as processes compute, rendering analyses by global state space exploration in general impossible. An alternative which was explored in [5] for CCS is to take a compositional, proof-based approach. In this paper we extend this approach to the π-calculus and show how a compositional proof system can be built with the scope of verifying quite general properties of π-calculus processes.

It is important to note that this is an ambitious and difficult task. There are few approaches to verification around that can deal satisfactorily with this kind of problem even in settings that are computationally simpler than that of the π-calculus. We do not claim to give the definite answer to this problem in this paper – much more work will be required before stable methods and criteria for measuring their usefulness have been found. What we do claim, however, is to present *one* possible approach that does, according to some set of criteria (soundness, weak forms of completeness, non-trivial examples) address and adequately solve the problem.

The investigation is cast in terms of judgments of the form $\Gamma \vdash^s E : \phi$ where E is an open π-calculus term, ϕ is a general temporal formula, Γ is a sequence of assumptions governing agent variables free in E, and s is a set of channel names local to E. Verification in such a setting must be inherently compositional, since the behaviour of E is defined only up to properties of its constituent parts, as determined by Γ. Temporal specifications are given in an extension of the modal μ-calculus with name equality, inequality, first-order universal and existential quantification, and primitives for name input and output along the lines of [4]. We present a proof system for judgments which is sound and weakly complete for recursion-free formulas, and for general formulas we show that the proof system is sound in general and weakly complete for finite control processes. As the CCS-based proof system of [5] the proof system consists of a (rather large) number of proof rules that account, roughly, for the modal fragment, plus a single rule of discharge to handle fixed points.

We illustrate the use and scope of the proof system by means of two examples. First an example based on Milner's encoding of data types into the π-calculus shows how the type of natural numbers can be encoded. While we have no inherent interest in the natural numbers the example is of interest since it exercises all important aspects of both the π-calculus (dynamic process creation, name generation, scope extrusion, and communication), and of the temporal logic (fixed points, alternation, quantification, equality and inequality, input and output). Moreover, it illustrates well an important point of the proof system, namely its capacity to uniformly prove properties pertaining to infinite collections of essentially distinct agents. Finally the example is surprisingly subtle: It is not at all trivial to arrive at a suitable encoding of the property of "representing a natural number", let alone to formally prove type correctness properties such as those we consider.

As a second example we consider bounded and (in particular) unbounded buffers and show, as the main example, absence of spurious output from an unbounded directional buffer which is "garbage-collecting" in the sense that buffer elements which become empty are terminated.

The chosen setting for the π-calculus is introduced in Section 2. We work with a largely standard version of the polyadic π-calculus [10], using recursive process definitions and incorporating a standard conditional. We then proceed to introduce the "π-μ-calculus", a version of first-order μ-calculus in the style of Park [14], extended with π-calculus-specific primitives. In Section 4 we present a few example specifications, notably some examples of specifications addressing basic buffer properties (order preservation, absence of spurious output, absence of message loss) and a formalisation of the "type of natural numbers". Then we proceed to present the modal fragment of the proof system, the shape of judgments, their semantics, and the proof rules along with a few example derivations. The modal fragment of the proof system is grouped naturally into several collections of rules. Some (the logical rules, governing the first-order connectives) are largely standard. Others (the structural rules and – to a slightly less extent – the rules for equality and inequality) are somewhat interesting in the way private names are handled. Two further collections of rules are the rules for transition modalities and the rules for modalities reflecting name input and output, the io-modalities. The modal fragment of the proof system is shown to be sound and weakly complete in Section 6. We then proceed to consider fixed points. In Section 7 the semantics and proof rules for fixed point formulas is introduced. The handling of fixed points follows a novel approach, introduced first in [6]. This approach exploits approximation ordinals in an explicit way paving the way for a logical account which is far simpler and more elegant than the story given in [5]. The proof rules are grouped in two: First is a collection of local rules, providing support for fixed point unfolding, and introduction and unfolding of approximated formulas. The second and final part of the proof system consists of a single rule of discharge, providing, roughly, a formal correlate of well-founded induction. In Section 8, then, we go on to prove a weak form of completeness, by reducing proofs in a "global", model-checking oriented version of the proof system which is known to be complete, to proofs in the compositional system. Sections 9 and 10 contain verification examples pertaining to the natural number and buffer examples and Section 11, finally, contains our conclusions and pointers for future work. Proofs of the main completeness results have been deferred to appendices.

2. Preliminaries on the π-calculus

In this section we introduce the π-calculus, its syntax and operational semantics.

Syntax. Assume denumerable sets of *agent (abstraction) identifiers* D and of (channel) *names* $a, b, c \in \mathcal{G}$. We use $\tilde{a}$ to denote vectors $a_1, \ldots, a_n$. *Processes* $P, Q \in \mathcal{P}$ along with *abstractions*, $A \in \mathcal{A}$, and *concretions*, $C \in \mathcal{C}$, are generated by the following abstract syntax:

$$\begin{aligned} P &::= 0 \,\big|\, P + P \,\big|\, \tau.P \,\big|\, a.A \,\big|\, \overline{a}.C \,\big|\, P \mid P \,\big|\, \textbf{if } a = b \textbf{ then } P \textbf{ else } P \,\big| \\ &\quad\;\; \nu a.P \,\big|\, D(\tilde{a}) \\ A &::= (a)A \,\big|\, P \\ C &::= \langle a \rangle C \,\big|\, \nu a.C \,\big|\, P \end{aligned}$$

The notation is largely standard. We use a standard if-then-else notation for the conditional in place of the matching/mismatching notations $[a = b]P$ and $[a \neq b]P$ more common in π-calculus contexts. This is to avoid excessive proliferation of the "box" notation which is used later also for one of the modal operators.

Agents are interpreted relative to an environment which determines, for each agent identifier D, a defining equation $D(\tilde{a}) \stackrel{\Delta}{=} P$. An alternative to such recursive definitions is to use the "bang" operator $!P$, easily definable by the identity $!P = P \mid !P$. It is also possible, though a little more involved, to derive recursively defined processes using the "bang".

Binding. The calculus has two binding operators:

- Abstractions $(a)A$ binds a in A

- Restrictions $\nu a.P$ ($\nu a.C$) binds a in P (in C).

Agent expressions are considered only up to renaming of bound names. $fn(P)$, $bn(P)$, and $n(P)$ are the free names, bound names, and names (free or bound) of P, respectively. We use $P\{a_1/b_1, \ldots, a_n/b_n\}$, or in vectorized form, $P\{\tilde{a}/\tilde{b}\}$, as the notation for uniform and simultaneous substitution, here of names.

Actions. An *action*, α, is either the internal action τ, an *input action* $a(\tilde{b})$, or an *output action* of the shape $\nu\tilde{b}.\overline{a}\langle\tilde{c}\rangle$ where $\tilde{b} \subseteq \tilde{c}$ and $a \notin \tilde{b}$. Actions are used as derived notation, by the clauses

$$
\begin{aligned}
a(b_1, \ldots, b_n).P &\triangleq a.(b_1) \cdots (b_n)P \\
\nu b_1, \ldots, b_n.\overline{a}\langle c_1, \ldots, c_m\rangle.P &\triangleq \nu b_1. \cdots .\nu b_n.\overline{a}.\langle c_1\rangle \cdots \langle c_m\rangle P
\end{aligned}
$$

By means of the identifications

$$
\begin{aligned}
\nu a.\nu b.C &= \nu b.\nu a.C \\
\nu a.\langle b\rangle C &= \langle b\rangle \nu a.C \quad (\text{if } a \neq b) \\
\overline{a}.\nu b.C &= \nu b.\overline{a}.C \quad (\text{if } a \neq b)
\end{aligned}
$$

it is possible to rewrite each process $a.A$ or $\overline{a}.C$ into one of the shape $\alpha.P$. Given the binding conventions, the functions fn, bn and n are extended to actions in the obvious way.

Transition semantics. The intended semantics is quite well known from CCS [9] and the original π-calculus papers [12]:

- 0 is the inert process, incapable of performing transitions.
- $P + Q$ is the CCS choice construction: A transition of $P + Q$ is a transition of either P or of Q.
- $\alpha.P$ is an agent offering just one transition, labelled with α, with P as the resulting next state.
- Actions of the shape τ represent internal, or unobservable, actions. Actions of the shape $a(\tilde{b})$ represents input actions, and actions of the shape $\nu\tilde{b}.\overline{a}\langle\tilde{c}\rangle$ represents output actions.
- In the case $\alpha = \nu b.\overline{a}.\langle b\rangle$ the process concerned is emitting a private name b to the receiver, causing the scope of that b to be *extruded*, i.e. extended, possibly involving alpha conversions, to encompass the receiving process.

$$\textsc{Sum}\ \frac{P \xrightarrow{\alpha} P'}{P + Q \xrightarrow{\alpha} P'} \qquad \textsc{Pre}\ \frac{\cdot}{\alpha.P \xrightarrow{\alpha} P}$$

$$\textsc{Com}\ \frac{P \xrightarrow{\nu\tilde{b}.\bar{a}\langle\tilde{b'}\rangle} P' \quad Q \xrightarrow{a(\tilde{c})} Q'}{P \mid Q \xrightarrow{\tau} \nu b.(P' \mid (Q'\{\tilde{b'}/\tilde{c}\}))} \qquad \textsc{Par}\ \frac{P \xrightarrow{\alpha} P'}{P \mid Q \xrightarrow{\alpha} P' \mid Q}$$

$$\textsc{If}_1\ \frac{P \xrightarrow{\alpha} P'}{\textbf{if } a = a \textbf{ then } P \textbf{ else } Q \xrightarrow{\alpha} P'}$$

$$\textsc{If}_2\ \frac{Q \xrightarrow{\alpha} Q'}{\textbf{if } a = b \textbf{ then } P \textbf{ else } Q \xrightarrow{\alpha} Q'}\ (a \neq b)$$

$$\textsc{Res}\ \frac{P \xrightarrow{\alpha} P'}{\nu a.P \xrightarrow{\alpha} \nu a.P'}\ (a \notin n(\alpha))$$

$$\textsc{Open}\ \frac{P \xrightarrow{\nu\tilde{b}.\bar{a}\langle\tilde{b'}\rangle} P'}{\nu c.P \xrightarrow{\nu\tilde{b}.c.\bar{a}\langle\tilde{b'}\rangle} P'}\ (c \neq a, c \in \tilde{b'}\backslash\tilde{b})$$

$$\textsc{Id}\ \frac{P\{\tilde{b}/\tilde{a}\} \xrightarrow{\alpha} P'}{D(\tilde{b}) \xrightarrow{\alpha} P'}\ (D(\tilde{a}) \triangleq P)$$

Table 4.1. Transition rules

- $P \mid Q$ is parallel composition offering the transitions of P and Q separately, as well as internal transitions arising from (synchronous) communication between P and Q. Communications cause input and output actions to be matched.
- **if** $v_1 = v_2$ **then** P_1 **else** P_2 is the conditional.
- $\nu a.P$ declares a new name a to be used by P.
- $D(\tilde{a})$ is invocation of the process defined by D.

Example 1 The agent D defined by

$$D(a) = (\nu b).\bar{a}\langle b\rangle.b(c).D(c)$$

is a recursive agent, parametrised on a, which declares a new local name b which is passed along a to the environment whereupon a name c can be received along b, resulting in the agent $D(c)$.

The transition rules are given in Table 4.1. The table does not include symmetric versions of the rules Sum, Com, and Par. Those should be assumed. Since terms are identified up to alpha-conversion many side conditions intending to prevent confusion of scope can be avoided.

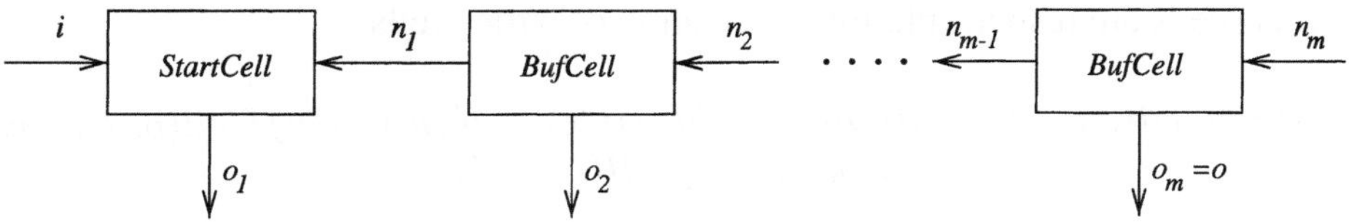

Figure 4.1. Buffer with reclaimable cells

Example 2 Let $P = a(b).P_1$ and $Q = \nu c.\overline{a}\langle c\rangle Q_1$. Up to choice of bound names there are three transitions enabled from $P \mid Q$, namely

- $P \mid Q \overset{a(d)}{\to} (P_1\{d/b\} \mid Q)$,
- $P \mid Q \overset{\nu d.\overline{a}\langle d\rangle}{\to} (P \mid Q_1\{d/c\})$, and
- $P \mid Q \overset{\tau}{\to} \nu d.(P_1\{d/b\} \mid Q_1\{d/c\})$

We conclude the section by introducing the two running examples of the paper.

Example 3 (Buffers) A 1-place buffer reads a name from an input port i and delivers it to an output port o:

$$Buf_1(i,o) = i(a).\overline{o}\langle a\rangle.Buf_1(i,o). \tag{4.1}$$

Inductively, if Buf_{n_1} (Buf_{n_2}) is an n_1-place (n_2-place) buffer abstraction similar to Buf_1 then

$$Buf_{n_1+n_2}(i,o) = \nu m.(Buf_{n_1}(i,m) \mid Buf_{n_2}(m,o)) \tag{4.2}$$

is an $n_1 + n_2$-place buffer abstraction. Unbounded buffers need to be able to allocate new buffer cells dynamically:

$$Buf(i,o) = i(a).\nu m.(Buf(i,m) \mid \overline{o}\langle a\rangle.Buf(m,o)). \tag{4.3}$$

Whenever a data item is input by Buf a new buffer cell is allocated, never to be deallocated. This is clearly wasteful. We may also consider an unbounded buffer with reclaimable cells, organised as a linked list as shown in Fig. 4.1. The main component is the buffer cell $BufCell(o,d,n_l,n_r)$ which holds the value d and waits to either output d along o and then terminate and communicate o and n_r "upstream" along n_l, or else receive a new output port o' and a new "down-link" n' along n:

$$BufCell(o,d,n_l,n_r) = \overline{o}\langle d\rangle.\overline{n_l}\langle o,n_r\rangle.0 + n_r(o',n).BufCell(o',d,n_l,n) \tag{4.4}$$

Start cells are responsible for the creation of buffer cells:

$$\begin{aligned}StartCell(i,o,n) = {} & i(d).\nu o'.\nu n'.(StartCell(i,o',n') \mid BufCell(o,d,n',n)) \\ & + n(o',n').StartCell(i,o',n')\end{aligned}$$

A "garbage collecting" unbounded buffer then consists initially of just a start cell:

$$GCBuf(i,o) = \nu n.StartCell(i,o,n) \tag{4.5}$$

Example 4 (Natural numbers) We consider an encoding of natural numbers based on Milner's encoding of data types in the polyadic π-calculus (cf. [10]). The idea is to represent a natural number "located" at a name n as a process which expects two names, say s and z, along n, and then proceed to either synchronize on z to signal to the receiver that the "value" of n is 0, or else to output the location n' of another natural number along s to signal that the value of n is $n' + 1$.

Define the constants $ZERO$ and $SUCC$ in the following way:

$$\begin{aligned}ZERO(n) &= n(s,z).\overline{z}\langle\rangle.0 \\ SUCC(n,m) &= m(s,z).\overline{s}\langle n\rangle.0.\end{aligned}$$

Given a location n, $ZERO(n)$ inputs two names along n to attempt an output along the second of the two, the "zero" channel. Similarly, $SUCC(n,m)$ receives two names along m to attempt an output along the first, the "successor" channel, passing along it the location of n. Thus we can represent, eg., the natural number 1 as the agent

$$ONE(one) = \nu zero.(SUCC(zero, one) \mid ZERO(zero)).$$

A variety of operations on naturals can be defined, including addition, multiplication, and as a very basic one the "copying" operation

$$\begin{aligned}COPY(n,m) = {} & \nu s_1.\nu z_1.\overline{n}\langle s_1, z_1\rangle. \\ & z_1().ZERO(m) + \\ & s_1(n_1).\nu m_1.(SUCC(m_1,m) \mid COPY(n_1,m_1))\end{aligned}$$

that turns a natural at location n into a natural at location m. Addition can be defined as follows:

$$\begin{aligned}ADD(n_1,n_2,m) = {} & \nu s_1.\nu z_1.\overline{n_1}\langle s_1, z_1\rangle.(z_1().COPY(n_2,m)) + \\ & (s_1(n_{11}).\nu m_1.(SUCC(m_1,m) \mid ADD(n_{11},n_2,m_1)))\end{aligned}$$

3. A π-μ-calculus

In this section we introduce the logic which we use for specifying properties of agents. The logic is a first-order version of the modal μ-calculus extended with operations that describe input/output behaviour.

Syntax. Assume a denumerable set of *predicate variables* X. *Formulas* $\phi, \psi \in \mathcal{F}$ in the π-μ-calculus are generated in the following way:

$$\begin{aligned}\phi \; ::= \; & a = b \,\big|\, a \neq b \,\big|\, \forall a.\phi \,\big|\, \exists a.\phi \,\big|\, \phi \wedge \phi \,\big|\, \phi \vee \phi \,\big|\, <l_\tau>\phi \,\big|\, [l_\tau]\phi \\ & \big|\, a \rightarrow \phi \,\big|\, a \leftarrow \phi \,\big|\, \nu a \rightarrow \phi \,\big|\, \nu a \leftarrow \phi \,\big|\, X(\vec{a}) \,\big|\, (\nu X(\vec{a}).\phi)(\vec{b}) \\ & \big|\, (\mu X(\vec{a}).\phi)(\vec{b})\end{aligned}$$

In this grammar we use l to range over names, a, and co-names, $\overline{a}$, and we use l_τ to range over names, co-names, and τ. A *monadic* formula is one with strict alternation between "transition" modalities (of the shape $<l>\phi$ or $[l]\phi$) and "io" modalities (of the shape $a \rightarrow \phi$, $a \leftarrow \phi$, $\nu a \rightarrow \phi$, or $\nu a \leftarrow \phi$), where we also require that any outermost modal operator be a transition modality.

The connectives. The language is based on a first-order modal μ-calculus with name equality and inequality, universal and existential name quantification, boolean connectives "and" ($\wedge$) and "or" ($\vee$), modal operators $<l>$ and $[l]$, and (parametrised) least (μ) and greatest (ν) fixed points. A *modal* formula is a formula in the fragment with neither fixed points nor predicate variables. An *elementary* formula is a formula in the first-order language of equality (over names). It is often convenient to treat elementary formulas differently from other formulas as they do not depend on the agent being predicated.

The connectives $a \rightarrow$, $a \leftarrow$, $\nu a \rightarrow$, and $\nu a \leftarrow$ are used for input, free output, input of a fresh name, and bound output, respectively. For instance, $a \rightarrow \phi$ predicates an abstraction and is taken to mean that, when applied to the name a, the resulting agent satisfies ϕ. Observe that a is not bound in $a \rightarrow \phi$. Similarly $a \leftarrow \phi$ predicates a concretion term, and states that the term is of the shape $\langle b \rangle \phi$ such that a and b are equal and ϕ holds of the continuation. For bound outputs the operation $\nu a \leftarrow \phi$ is used. The operation $\nu a \rightarrow \phi$ was introduced in a slightly different shape in [1]. This connective predicates abstractions and means that whenever the agent being predicated inputs a name a which is fresh then ϕ holds of the continuation. Both the modalities $\nu a \rightarrow \phi$ and $\nu a \leftarrow \phi$ bind a in ϕ. The syntax presented here divorces handling of transition labels, or subjects, from handling of the parameters, or objects. An alternative which we return to in more detail below, is to devise connectives in the style of those used by Milner et al [13] which combine the two, in modalities which reflect action capabilities more directly.

Example 5 The formula

$$(\nu X(a).<\bar{a}>\nu b \leftarrow [b]\forall c.c \rightarrow X(c))(a)$$

expresses of an agent that it can perform a bound output of a b along channel a. After this whenever a c is received along b the continuation satisfies $X(c)$.

Bindings, sentences. We use σ to range over $\{\nu, \mu\}$, and (l) to range over the modalities $<l>$ and $[l]$. Formulas are considered only up to alpha-renaming of bound predicate and name variables. Name binders are the first-order quantifiers, the bound output connective, and the fixed point operators. For instance, in $(\nu X(\tilde{a}).\phi)(\tilde{b})$, names in $\tilde{b}$ occur freely and names in $\tilde{a}$ bind their occurrences in ϕ. We can without loss of generality assume that fixed point formulas do not contain free names. A *sentence* is a formula that do not contain free occurrences of predicate variables. Unless otherwise specified we restrict attention to sentences.

Predicate and name interpretations. Names serve a double role, both as constants (two distinct names occurring freely at the top level of a process term are regarded as distinct) and as variables (since names can be bound and instantiated). A *name interpretation* is a mapping $\eta : \mathcal{G} \rightarrow \mathcal{G}$ which assigns names (as values) to names (as variables). We require of name interpretations η that $\mathcal{G} - range(\eta)$ is an infinite set. A name $b \in \mathcal{G}$ is said to be *fresh for* η if $b \notin range(\eta)$, and $\eta\{b/a\}$ is the update of η that maps a to b and otherwise acts as η. Let $\nu b.\eta = \eta\{b/b\}$, and if $s = \{a_1, \ldots, a_n\}$ then $\nu s.\eta = \nu a_1. \cdots .\nu a_n.\eta$. Observe that the operation $\nu b.-$ is generally only applied to η for which b is fresh. Name interpretations are extended to actions by the clause $\eta(\tau) = \tau$.

Predicate variables depend for their semantics upon a list of argument names and a name interpretation. Thus, a predicate variable interpretation can be taken to be a mapping $\rho(X)(\eta, \tilde{a}) \subseteq \mathcal{P}$. Maps $f : (\eta, \tilde{a}) \mapsto S \subseteq \mathcal{P}$ are ordered by $\leq$ defined as subset containment lifted pointwise to functions.

Semantics, validity. The semantics is given in terms of a mapping $\|\phi\|\rho\eta \subseteq \mathcal{P}$ where ρ is a predicate variable interpretation and η is a name interpretation. The definition of $\|\cdot\|$ is given in Table 4.2. In the clauses for bound input and bound output the notation *c fresh* means that c does not occur freely in neither A nor ϕ, and c is fresh for η. For sentences ϕ the ρ is immaterial, and we abbreviate $A \in \|\phi\|\rho\eta$ by $A \in \|\phi\|\eta$. Sometimes we write $\models_\eta A : \phi$ in place of $A \in \|\phi\|\eta$. A sentence ϕ is said to be *valid* if $\|\phi\|\eta = \mathcal{P}$ for all name interpretations η. Sentences ϕ and ψ are *equivalent*, written $\phi \equiv \psi$, if for all η, $\|\phi\|\eta = \|\psi\|\eta$.

$$
\begin{aligned}
\|a = b\|\rho\eta &= \{A \in \mathcal{P} \cup \mathcal{A} \cup \mathcal{C} \mid \eta(a) = \eta(b)\} \\
\|a \neq b\|\rho\eta &= \{A \in \mathcal{P} \cup \mathcal{A} \cup \mathcal{C} \mid \eta(a) \neq \eta(b)\} \\
\|\forall a.\phi\|\rho\eta &= \bigcap\{\|\phi\|\rho\eta\{b/a\} \mid b \in \mathcal{G}\} \\
\|\exists a.\phi\|\rho\eta &= \bigcup\{\|\phi\|\rho\eta\{b/a\} \mid b \in \mathcal{G}\} \\
\|\phi \wedge \psi\|\rho\eta &= \|\phi\|\rho\eta \cap \|\psi\|\rho\eta \\
\|\phi \vee \psi\|\rho\eta &= \|\phi\|\rho\eta \cup \|\psi\|\rho\eta \\
\|{<}\tau{>}\phi\|\rho\eta &= \{P \in \mathcal{P} \mid \exists Q.P \stackrel{\tau}{\rightarrow} Q \text{ and } Q \in \|\phi\|\rho\eta\} \\
\|{<}a{>}\phi\|\rho\eta &= \{P \in \mathcal{P} \mid \exists c, \vec{b}, Q, P \stackrel{c(\vec{b})}{\rightarrow} Q, c = \eta(a), \text{and } (\tilde{b})Q \in \|\phi\|\rho\eta\} \\
\|{<}\overline{a}{>}\phi\|\rho\eta &= \{P \in \mathcal{P} \mid \exists \vec{b_1}, c, \vec{b_2}, Q.P \stackrel{\nu\vec{b_1}.c\langle\vec{b_2}\rangle}{\rightarrow} Q, c = \eta(a), \\
&\qquad \text{and } \nu\vec{b_1}.\langle\vec{b_2}\rangle Q \in \|\phi\|\rho\eta\} \\
\|[\tau]\phi\|\rho\eta &= \{P \in \mathcal{P} \mid \forall Q.P \stackrel{\tau}{\rightarrow} Q \text{ implies } Q \in \|\phi\|\rho\eta\} \\
\|[a]\phi\|\rho\eta &= \{P \in \mathcal{P} \mid \forall c, \vec{b}, Q, P \stackrel{c(\vec{b})}{\rightarrow} Q, c = \eta(a) \text{ implies } (\tilde{b})Q \in \|\phi\|\rho\eta\} \\
\|[\overline{a}]\phi\|\rho\eta &= \{P \in \mathcal{P} \mid \forall \vec{b_1}, c, \vec{b_2}, Q.P \stackrel{\nu\vec{b_1}.c\langle\vec{b_2}\rangle}{\rightarrow} Q, \\
&\qquad c = \eta(a) \text{ implies } \nu\vec{b_1}.\langle\vec{b_2}\rangle Q \in \|\phi\|\rho\eta\} \\
\|a \rightarrow \phi\|\rho\eta &= \{(b)A \mid A\{\eta(a)/b\} \in \|\phi\|\rho\eta\} \\
\|a \leftarrow \phi\|\rho\eta &= \{\langle b\rangle C \mid \eta(a) = b \text{ and } C \in \|\phi\|\rho\eta\} \\
\|\nu a \rightarrow \phi\|\rho\eta &= \{(b)A \mid \forall c.c \text{ fresh implies } A\{c/b\} \in \|\phi\|\rho(\nu c.\eta)\} \\
\|\nu a \leftarrow \phi\|\rho\eta &= \{\nu b.\langle b\rangle C \mid \exists c.c \text{ fresh and } C\{c/b\} \in \|\phi\|\rho\eta\{c/a\}\} \\
\|X(\tilde{a})\|\rho\eta &= \rho(X)(\eta, \tilde{a}) \\
\|(\nu X(\tilde{a}).\phi)(\tilde{b})\|\rho\eta &= (\sqcup\{f \mid f \leq M(f)\})(\eta, \tilde{b}),\ (M(f)(\eta, \tilde{c}) = \|\phi\|\rho\{f/X\}\eta\{\tilde{c}/\tilde{a}\}) \\
\|(\mu X(\tilde{a}).\phi)(\tilde{b})\|\rho\eta &= (\sqcap\{f \mid M(f) \leq f\})(\eta, \tilde{b}),\ (M \text{ as above})
\end{aligned}
$$

Table 4.2. Formula semantics

The clauses for the modal operators lead to derived notation for the transition relation:

$$P \stackrel{a}{\to} (\vec{b})Q \quad \text{iff} \quad P \stackrel{a(\vec{b})}{\to} Q$$

$$P \stackrel{\overline{a}}{\to} \nu\vec{b_1}.\langle\vec{b_2}\rangle Q \quad \text{iff} \quad P \stackrel{\nu\vec{b_1}.\overline{a}.\langle\vec{b_2}\rangle}{\to} Q$$

We can thus simplify the transition modality clauses in the following style:

$$\|{<}a{>}\phi\|\rho\eta = \{P \in \mathcal{P} \mid \exists c, A.P \stackrel{c}{\to} A, c = \eta(a), \text{ and } A \in \|\phi\|\rho\eta\}.$$

Abbreviations. We introduce the following derived forms:

$$\top \triangleq \forall a.a = a \qquad \bot \triangleq \forall a.a \neq a$$

$$isP \triangleq [\tau]\top$$

$$isA \triangleq \forall a.a \to \top \qquad isCf \triangleq \exists a.a \leftarrow \top \qquad isCb \triangleq \nu a \leftarrow \top$$

Observe that *isP* holds of processes, *isA* of (non-trivial) abstractions, *isCf* of free outputs and *isCb* of bound outputs. Let $\mathcal{S}$ range over these four "typing formulas". Two typing formulas are said to be *complementary* if one of them is *isA* and the other either *isCf* or *isCb*.

An alternative to the definitions of $\top$ and $\bot$ is to set $\top \triangleq (\nu X().X())()$ and similarly $\bot \triangleq (\mu X().X())()$. These definitions are easily shown to be equivalent to the original ones in the proof system below.

The logic is closed under negation, as we can define the operation $\neg$ by the usual De-Morgan rules plus:

$$\begin{aligned}
\neg{<}\alpha{>}\phi &= ([\alpha]\neg\phi) \vee isA \vee isCf \vee isCb \\
\neg[\alpha]\phi &= ({<}\alpha{>}\neg\phi) \vee isA \vee isCf \vee isCb \\
\neg(a \to \phi) &= (a \to \neg\phi) \vee isP \vee isCf \vee isCb \\
\neg(a \leftarrow \phi) &= (\exists b.a \neq b \wedge b \leftarrow \top) \vee (a \leftarrow \neg\phi) \vee isP \vee isA \vee isCb \\
\neg(\nu a \to \phi) &= (\nu a \to \neg\phi) \vee isP \vee isCf \vee isCb \\
\neg(\nu a \leftarrow \phi) &= (\nu a \leftarrow \neg\phi) \vee isP \vee isA \vee isCf
\end{aligned}$$

With classical negation a number of standard abbreviations like $\phi \supset \psi \triangleq \neg\phi \vee \psi$ becomes available (but observe that recursion variables can in general only be negated an even number of times). There would in principle be no problem in taking negation as primitive. We choose not to do so, mainly as a matter of convenience, as otherwise the proof rules, which are cluttered up enough as they stand, would become even harder to read.

π-calculus modalities. Comparing with earlier attempts to define modal and temporal logics for the π-calculus we consider first the modal logic of Milner, Parrow, and Walker [13]. Their basic modalities are rendered in our framework as follows:

$$\begin{aligned}
<\overline{a}b>\phi &\triangleq <\overline{a}>b \leftarrow \phi \\
<ab>\phi &\triangleq <a>b \rightarrow \phi \\
<\overline{a}(b)>\phi &\triangleq <\overline{a}>\nu b \leftarrow \phi \\
<a(b)>\phi &\triangleq <a>\exists b.b \rightarrow \phi \\
<a(b)>^{\mathrm{L}}\phi &\triangleq <a>\forall b.b \rightarrow \phi \\
<a(b)>^{\mathrm{E}}\phi &\triangleq \forall b.<a>b \rightarrow \phi
\end{aligned}$$

We refer to these operators as the "MPW modalities". We leave aside the issue of whether a reduction in the other direction exists (of modal monadic formulas to formulas in the logic of [13]). Trivially this is not so since the logic of [13] lacks a modality for bound input. With the addition of such a modality the matter is less easily resolved, however.

Other logics for π-calculus have been proposed in work by Milner [10], Dam [4], and Amadio and Dam [1]. In [10] dependent sum and product constructions were proposed, used also in [4]:

$$\begin{aligned}
\Sigma a.\phi &\triangleq (\exists a.a \leftarrow \phi) \vee (\nu a \leftarrow \phi) \\
\forall' a.\phi &\triangleq \forall a.a \rightarrow \phi \\
\exists' a.\phi &\triangleq \exists a.a \rightarrow \phi
\end{aligned}$$

We use "primed" versions of the quantifiers here so as not to confuse with the standard first-order quantifiers. Observe that, even in the absence of bound input modalities, the logics of [10, 4] are strictly less expressive than that of [13] as the former does not separate free and bound output. The modalities of [1], finally, are easily derivable as well. We leave out the details.

Logical characterisation. An important property of a modal logic such as the one considered here is its capability to separate models. In [13] it was shown that the MPW modalities could be used to characterise a particular, quite strong, process equivalence called late bisimulation equivalence (cf. [12]).

Definition 6 (Late bisimulation equivalence) A binary relation $\mathcal{R}$ on $\mathcal{P}$ is a *late simulation* if $P_1 \mathcal{R} P_2$ implies:

1 If $P_1 \xrightarrow{\alpha} Q_1$ and α is not an input action then there is some Q_2 such that $P_2 \xrightarrow{\alpha} Q_2$ and $Q_1 \mathcal{R} Q_2$.

2 If $P_1 \stackrel{a(\vec{b})}{\rightarrow} Q_1$ then for some Q_2, $P_2 \stackrel{a(\vec{b})}{\rightarrow} Q_2$ and
for all $\vec{c}$, $Q_1\{\vec{c}/\vec{b}\}\mathcal{R}Q_2\{\vec{c}/\vec{b}\}$.

The relation $\mathcal{R}$ is a *late bisimulation* if both $\mathcal{R}$ and $\mathcal{R}^{-1}$ are late simulations. P_1 and P_2 are *late bisimilar*, $P_1 \dot{\sim} P_2$, if $P_1\mathcal{R}P_2$ for some late bisimulation $\mathcal{R}$.

In the second clause of Definition 6 the arities of vectors $\vec{b}$ and $\vec{c}$ are required to coincide. Observe that the simplicity of the definition, both for input and output actions, relies heavily on alpha-conversion to avoid accidental capture of bound names.

Other equivalences appear in the π-calculus literature. *Early bisimulation* ([12]) is strictly weaker than late bisimulation. It is obtained by swapping the quantifications over Q_2 and $\vec{c}$ in clause 2 of Definition 6. *Open bisimulation* [15] is strictly stronger than late bisimulation and requires the bisimulation relation to be closed under substitution. *Ground bisimulation*, finally, is standard (CCS) bisimulation equivalence [9] applied to the π-calculus. Thus ground bisimulation avoids quantification entirely in its defining clauses, and it is strictly weaker than early bisimulation equivalence. For a substantially constrained version of the π-calculus, eliminating the conditional, matching (a conditional of the special form **if** $a = b$ **then** P **else** 0), and continuation under output prefix, it can be shown, however, that all four equivalences coincide [15].

We obtain the following logical characterisation result:

Proposition 7 (Logical characterisation) *Two processes P and Q are late bisimilar if and only if for all sentences ϕ and all name interpretations η, $\models_\eta P : \phi$ implies $\models_\eta Q : \phi$.*

PROOF. In [13] the logical characterisation result is proved for the logic constructed using equations, inequations, conjunction, disjunction, and the MPW modalities. The adaptation of this result to the present setting is easy and left out. It thus suffices to show that all formulas in our language respect bisimulation equivalence in the sense that if $\models_\eta P : \phi$ and $P \dot{\sim} Q$ then $\models_\eta Q : \phi$ as well. To cater for fixed points the assertion need to be generalised: Say a predicate interpretation ρ *respects* $\dot{\sim}$ if whenever $P \in \rho(X)(\eta, \tilde{a})$ and $P \dot{\sim} Q$ then $Q \in \rho(X)(\eta, \tilde{a})$ too. We show by induction in the structure of ϕ that if ρ respects $\dot{\sim}$, $P \in \|\phi\|\rho\eta$, and $P \dot{\sim} Q$ then $Q \in \|\phi\|\rho\eta$. The details of this induction are not difficult and left to the reader. ∎

By constraining the nesting of transition and io modalities a similar characterisation of early bisimulation can be given. In effect the formation of the late input MPW modality $<a(b)>^{\mathrm{L}}$ needs to be prevented, as shown by [13].

4. Example specifications

In this section we give some examples of agent properties that can be specified using the above logic.

Example 8 (Weak modalities) One can easily derive modalities which are insensitive to the number of initial τ-transitions.

$$
\begin{aligned}
<<>>\phi &\triangleq \mu X.\phi \vee <\tau>X \\
<<l>>\phi &\triangleq <<>><l>\phi \\
[[]]\phi &\triangleq \nu X.\phi \wedge [\tau]X \\
[[l]]\phi &\triangleq [[]][l]\phi
\end{aligned}
$$

Example 9 (Wildcard input and output) Often one wishes to ignore the identity of names being input or output:

$$
\begin{aligned}
\_ \rightarrow \phi &= \forall a.a \rightarrow \phi \\
\_ \leftarrow \phi &= (\exists a.a \leftarrow \phi) \vee \nu a \leftarrow \phi
\end{aligned}
$$

This definition presupposes that a does not appear freely in ϕ.

Example 10 (Buffer properties) We consider some properties one might like to impose of a π-calculus buffer taking input along the channel i and producing output to the channel o. The formulas in this example are due to Joachim Parrow (personal communication).

- **Order preservation (of first data item).**

$$
\begin{aligned}
&FirstOut(i, o, d) = \\
&\quad (\nu X(d).[\tau]X(d) \wedge \forall a. \\
&\quad\quad ([\overline{a}](a = o \wedge d \leftarrow true)) \wedge \\
&\quad\quad ([a](a = i \wedge \_ \rightarrow X(d))))(d)
\end{aligned}
$$

$$
\begin{aligned}
&OrdPres(i, o) = \\
&\quad \nu X.[\tau]X \wedge \forall c. \\
&\quad\quad ([\overline{c}](c = o \wedge \_ \leftarrow X)) \wedge \\
&\quad\quad ([c](c = i \wedge \forall d.d \rightarrow FirstOut(i, o, d)))
\end{aligned}
$$

The idea is quite simple: $OrdPres(i, o)$ holds until something (the d) is input along i. Then $FirstOut(i, o, d)$ takes effect. This fixed point holds invariantly (along τ and i transitions) until something is output. That something must be d.

- **No spurious output.** Consider the following formula:

$$
\begin{aligned}
&NoSpuOut'(i,o,a) = \\
&\quad (\nu X(b).[\tau]X(b) \wedge \forall c. \\
&\qquad ([\overline{c}](c = o \wedge ((\nu d \leftarrow X(b)) \vee (\exists d.d \neq b \wedge X(b))))) \wedge \\
&\qquad ([c](c = i \wedge \forall d.d \rightarrow (b = d \vee X(b)))))(a) \\
&NoSpuOut(i,o) = \forall a.NoSpuOut'(i,o,a)
\end{aligned}
$$

  The formula expresses, roughly, that an a must be input before it can be output. Observe that in the context of the π-calculus this gives an operational meaning to the notion of *authenticity:* it can be trusted that information emitted on o really was received along i.

- **No lost input.**

$$
\begin{aligned}
&EvOut(i,o,a) = \\
&\quad (\mu X(b).(<\tau>\top \vee <\overline{o}>\top \vee <i>\top) \wedge [\tau]X(b) \wedge \forall e. \\
&\qquad ([\overline{e}](e = o \wedge ((\nu c \leftarrow X(b)) \vee (\exists c.c \leftarrow X(b)) \vee (b \leftarrow \top)))) \wedge \\
&\qquad ([e](e = i \wedge \_ \rightarrow X(b))))(a)
\end{aligned}
$$

$$
\begin{aligned}
&NoLostInput(i,o) = \\
&\quad \nu X.[\tau]X \wedge \forall c. \\
&\qquad ([\overline{c}](c = o \wedge \_ \leftarrow X)) \wedge \\
&\qquad ([c](c = i \wedge \forall a.a \rightarrow X \wedge EvOut(i,o,a)))
\end{aligned}
$$

  For $NoLostInput(i,o)$ X must hold invariantly, and whenever an a is input then $EvOut(i,o,a)$ holds. The latter property expresses that some transition is enabled, and whatever transition is taken (among τ, i and o), either the transition was an output of a along o or else $EvOut(i,o,a)$ continues to hold. However, since we use a least fixed point eventually the former case must hold, so a is eventually output.

Example 11 (Natural numbers) We wish to define the property $Nat(n)$ of "possessing a natural number object located at n". Keeping in mind the concrete representations of Example 4, $Nat(n)$ should be expected to hold of a process P under the following circumstances:

- It should be possible to force P to synchronize along n, if not immediately then after an initial number of τ steps.
- No sequence of τ-labelled steps can disable an n-transition.
- All n-transitions take two arguments.

- Whenever fresh names s and z are provided along n the following properties will hold:
- At least one outgoing synchronisation along s or z is possible, but not both.
- Only unary outgoing synchronisations along s, or along z are possible.
- No sequence of internal steps can disable an s- or z-transition.
- Whenever a z transition takes place, the result is a process.
- Whenever an s-transition takes place a new name is output, serving as the location for the "predecessor of n".

We do not claim that these points unambiguously pin down the intended behaviour of "a natural number object" – in fact most of the points above leave room for debate. Resolving this we formalize the intuition in the π-μ-calculus.

We describe the behaviour of natural number objects as a state machine wrapped inside a least fixed point reflecting the well-foundedness property of natural numbers. We propose the following definition:

$$\begin{aligned} \mathit{Nat}_0(n,\phi) &\triangleq \mu Y.{<<}n{>>}\top \wedge \\ &\quad [\tau]Y \wedge \\ &\quad \forall b.[b](b = n \wedge \nu s \to \nu z \to \phi(s,z)) \wedge \\ &\quad \forall b.[\bar{b}]\bot \\ \mathit{Nat}_1(X)(s,z) &\triangleq \mu Z.\neg({<<}\bar{s}{>>}\top \wedge {<<}\bar{z}{>>}\top) \wedge \\ &\quad ({<<}\bar{s}{>>}\top \vee {<<}\bar{z}{>>}\top) \wedge \\ &\quad [\tau]Z \wedge \\ &\quad \forall b.[b]\bot \wedge \\ &\quad \forall b.[\bar{b}]((b = z \wedge \mathit{isP}) \vee (b = s \wedge \nu m \leftarrow X(m))) \\ \mathit{Nat} &\triangleq \mu X(n).\mathit{Nat}_0(n, \mathit{Nat}_1(X)) \end{aligned} \quad (4.6)$$

The definition uses higher-order parameters in a manner which goes beyond the syntax as presented in the start of this section. However, at the expense of a more monolithic and less readable notation it is quite easy to rewrite the definition to eliminate the higher-order abbreviations.

The idea of the definition is the following: We describe the behaviour of a natural number object as a sort of state machine formalizing a sort of natural number protocol. The state machine has two states, Nat_0 and Nat_1. Outside the state description is a least fixed point, used to reflect progress of entire protocol runs. The description of each state must bring out what actions are enabled, and how, and it must describe, since it will be used in a compositional manner, for each type of action, what the effect of performing such an action will be. In several case studies we have performed over the years we have found this sort of abstract state description very useful for proving properties of infinite state systems, and more examples will be given toward the end of this paper.

Now, suppose that P possesses a natural number object located at n. Then n is enabled and can be supplied with first a fresh s and then a fresh z. If n is supplied with a z or an s which is *not* fresh we can say little about the behaviour of the resulting system, as unintended internal communications can result. This is an important point, and it is the main reason why we have chosen to include fresh input as a primitive. Continuing the protocol either s or z is offered, but not both. If z is offered the result is a process and the protocol is terminated. If s is offered a new location m is output, and the agent will continue to behave as a natural number object located at m. Moreover, because of the least fixed point, the first option must eventually be selected, so the natural number object is ensured to be well-founded.

The natural number type given here is a candidate for the type of "one-shot", or ephemeral natural numbers. Expressing a property such as "n is the location of a natural number object from now until some future event takes place" is an easy embedding of Nat into an invariant. Many variations on the definition of Nat are possible. For instance we might not insist on well-foundedness, or Nat might be permitted to diverge.

5. Proof system, modal fragment

A *closed correctness assertion* is an assertion of the shape $\vdash P : \phi$ where P is an arbitrary process and ϕ is a π-μ-calculus sentence. For certain so-called finite-control agents that refrain from creating new processes dynamically (by avoiding parallel compositions in recursive contexts, or the replication operator) the problem of deciding validity of closed correctness assertions is decidable [4]. This is due to the fact that, up to choice of names and a little garbage collection, the state spaces of finite-control agents are finite [4]. Many interesting agents, however, fall outside the class of finite-control agents, including the natural number agents of Example 4 and the unbounded buffers of Example 3, because processes are created dynamically. In fact it is just this dynamic process creation capability that in conjunction with the capability of creating and communicating new names gives the π-calculus its remarkable expressive power, and it is also the feature that makes verification difficult.

Limits of global state exploration. Approaches to verification which merely chase transitions and global states are unlikely to be very successful in proving interesting properties of non-finite-control agents. Consider for instance the unbounded buffer Buf of Example 3. A temporal property that depends on Buf being continually able to input new data items will give rise to an unbounded state space quite trivially, as each input action gives rise to the creation of a new component process. But it may also be that despite being non-finite-control the state space is in fact bounded. As an example we may consider any process of the shape

$$P_n(b) = \nu a_n.COPY(a_n, b) \mid \nu a_{n-1}.\cdots \mid \nu a_0.SUCC(a_0, a_1) \mid ZERO(a_0) \tag{4.7}$$

which can only perform a bounded number of actions despite the dynamic process creation involved in the definition of $COPY$. However, the interesting correctness property of $COPY$ is less the collection of n facts that (for instance) $\nu b.(COPY(b,c) \mid P_n(b))$ represents a natural number located at c, but rather the fact that $COPY$ is type correct, ie. that for *any* x which represents a natural number located at b, $\nu b.(COPY(b,c) \mid x)$ represents a natural number located at c. But this latter assertion is not a closed correctness assertion, and it is not within the scope of model checking techniques such as that of [4] that explore global state spaces since the agent expression being predicated is open.

Open correctness assertions. We thus need to address more general *open correctness assertions* that allow correctness properties $P : \phi$ to be made conditional upon properties of the parameters of P (such as: x represents a natural number located at b). This can give a handle on dynamic process creation in the following way: Suppose the goal is to prove a correctness assertion of the shape

$$\vdash P : \phi. \tag{4.8}$$

Assume that P involves dynamic process creation and that through a number of steps the proof goal (4.8) is "reduced" to one the shape

$$\vdash P \mid Q : \psi \tag{4.9}$$

For instance P may be the agent $Buf\ i\ o$ of Example 3. The idea now is to apply a cut, guessing a property γ to hold of P, and then reduce (4.9) to the proof goals

$$\vdash P : \gamma \tag{4.10}$$

and

$$x : \gamma \vdash x \mid Q : \psi \tag{4.11}$$

representing the assertion that ψ holds on $x \mid Q$ conditional upon γ holding on x. Say, for instance, that we can choose $\gamma = \phi$ which turns out to be the case in many examples. Then the proof goal (4.10) is an instance of (4.8) and it may consequently be that the subgoal (4.10) for this reason can be discharged. If in turn Q does not involve dynamic process creation the problem (of dealing with this feature) may have been resolved, and if it does we will want to iterate the approach, keeping in mind that we now have to deal with more general open correctness assertions.

5.1 Basic judgments

In [5] this idea was worked out for CCS. A delicate issue in generalising the approach to the π-calculus is how to deal with private names, name generation, and scope extrusion. In general one will wish to verify properties of a process $\nu a.P(x)$ relative to a property ψ of x. ψ must be allowed to depend on a. Consider for instance the property that $\nu b.(COPY\ b\ c \mid x)$ is a natural number located at c given that x is a natural number located at b. But if we take alpha-conversion for granted then $\nu b.(COPY\ b\ c \mid x)$ should be indistinguishable from $\nu b'.(COPY\ b'\ c \mid x)$ provided no name clashes arise, and the dependency upon b is lost. We thus need a mechanism to "freeze" b to extend its scope to cover also formulas to the left of the turnstile. In [1] we suggested annotating the turnstile with a "restriction set" s for this purpose. We are following this suggestion here. Thus judgments take the more general form $x : \psi \vdash^s P : \phi$ where s is a finite set of names with a scope extending over both P and ψ, but *not* over ϕ. In fact a very similar annotation was introduced already by Stirling [18]. Here, however, the annotation has a somewhat deeper function, due to the richer name discipline of the π-calculus.

Term variables and open terms. Before defining formally the notion of judgment and its semantics observe that we need to extend the basic syntax of the π-calculus to open terms. We use x, y, and z as term variables (to range over processes, abstractions, and concretions). Terms in $\mathcal{P}$, $\mathcal{A}$, or $\mathcal{C}$ may from now on be open, i.e. involve term variables. A *closed term* will be a term which does not contain term variables (but it may contain free names). We use E to range over $\mathcal{P} \cup \mathcal{A} \cup \mathcal{C}$.

The consideration of open terms also leads us to extend the π-calculus syntax slightly, by allowing the parallel composition operator to be applied not only to processes but also to abstractions and concretions (cf. [10]). This is done by rewriting (where symmetric versions are assumed to be added implicitly):

$$\begin{aligned} ((\vec{b})A) \mid P &\rightarrow (\vec{b})(A \mid P) \\ (\nu\vec{c_1}.\langle\vec{c_2}\rangle C) \mid P &\rightarrow \nu\vec{c_1}.\langle\vec{c_2}\rangle(C \mid P) \\ ((\vec{b})A) \mid (\nu\vec{c_1}.\langle\vec{c_2}\rangle C) &\rightarrow \nu\vec{c_1}.(A\{\vec{c_2}/\vec{b}\} \mid C) \end{aligned}$$

We assume that symmetric versions of these rewrite rules are added implicitly, and appeal to alpha-conversion to prevent variable capture as ever.

Definition 12 (Basic judgments, validity) A *basic judgment* is an expression of the form

$$\Gamma \vdash^s E : \Delta \tag{4.12}$$

where

1 $\Gamma = \{x_1 : \phi_1, \ldots, x_n : \phi_n\}$ is a finite set (of *assumptions*) such that the ϕ_i are sentences, $1 \leq i \leq n$,

2 $s = \{a_1, \ldots, a_k\}$ is a finite set of names, the *restriction set*,

3 $\Delta = \{\psi_1, \ldots, \psi_m\}$ is a finite set of sentences, and

4 E is an open term.

An agent variable x occurs freely in $\Gamma \vdash^s E : \Delta$ if either x occurs freely in E or Γ contains an assumption on x, ie. an assumption of the shape $x : \phi$.

For the semantics name interpretations η are extended to general substitutions by mapping both names to names and term variables to closed terms. In doing so the sets of names and of term variables are assumed to be distinct. We can then extend $\nu s.\eta$ to substitutions by restricting η to names. Then, the judgment $\Gamma \vdash^s E : \Delta$ is *valid* (or *true*, written $\Gamma \models^s E : \Delta$) if for all substitutions η, if $\models_{\nu s.\eta} \eta(x_i) : \phi_i$ for all $i : 1 \leq i \leq n$, then $\models_\eta \nu s.(E\eta) : \psi_j$ for some $j : 1 \leq j \leq m$.

Notation. For sets such as Γ, s, and Δ we use a standard sequence-like notation, writing eg. $\Gamma, x_1 : \phi_1, x_2 : \phi_2$ in place of $\Gamma \cup \{x_1 : \phi_1, x_2 : \phi_2\}$, or $\Gamma \vdash^{()} A : \Delta_1, \Delta_2$ in place of $\Gamma \vdash^{\emptyset} A : \Delta_1 \cup \Delta_2$. For a basic judgment $\Gamma, x : \phi \vdash^s E : \Delta$, if ϕ is elementary (so that the holding of ϕ does not depend on x), we allow the judgment to be abbreviated by $\Gamma, \phi \vdash^s E : \Delta$.

Restrictions sets and scoping. The point already made concerning scope of restriction sets deserves to be reiterated. Observe that in a judgment of the shape $x : \phi \vdash^a E : \psi$ both ϕ, E and ψ may mention a. Let η be an arbitrary name interpretation, and suppose that $\models_{\nu a.\eta} \eta(x) : \phi$. Then occurrences of a in ϕ refer, because of the use of the name interpretation $\nu a.\eta$, to occurrences of a in $\eta(x)$. Moreover, no other name occurring in ϕ can be confused with a in $\eta(x)$. In forming $\nu a.(E\eta)$ the a's in E and the a's in $\eta(x)$ are identified (they are equal). An a occurring in ψ, on the other hand, can, through η, be identified with or distinguished from any name occurring freely in $\nu a.(E\eta)$, but in that agent a itself is bound. So the scope of a in $x : \phi \vdash^a E : \psi$ extends to E and ϕ, but *not* to ψ. This is reflected in the proof system below by the rule (ALPHA).

5.2 A proof system for the modal fragment

We now turn to the problem of proving validity of basic judgments, restricting attention for now to the modal fragment. A proof system will consist of a number of clearly discernible parts:

1 A group of structural rules governing aspects like the introduction and use of assumptions.

2 A group of logical rules to deal with connectives like conjunction, disjunction, and the quantifiers.

3 A group of rules for name equality and inequality.

4 A group of rules for the process modalities $<\alpha>$ and $[\alpha]$.

5 A group of rules for the input/output modalities $a \rightarrow \phi$, $a \leftarrow \phi$, $\nu a \rightarrow \phi$ and $\nu a \leftarrow \phi$.

The first three groups of rules are based on a standard sequent-style formalisation of first-order logic with equality. The adaptation is not completely trivial, however, due to the presence of agent terms and variables, and restriction sets. An important and delicate issue concerns the choice of rules to include as primitive. Our strategy here is to include only rules that are needed in a completeness argument, but it is probably useful to bear in mind that this is far less clearcut and definitive a criterion than may be thought at first glance, and other less tangible criteria like "elegance", or "orthogonality" are important too in the detailed formulation of rules.

We go through each group of rules in turn.

5.3 Structural rules

We first have a minimal set of rules for introducing and applying assumptions, namely the identity, weakening, and the cut, bearing in mind that in general these rules are affected by the presence of restriction sets:

$$\text{(I)} \quad \frac{\cdot}{\Gamma, x : \phi \vdash^{()} x : \phi, \Delta}$$

$$\text{(W-L)} \quad \frac{\Gamma \vdash^{s} E : \Delta}{\Gamma, x : \phi \vdash^{s} E : \Delta} \qquad \text{(W-R)} \quad \frac{\Gamma \vdash^{s} E : \Delta}{\Gamma \vdash^{s} E : \psi, \Delta}$$

As the Γ, s, and Δ are sets no rules for contraction and permutation are needed. We comment on the need for weakening later. The cut rule comes in three flavours. The first, (CUT-1), is essential to accomodate reasoning on agent structure and it is not in general eliminable.

$$\text{(CUT-1)} \quad \frac{\Gamma, s\ \mathit{fresh} \vdash^{()} P : \phi \quad \Gamma, x : \phi \vdash^{s} E : \Delta}{\Gamma \vdash^{s} E\{P/x\} : \Delta} \quad (x \notin \Gamma)$$

We here introduce two new pieces of shorthand, writing $x \notin \Gamma$ for the condition that Γ does not contain an assumption on x, and $s\ fresh$ for an elementary formula expressing that all names in s are distinct from each other, and from any name occurring in Γ or Δ.

To illustrate the complications involved in the first cut rule assume that the judgments $\Gamma, s\ fresh \vdash^{()} P : \phi$ and $\Gamma, x : \phi \vdash^{s} E : \Delta$ are valid, and that a substitution η is given such that all assumptions in Γ are validated for the name interpretation $\nu s.\eta$. The substitution $\nu s.\eta$ validates also the condition $s\ fresh$, so we can conclude that $\models_{\nu s.\eta} P\eta : \phi$. Now x does not occur in Γ so η can be extended to the substitution $\eta' = \eta\{P\eta/x\} = \{P/x\}\eta$ which validates both Γ and $x : \phi$ for the restriction set s. But then $\models_{\eta} \nu s.(E\{P/x\})\eta) : \Delta$ as required.

The second cut rule will be needed when we come to consider logical fixed points and discharge by loop termination by providing garbage collection of restrictions that are no longer used. For the modal fragment, however, (CUT-2) is admissible, as can be seen from the completeness proof below.

$$(\text{CUT-2}) \quad \frac{\Gamma, s_2\ fresh \vdash^{s_1} E : \phi \quad \Gamma, x : \phi \vdash^{s_2} x : \Delta}{\Gamma \vdash^{s_1, s_2} E : \Delta} \quad (x \notin \Gamma, s_1 \cap s_2 = \emptyset)$$

Here $s_2\ fresh$ is used in the sense of requiring also names in s_2 to be distinct from names in s_1. To see the soundness of this rule assume $\Gamma, s_2\ fresh \models^{s_1} E : \phi$, $\Gamma, x : \phi \models^{s_2} x : \psi$, and $s_1 \cap s_2 = \emptyset$. Let η be given such that $\eta(x_i)$ is a process for all $i : 1 \leq i \leq n$.. Assume that $\eta(x) \in \|\phi_i\|(\nu s_1.\nu s_2.\eta)$ whenever $x = x_i$. By the first assumption, $\nu s_1.(P\eta) \in \|\phi\|(\nu s_2.\eta)$. But then $\nu s_2.\nu s_1.(P\eta) = \nu s_1.\nu s_2.(P\eta) \in \|\psi\|\eta$ as desired.

The third cut rule is the following:

$$(\text{CUT-3}) \quad \frac{\Gamma \vdash^{s} E : \phi, \Delta \quad \Gamma, y : \phi \vdash^{()} y : \Delta}{\Gamma \vdash^{s} E : \Delta}$$

where y does not occur in Γ. For the modal fragment (CUT-3) is admissible. We conjecture that this is not so in general. However for $s = ()$ the rule (CUT-3) is derivable quite easily using the logical rules introduced in the following section.

Finally we need the following rule to reflect the scoping rule for restriction sets:

$$(\text{ALPHA}) \quad \frac{\Gamma\sigma \vdash^{s\sigma} E\sigma : \Delta}{\Gamma \vdash^{s} E : \Delta}$$

where $\sigma : s \to \mathcal{G}$ is injective, the range of σ is disjoint from $(fn(\Gamma) \cup fn(E) \cup fn(\Delta)) - s$, and the postfixing of σ is the extension of σ to agents, restriction sets, and assumption sets Γ. Other substitution rules such as injective substitutions of unrestricted names, or of term variables, are admissible.

5.4 Logical rules

As a first approximation we require standard rules for introducing elementary connectives to the left and to the right.

$$(\wedge\text{-L})\quad \frac{\Gamma, x:\phi, x:\psi \vdash^s E:\Delta}{\Gamma, x:\phi\wedge\psi \vdash^s E:\Delta} \qquad (\wedge\text{-R})\quad \frac{\Gamma \vdash^s E:\phi,\Delta \quad \Gamma\vdash^s E:\psi,\Delta}{\Gamma\vdash^s E:\phi\wedge\psi,\Delta}$$

$$(\vee\text{-L})\quad \frac{\Gamma, x:\phi \vdash^s E:\Delta \quad \Gamma, x:\psi\vdash^s E:\Delta}{\Gamma, x:\phi\vee\psi \vdash^s E:\Delta}$$

$$(\vee\text{-R})\quad \frac{\Gamma\vdash^s E:\phi,\psi,\Delta}{\Gamma\vdash^s E:\phi\vee\psi,\Delta}$$

$$(\forall\text{-L})\quad \frac{\Gamma, x:\phi\{b/a\}\vdash^s E:\Delta}{\Gamma, x:\forall a.\phi\vdash^s E:\Delta} \qquad (\forall\text{-R})\quad \frac{\Gamma\vdash^s E:\phi,\Delta}{\Gamma\vdash^s E:\forall a.\phi,\Delta}\quad (a \text{ fresh})$$

$$(\exists\text{-L})\quad \frac{\Gamma, x:\phi\vdash^s E:\Delta}{\Gamma, x:\exists a.\phi\vdash^s E:\Delta}\quad (a \text{ fresh}) \qquad (\exists\text{-R})\quad \frac{\Gamma\vdash^s E:\phi\{b/a\},\Delta}{\Gamma\vdash^s E:\exists a.\phi.\Delta}$$

In the rules ∀-R and ∃-L we use a fresh as a shorthand for the condition that a is not free in the conclusion judgment and, in the case of ∃-L, neither does a occur in s.

Since the logic is closed under classical negation we further need rules e.g. to reflect that contradictory assumptions can be made governing the same agent variable. We suggest the following two rules, left and right dilemma:

$$(\text{D-L})\quad \frac{\Gamma, x:\phi\vdash^s E:\Delta \quad \Gamma, x:\neg\phi\vdash^s E:\Delta}{\Gamma\vdash^s E:\Delta}$$

$$(\text{D-R})\quad \frac{\Gamma\vdash^s E:\phi,\Delta \quad \Gamma\vdash^s E:\neg\phi,\Delta}{\Gamma\vdash^s E:\Delta}$$

Two further rules are needed to reflect the fact that elementary formulas do not depend on the agent term being predicated:

$$(\text{E-L})\quad \frac{\Gamma, y:\phi\vdash^s E:\Delta}{\Gamma, x:\phi\vdash^s E:\Delta} \qquad (\text{E-R})\quad \frac{\Gamma\vdash^s E:\phi}{\Gamma\vdash^s E':\phi}$$

where both rules require that ϕ is elementary.

Lemma 13 *The following rules are derivable:*

$$\neg\text{-L}\quad \frac{\Gamma\vdash^{()} x:\phi,\Delta}{\Gamma, x:\neg\phi\vdash^{()} x:\Delta} \qquad \neg\text{-R}\quad \frac{\Gamma, x:\phi\vdash^{()} x:\Delta}{\Gamma\vdash^{()} x:\neg\phi,\Delta}$$

PROOF. Easy derivations using weakening and dilemma. ∎

5.5 Rules for equality and inequality

For equality and inequality we suggest four axioms and one rule of inference. The four axioms, first, express the following properties:

1 Names are equal to themselves.

2 Even possibly restricted names are equal to themselves.

3 A name can not be identified with a name in the restriction set unless it is textually identical.

4 There are infinitely many names.

The addition is a rule of substitution of equals for equals.

$$(\text{REFL}) \quad \frac{}{\Gamma \vdash^s E : a \dot{=} a, \Delta}$$

$$(\text{IRR}) \quad \frac{}{\Gamma, a \dot{\neq} a \vdash^s E : \Delta}$$

$$(\text{NEW1}) \quad \frac{}{\Gamma, x : a \dot{=} b \vdash^{s,a} E : \Delta} \quad (a \neq b)$$

$$(\text{INFTY}) \quad \frac{}{\Gamma \vdash^s E : \exists b. b \dot{\neq} a_1 \wedge \cdots \wedge b \neq a_n, \Delta}$$

$$(\text{SUBST}) \quad \frac{\Gamma\{b/c\} \vdash^s E\{b/c\} : \Delta\{b/c\}}{\Gamma\{a/c\}, a = b \vdash^s E\{a/c\} : \Delta\{a/c\}} \quad (a, b \notin s)$$

To see the soundness of (NEW1) observe that it can not simultaneously be the case that $\models_{\nu s.\nu a.\eta} \eta(x) : a = b$ and that a and b are distinct names. Observe also that the rule (IRR) is needed only for $a \in s$. We note the derivability of a number of useful proof rules.

Lemma 14 *The following rules are derivable:*

$$(\text{EQ-I}) \quad \frac{}{\Gamma, a = b \vdash^s E : a \dot{=} b, \Delta} \quad (a, b \notin s)$$

$$(\text{ELEM-I}) \quad \frac{}{\Gamma, \phi \vdash^s E : \dot{\phi}, \Delta} \quad (\phi \text{ elementary}, fn(\phi) \cap s = \emptyset)$$

$$(\text{INEQ}) \quad \frac{\Gamma\{b/a\} \vdash^s E\{b/a\} : \Delta\{b/a\}}{\Gamma \vdash^s E : a \neq b, \Delta} \quad (a, b \notin s)$$

$$(\text{SYM1}) \quad \frac{\Gamma \vdash^s E : b = a, \Delta}{\Gamma \vdash^s E : a = b, \Delta}$$

$$(\text{TR}) \quad \frac{\Gamma \vdash^s E : a = b, \Delta \quad \Gamma \vdash^s E : b = c, \Delta}{\Gamma \vdash^s E : a = c, \Delta}$$

$$(\text{SYM2}) \quad \frac{\Gamma \vdash^s E : a \neq b, \Delta}{\Gamma \vdash^s E : b \neq a, \Delta}$$

$$(\text{DIST}) \quad \frac{\Gamma \vdash^s E : a \neq b, \Delta \quad \Gamma \vdash^s E : b = c, \Delta}{\Gamma \vdash^s E : a \neq c, \Delta}$$

PROOF. (EQ-I): Use (SUBST) and (REFL).

(ELEM-I): Use EQ-I) and the dilemma rules along with the logical rules and structural induction in ϕ.

(INEQ): Use (D-L), weakening, (ELEM-I), and (SUBST).

(SYM1): This case is slightly more delicate than the previous ones. We prove $\Gamma \vdash^s E : b = a, \Delta$ given a proof of $\Gamma \vdash^s E : a = b, \Delta$. First observe that $\Gamma \vdash^s a = b, b = a, \Delta$ by weakening. Then observe that $\Gamma \vdash^{()} b = b, \Delta$ so that $\Gamma, a = b \vdash^{()} b = a, \Delta$ by (SUBST). Then the proof is complete by (CUT-3).

The remaining derivations are easy exercises. ∎

5.6 Rules for transition modalities

We then arrive at the rules for the modalitites $<l>$ and $[l]$. Proof goals in this case have the following shape:

$$\Gamma \vdash^{s} E : (l)\phi \tag{4.13}$$

where (l) is used as a wildcard ranging over $<l>$ and $[l]$. Observe that this shape is not quite as general as we would wish it to be. Rather we would want to permit sets of modal formulas to the right of the turnstile instead of the single formula permitted here. The restricted format (4.13) is chosen for the following reasons:

1 Modal proof rules for more general multi-conclusioned judgments would become unduly complicated.

2 The restricted format suffices for both the examples and the weak completeness results which we present here.

The transition capabilities of E in (4.13) are determined by the structure of E according the operational semantics given earlier, and according to the assumptions made in Γ on variables occurring freely in E. As the operational semantics is determined by induction in the structure of E it is hardly surprising that in general the number of rules governing proof goals of the shape $\Gamma \vdash^{s} E : (\alpha)\phi_1, \ldots, (\alpha)\phi_n$ depends on the number of primitive operators, and for each operator, the number of operational semantics rules of Section 2 determining its behaviour.

We then proceed by induction in the structure of the right hand agent term being predicated to give rules that show how (typically modal, but in some cases quite general) properties of an agent with a specific outermost connective can be inferred in terms of properties of its immediate constituents.

Term variables. The case for E a variable corresponds to the monotonicity rule familiar from sequent-style formalisations of modal logic:

$$(<l_\tau>\text{-MON}) \quad \frac{\Gamma, y : \phi, x : \phi_1, \ldots, y : \phi_n \vdash^{s} y : \psi_1, \ldots, \psi_m}{\Gamma, x : <l_\tau>\phi, x : [l_\tau]\phi_1, \ldots, x : [l_\tau]\phi_n \vdash^{s} x : <l_\tau>\psi_1, \ldots, <l_\tau>\psi_m}$$

$$([l_\tau]\text{-MON}) \quad \frac{\Gamma, y : \phi_1, \ldots, y : \phi_n \vdash^{s} y : \psi, \psi_1, \ldots, \psi_m}{\Gamma, x : [l_\tau]\phi_1, \ldots, x : [l_\tau]\phi_n \vdash^{s} x : [l_\tau]\psi, <l_\tau>\psi_1, \ldots, <l_\tau>\psi_m}$$

For both rules we require that $y \notin \Gamma$. With this proviso the rules are clearly sound.

Nil.

$$([L_\tau]\text{-NIL}) \quad \frac{}{\Gamma \vdash^{s} 0 : [l_\tau]\phi}$$

Summation.

$$<l_\tau>\text{-PLUS} \quad \frac{\Gamma \vdash^s P_i : <l_\tau>\phi}{\Gamma \vdash^s P_1 + P_2 : <l_\tau>\phi, \Delta} \quad (i \in \{1, 2\})$$

$$([L_\tau]\text{-PLUS}) \quad \frac{\Gamma \vdash^s P_1 : [l_\tau]\phi \quad \Gamma \vdash^s P_2 : [l_\tau]\phi}{\Gamma \vdash^s P_1 + P_2 : [l_\tau]\phi}$$

Prefixing.

$$((\tau)\text{-TAU}) \quad \frac{\Gamma \vdash^s P : \phi}{\Gamma \vdash^s \tau.P : (\tau)\phi} \qquad ([\tau]\text{-ACT}) \quad \frac{\cdot}{\Gamma \vdash^s l.E : [\tau]\phi}$$

$$([L]\text{-TAU}) \quad \frac{\cdot}{\Gamma \vdash^s \tau.P : [l]\phi} \qquad ([A]\text{-}\overline{B}) \quad \frac{\cdot}{\Gamma \vdash^s \overline{b}.C : [a]\phi}$$

$$([\overline{A}]\text{-}B) \quad \frac{\cdot}{\Gamma \vdash^s b.A : [\overline{a}]\phi}$$

$$(<l>\text{-ACT}) \quad \frac{\Gamma \vdash^s E : \phi \quad \Gamma \vdash^s E : a_1 = a_2}{\Gamma \vdash^s \hat{a_1}.E : <\hat{a_2}>\phi} \quad (l_1 \notin s)$$

$$([l]\text{-ACT}) \quad \frac{\Gamma, a_1 = a_2 \vdash^s E : \phi}{\Gamma \vdash^s \hat{a_1}.E : [\hat{a_2}]\phi} \quad (a_2 \notin s)$$

In the two last rules the notation $\hat{a}$ is used to indicate overbarring which is optional in the sense that either both transition labels indicated are overbarred, or else none is.

The rule ([l]-ACT) reveals where (ALPHA) is required: To show $\vdash^a a.A : [a]\bot$ (which is a valid judgment for any A) first we need to use (ALPHA) to rename the restricted a, then use ([l]-ACT) to reduce to a goal of the shape $b = a \vdash^b A : \bot$, which is then resolved by (NEW1).

Parallel composition. The rules for parallel composition are shown on Table 4.3. The rule (<l>-PAR) comes with a symmetric version. All rules for $|$ are marked $*$ indicating that they are subject to the side-condition that x and y are fresh (ie. do not appear free in the conclusion judgment), and the typing formulas S_1 and S_2 (where they appear) are complementary. The typing formulas have the important role of matching input and output. This is evident in the rule (<τ>-PAR). In ([τ]-PAR) the role is implicit: The rule requires an ancillary rule schema

$$(\text{AR}) \quad \frac{\Gamma \vdash^s E : S_1 \quad \Gamma \vdash^s E : S_2}{\Gamma \vdash^s F : \phi} \quad (S_1 \neq S_2)$$

to ensure that inputs are matched with outputs of the proper sort.

To see the soundness of eg. (<τ>-PAR) assume that the antecedents of that rule are valid. Assume also that a substitution η is given such that $\models_{\nu s.\eta} \eta(x_i) : \phi_i$ for all $i : 1 \leq i \leq n$. Then $\models E\eta : <l>\phi$ and $\models F\eta : <\bar{l}>\psi$. Thus $E\eta$ and $F\eta$ are processes, and there are E' and F' such that $E\eta \xrightarrow{l} E'$ and $F\eta \xrightarrow{\bar{l}} F'$. Moreover we can assume that E' and F' have complementary arities by the

(<l>-PAR)*:

$$\frac{\Gamma, s\ \mathit{fresh} \vdash^{()} F : \mathit{isP} \quad \Gamma, s\ \mathit{fresh} \vdash^{()} E : \langle l \rangle \phi \quad \Gamma, x : \phi \vdash^{s} x \mid F : \gamma}{\Gamma \vdash^{s} E \mid F : \langle l \rangle \gamma}$$

(<τ>-PAR)*:

$$\frac{\begin{array}{c}\Gamma, s\ \mathit{fresh} \vdash^{()} E : \langle l \rangle \phi \quad \Gamma, s\ \mathit{fresh} \vdash^{()} F : \langle \bar{l} \rangle \psi \\ \Gamma, s\ \mathit{fresh}, x : \phi \vdash^{()} x : \mathcal{S}_1 \quad \Gamma, s\ \mathit{fresh}, y : \phi \vdash^{()} y : \mathcal{S}_2 \\ \Gamma, x : \phi, y : \psi \vdash^{s} x \mid y : \gamma\end{array}}{\Gamma \vdash^{s} E \mid F : \langle \tau \rangle \gamma}$$

([l]-PAR)*:

$$\frac{\begin{array}{c}\Gamma, s\ \mathit{fresh} \vdash^{()} E : \mathit{isP} \quad \Gamma, s\ \mathit{fresh} \vdash^{()} F : \mathit{isP} \\ \Gamma, s\ \mathit{fresh} \vdash^{()} E : [l]\phi_1 \ \cdots \ \Gamma, s\ \mathit{fresh} \vdash^{()} E : [l]\phi_n \\ \Gamma, s\ \mathit{fresh} \vdash^{()} F : [l]\psi_1 \ \cdots \ \Gamma, s\ \mathit{fresh} \vdash^{()} F : [l]\psi_m \\ \Gamma, x : \phi_1, \ldots, x : \phi_n \vdash^{s} x \mid F : \gamma \quad \Gamma, y : \psi_1, \ldots, y : \psi_n \vdash^{s} E \mid y : \gamma\end{array}}{\Gamma \vdash^{s} E \mid F : [l]\gamma}$$

([τ]-PAR)*:

$$\frac{\begin{array}{c}\Gamma, s\ \mathit{fresh} \vdash^{()} E : \mathit{isP} \quad \Gamma, s\ \mathit{fresh} \vdash^{()} F : \mathit{isP} \\ \Gamma, s\ \mathit{fresh} \vdash^{()} E : [\tau]\phi_{1,1} \ \cdots \ \Gamma, s\ \mathit{fresh} \vdash^{()} E : [\tau]\phi_{1,n_1} \\ \Gamma, s\ \mathit{fresh} \vdash^{()} F : [\tau]\psi_{1,1} \ \cdots \ \Gamma, s\ \mathit{fresh} \vdash^{()} F : [\tau]\psi_{1,m_1} \\ \Gamma, s\ \mathit{fresh} \vdash^{()} E : \forall b.[b]\phi_{2,1} \ \cdots \ \Gamma, s\ \mathit{fresh} \vdash^{()} E : \forall b.[b]\phi_{2,n_2} \\ \Gamma, s\ \mathit{fresh} \vdash^{()} E : \forall b.[\bar{b}]\phi_{3,1} \ \cdots \ \Gamma, s\ \mathit{fresh} \vdash^{()} E : \forall b.[\bar{b}]\phi_{3,n_3} \\ \Gamma, s\ \mathit{fresh} \vdash^{()} F : \forall b.[b]\psi_{2,1} \ \cdots \ \Gamma, s\ \mathit{fresh} \vdash^{()} F : \forall b.[b]\psi_{2,m_2} \\ \Gamma, s\ \mathit{fresh} \vdash^{()} F : \forall b.[\bar{b}]\psi_{3,1} \ \cdots \ \Gamma, s\ \mathit{fresh} \vdash^{()} F : \forall b.[\bar{b}]\psi_{3,m_3} \\ \Gamma, x : \phi_{1,1}, \ldots, x : \phi_{1,n_1} \vdash^{s} x \mid F : \gamma \\ \Gamma, y : \psi_{1,1}, \ldots, y : \psi_{1,m_1} \vdash^{s} E \mid y : \gamma \\ \Gamma, x : \phi_{2,1}, \ldots, x : \phi_{2,n_2}, x : \mathit{isA}, y : \psi_{3,1}, \ldots, y : \psi_{3,m_3}, y : \mathit{isCf} \vdash^{s} x \mid y : \gamma \\ \Gamma, x : \phi_{2,1}, \ldots, x : \phi_{2,n_2}, x : \mathit{isA}, y : \psi_{3,1}, \ldots, y : \psi_{3,m_3}, y : \mathit{isCb} \vdash^{s} x \mid y : \gamma \\ \Gamma, x : \phi_{3,1}, \ldots, x : \phi_{3,n_3}, x : \mathit{isCf}, y : \psi_{2,1}, \ldots, y : \psi_{2,m_2}, y : \mathit{isA} \vdash^{s} x \mid y : \gamma \\ \Gamma, x : \phi_{3,1}, \ldots, x : \phi_{3,n_3}, x : \mathit{isCb}, y : \psi_{2,1}, \ldots, y : \psi_{2,m_2}, y : \mathit{isA} \vdash^{s} x \mid y : \gamma\end{array}}{\Gamma \vdash^{s} E \mid F : [\tau]\gamma}$$

Table 4.3. Proof rules for parallel composition vs. transition modalities

third and fourth antecedents. Moreover, $\models_{\nu s.\eta} \nu s.(x \mid y\eta\{E'/x\}\{F'/y\}) : \gamma$. But then $\models_{\nu s.\eta} E \mid F\eta : <\tau>\gamma$ as desired.

The rule ($[\tau]$-PAR) is at first sight ridiculously complex. On closer inspection, however, we argue that the rule merely brings out the quite complex modal behaviour of π-calculus parallel composition, and thus the complexity is inherent in the problem rather than due to the specificities of our formalisation. This is not to say that simpler formulations can not be found. In fact this may very well be possible, for instance by appealing directly to the operational semantics transition relation in a way which we have chosen *not* to do. However, we do believe quite strongly that a truly *compositional* and modal analysis of π-calculus parallel composition will have to perform the sort of quite convoluted case analysis brought out by the ($[\tau]$-PAR) rule.

Conditional.

$$(\text{COND}) \quad \frac{\Gamma, a = b \vdash^s P : \phi \quad \Gamma, a \neq b \vdash^s Q : \phi}{\Gamma \vdash^s \textbf{if } a = b \textbf{ then } P \textbf{ else } Q : \phi}$$

Restriction.

$$(\text{NEW2}) \quad \frac{\Gamma \vdash^{s,a} E : \phi}{\Gamma \vdash^s \nu a.E : \phi} \quad (a \text{ fresh})$$

Identifiers.

$$(\text{FIX}) \quad \frac{\Gamma \vdash^s P\{\vec{b}/\vec{a}\} : \phi}{\Gamma \vdash^s D(\vec{b}) : \phi} \quad D(\vec{a}) \triangleq P$$

Of these final three rules the rule (NEW2) is the least trivial. Assume that a is fresh for $\Gamma \vdash^s \nu a.E : \phi$, and that $\Gamma \models^{s,a} E : \phi$. Let η be given such that $\eta(x) \in \|\phi_i\|(\nu s.\eta)$ whenever $x = x_i$. Pick now a b which is fresh and not in free in any $\eta(x_i)$. Let $\eta'(x_i) = \eta(x)\{b/a\}$. Then $\eta'(x) \in \|\phi_i\|(\nu b.\nu s.\eta)$, so, by the assumption, and substituting b uniformly for a, $\nu s.\nu b.E\eta' \in \|\phi\|\eta$. But then $\nu s.(\nu a.E)\eta \in \|\phi\|\eta$ too.

Of all the rules for process modalities clearly the rule ($[\tau]$-PAR) is the most complex. It is nonetheless quite intuitive. To prove a property of the shape $[\tau]\phi$ of a parallel composition $A \mid B$ we have to describe in sufficient detail, by formulas $\phi_{i,j}$ and $\psi_{i,j}$, the properties of A and B after a transition has been performed. For almost all a, $\phi_{2,j}(a)$ and $\psi_{2,j}(a)$ will be false. The formulas $\phi_{i,j}$ and $\psi_{i,j}$ have to be shown to hold for A or B in the prescribed fashion, and they have to be shown to compose in the correct way, given that synchronisations must result in agents of complementary types.

Many variations can be played on the formulation of these rules. As an example we consider a version of ($[l]$-PAR) given in [5] (for CCS, so typing formulas and restriction sets were not needed there).

Proposition 15 *The following rule is derivable where x and y do not occur free in Γ:*

$$([l]\text{-PAR'}) \quad \frac{\Gamma, x : \phi_1, y : \psi_2 \vdash^s x \mid y : \gamma \qquad \Gamma, x : \phi_2, y : \psi_1 \vdash^s x \mid y : \gamma}{\Gamma, x : \phi_1 \wedge [l]\phi_2, y : \psi_1 \wedge [l]\psi_2 \vdash^s x \mid y : [l]\gamma}$$

PROOF. The proof goal is

$$\Gamma, x : \phi_1 \wedge [l]\phi_2, y : \psi_1 \wedge [l]\psi_2 \vdash^s x \mid y : [l]\gamma. \tag{4.14}$$

We get the following list of 6 subgoals:

$$\Gamma, s\ fresh, x : \phi_1 \wedge [l]\phi_2, y : \psi_1 \wedge [l]\psi_2 \vdash^{()} x : isP \tag{4.15}$$
$$\Gamma, s\ fresh, x : \phi_1 \wedge [l]\phi_2, y : \psi_1 \wedge [l]\psi_2 \vdash^{()} y : isP \tag{4.16}$$
$$\Gamma, s\ fresh, x : \phi_1 \wedge [l]\phi_2, y : \psi_1 \wedge [l]\psi_2 \vdash^{()} x : [l]\phi_2 \tag{4.17}$$
$$\Gamma, s\ fresh, x : \phi_1 \wedge [l]\phi_2, y : \psi_1 \wedge [l]\psi_2 \vdash^{()} y : [l]\psi_2 \tag{4.18}$$
$$\Gamma, x : \phi_1 \wedge [l]\phi_2, y : \psi_1 \wedge [l]\psi_2, x' : \phi_2 \vdash^s x' \mid y : \gamma \tag{4.19}$$
$$\Gamma, x : \phi_1 \wedge [l]\phi_2, y : \psi_1 \wedge [l]\psi_2, y' : \psi_2 \vdash^s x \mid y' : \gamma \tag{4.20}$$

To prove (4.15) and (4.16) we use a few logical and equational rules along with ($[l_\tau]$-MON). Subgoals (4.17) and (4.18) are trivial. To prove (4.19) and (4.20) apply weakening to eliminate the "old" version of x (or y), and then apply ($\wedge$-L) to arrive at the subgoal $\Gamma, x' : \phi_2, y : \psi_1 \vdash^s x' \mid y : [\tau]\gamma$ (for (4.19)) as desired. ∎

A small but important difference between the rules ($[l]$-PAR') and ($[l]$-PAR) is that in ($[l]$-PAR) (as elsewhere) we permit multiple assumptions on term variables. This is inessential for the modal fragment (the version with multiple assumptions can easily be derived), but when we come to consider recursive formulas the distinction will turn out to be more significant.

5.7 Rules for input/output modalities

Except for two rules needed for conversion between the free and bound input modalities, the rules for input and output modalities follow the pattern established for the process modalities in the previous section. As there introduced variables like x and y below are subject to the side condition that they do not appear in Γ.

Term variables. The rules for term variables are shown on Table 4.4. The side-condition a fresh in ($\nu \rightarrow$-MON) and ($\nu \leftarrow$-MON) means, as before, that a does not occur freely in the conclusion, nor is a a member of s.

The most delicate of the motonicity rules is ($\nu \leftarrow$-MON). To see that it is sound assume for simplicity that Γ is empty. Let η and C be given such

($\rightarrow$-MON):

$$\frac{\Gamma, x : \phi_1, \ldots, x : \phi_n \vdash^s x : \psi_1, \ldots, \psi_m}{\Gamma, x : a \rightarrow \phi_1, \ldots, x : a \rightarrow \phi_n \vdash^s x : a \rightarrow \psi_1, \ldots, a \rightarrow \psi_m} \quad (a \notin s)$$

($\leftarrow$-MON):

$$\frac{\Gamma, x : \phi_1, \ldots, x : \phi_n \vdash^s x : \psi_1, \ldots, \psi_m}{\Gamma, x : a \leftarrow \phi_1, \ldots, x : a \leftarrow \phi_n \vdash^s x : a \leftarrow \psi_1, \ldots, a \leftarrow \psi_m} \quad (a \notin s)$$

($\nu \rightarrow$-MON):

$$\frac{\Gamma, x : \phi_1, \ldots, x : \phi_n \vdash^s x : \psi_1, \ldots, \psi_m}{\Gamma, x : \nu a \rightarrow \phi_1, \ldots, x : \nu a \rightarrow \phi_n \vdash^s x : \nu a \rightarrow \psi_1, \ldots, \nu a \rightarrow \psi_m} \quad (a \text{ fresh})$$

($\nu \leftarrow$-MON):

$$\frac{\Gamma, x : \phi_1, \ldots, x : \phi_n \vdash^{s,a} x : \psi_1, \ldots, \psi_m}{\Gamma, x : \nu a \leftarrow \phi_1, \ldots, x : \nu a \leftarrow \phi_n \vdash^s x : \nu a \leftarrow \psi_1, \ldots, \nu a \leftarrow \psi_m} \quad (a \text{ fresh})$$

Table 4.4. Monotonicity rules for io-modalities

that $\models_{\nu s.\eta} C : \nu a \leftarrow \phi_i$ for all $i \in \{1, \ldots, n\}$. Then C can be written as $\nu b.[b]C'$, and $\models_{\nu c.\nu s.\eta} C'\{c/b\} : \phi_i\{c/a\}$ for all i where c is fresh. Assume that $x : \phi_1, \ldots, x : \phi_n \models^{s,a} x : \psi_1, \ldots, \psi_m$. Then $x : \phi_1\{c/a\}, \ldots, x : \phi_n\{c/a\} \models^{s,c} x : \psi\{c/a\}$ too, since c was fresh. Thus $\models_\eta \nu c.\nu s.C'\{c/b\} : \psi\{c/a\}$, showing that $\models_\eta \nu s.\nu b.[b]C' : \nu a \leftarrow \psi$ as required.

Parallel composition. The rules for parallel composition are shown on Table 4.5.

Input.

$$(\rightarrow\text{-IN}) \quad \frac{\Gamma, a = b \vdash^s A : \phi}{\Gamma \vdash^s (a)A : b \rightarrow \phi} \quad (a \text{ fresh}, b \notin s)$$

$$(\nu \rightarrow\text{-IN}) \quad \frac{\Gamma, a\ \mathit{fresh} \vdash^s A : \phi\{a/b\}}{\Gamma \vdash^s (a)A : \nu b \rightarrow \phi} \quad (a \text{ fresh})$$

In the context of a rule such as ($\nu \rightarrow$-IN) we use *a fresh* as abbreviation of the formula $\wedge\{a \neq c \mid c \text{ not fresh}\}$.

Output.

$$(\leftarrow\text{-OUT}) \quad \frac{\Gamma \vdash^s C : a = b \quad \Gamma \vdash^s C : \phi}{\Gamma \vdash^s \langle a \rangle C : b \leftarrow \phi}$$

$$(\nu \leftarrow\text{-OUT}) \quad \frac{\Gamma \vdash^s C : \phi}{\Gamma \vdash^{s,a} \langle a \rangle C : \nu a \leftarrow \phi}$$

Input modality conversion. Falling a little outside the patterns so far established we also need rule to convert between the free and bound input modalities:

$$(\nu \rightarrow\text{-}\rightarrow\text{-CONV}) \quad \frac{\cdot}{\Gamma, x : \nu a \rightarrow \phi \vdash^{()} x : a \rightarrow \phi} \quad (a \notin fn(\Gamma))$$

($\rightarrow$-PAR):

$$\frac{\begin{array}{c}\Gamma, s\ fresh \vdash^{()} F : isP \\ \Gamma, s\ fresh \vdash^{()} E : a \rightarrow \phi_1, \ldots, \Gamma, s\ fresh \vdash^{()} E : a \rightarrow \phi_n \\ \Gamma, x : \phi_1, \ldots, x : \phi_n \vdash^{s} x \mid F : \gamma\end{array}}{\Gamma \vdash^{s} E \mid F : a \rightarrow \gamma} \quad (a \notin s)$$

($\leftarrow$-PAR):

$$\frac{\begin{array}{c}\Gamma, s\ fresh \vdash^{()} F : isP \\ \Gamma, s\ fresh \vdash^{()} E : a \leftarrow \phi_1, \ldots, \Gamma, s\ fresh \vdash^{()} E : a \leftarrow \phi_n \\ \Gamma, x : \phi_1, \ldots, x : \phi_n \vdash^{s} x \mid F : \gamma\end{array}}{\Gamma \vdash^{s} E \mid F : a \leftarrow \gamma} \quad (a \notin s)$$

($\nu \rightarrow$-PAR):

$$\frac{\begin{array}{c}\Gamma, s\ fresh \vdash^{()} F : isP \\ \Gamma, s\ fresh \vdash^{()} E : \nu a \rightarrow \phi_1, \ldots, \Gamma, s\ fresh \vdash^{()} E : \nu a \rightarrow \phi_n \\ \Gamma, x : \phi_1, \ldots, x : \phi_n \vdash^{s} x \mid F : \gamma\end{array}}{\Gamma \vdash^{s} E \mid F : \nu a \rightarrow \gamma} \quad (a \text{ fresh})$$

($\nu \leftarrow$-PAR):

$$\frac{\begin{array}{c}\Gamma, s\ fresh \vdash^{()} F : isP \\ \Gamma, s\ fresh \vdash^{()} E : \nu a \leftarrow \phi_1, \ldots, \Gamma, s\ fresh \vdash^{()} E : \nu a \leftarrow \phi_n \\ \Gamma, x : \phi_1, \ldots, x : \phi_n \vdash^{s} x \mid F : \gamma\end{array}}{\Gamma \vdash^{s} E \mid F : \nu a \leftarrow \gamma} \quad (a \text{ fresh})$$

($\rightarrow$-$\leftarrow$-PAR):

$$\frac{\begin{array}{c}\Gamma, s\ fresh \vdash^{()} E : a \rightarrow \phi_1, \ldots, \Gamma, s\ fresh \vdash^{()} E : a \rightarrow \phi_n \\ \Gamma, s\ fresh \vdash^{()} F : a \leftarrow \psi_1, \ldots, \Gamma, s\ fresh \vdash^{()} F : a \leftarrow \psi_m \\ \Gamma, x : \phi_1, \ldots, x : \phi_n, y : \psi_1, \ldots, y : \psi_m \vdash^{s} x \mid y : \gamma\end{array}}{\Gamma \vdash^{s} E \mid F : \gamma}$$

($\nu \rightarrow$-$\nu \leftarrow$-PAR):

$$\frac{\begin{array}{c}\Gamma, s\ fresh \vdash^{()} E : \nu a \rightarrow \phi_1, \ldots, \Gamma, s\ fresh \vdash^{()} E : \nu a \rightarrow \phi_n \\ \Gamma, s\ fresh \vdash^{()} F : \nu a \leftarrow \psi_1, \ldots, \Gamma, s\ fresh \vdash^{()} F : \nu a \leftarrow \psi_m \\ \Gamma, x : \phi_1, \ldots, x : \phi_n, y : \psi_1, \ldots, y : \psi_m \vdash^{s,a} x \mid y : \gamma\end{array}}{\Gamma \vdash^{s} A \mid B : \gamma} \quad (a \text{ fresh})$$

Table 4.5. Rules for parallel composition vs. io-modalities

$$(\rightarrow\text{-}\nu \rightarrow\text{-CONV}) \quad \frac{\cdot}{\Gamma, x : a \rightarrow \phi \vdash^{()} x : \nu a \rightarrow \phi} \quad (a \notin fn(\Gamma))$$

As an example we show how to derive a rule matching bound output with free input.

Proposition 16 *The following rule is derivable:*

$(\rightarrow\text{-}\nu \leftarrow\text{-COM})$

$$\frac{\begin{array}{c}\Gamma, s\,\mathit{fresh} \vdash^{()} A : \forall a.a \rightarrow \phi_1, \ldots, \Gamma, s\,\mathit{fresh} \vdash^{()} A : \forall a.a \rightarrow \phi_n \\ \Gamma, s\,\mathit{fresh} \vdash^{()} C : \nu a \leftarrow \psi_1, \ldots, \Gamma, s\,\mathit{fresh} \vdash^{()} C : \nu a \leftarrow \psi_m \\ \Gamma, x : \phi_1, \ldots, x : \phi_n, y : \psi_1, \ldots, y : \psi_m \vdash^{s,a} x \mid y : \gamma\end{array}}{\Gamma \vdash^{s} A \mid C : \gamma} \quad (a\,\mathit{fresh})$$

PROOF. From the antecedents concerning A first fresh a's are introduced using ($\forall$-R), and then the resulting free inputs are converted to bound inputs using ($\rightarrow$-ν $\rightarrow$-CONV), so that (ν $\rightarrow$-ν $\leftarrow$-COM) can be used to yield the result. ∎

5.8 Examples

In this section we give a first little more substantial proof example. More proof examples are given later.

Example 17 Consider the processes

$$\begin{array}{rcl} P & = & \overline{a}.\nu b.\langle b\rangle b(c).0 \\ Q & = & a(d).\overline{d}\langle d\rangle.0 \end{array}$$

In $P \mid Q$ first b is passed as a private name from P to Q along a, then b is returned along itself back to P again. Clearly the judgment

$$\vdash P \mid Q : <\tau><\tau>[\tau]\bot \tag{4.21}$$

is valid. First CUT-1 is applied. Let

$$\begin{array}{rcl} \mathit{STOP} & = & [\tau]\bot \wedge \forall a.[a]\bot \wedge \forall a.[\overline{a}]\bot \\ \phi_1 & = & <\overline{a}>\nu b \leftarrow <b>\forall c.c \rightarrow \mathit{STOP} \\ \phi_2 & = & <a>\forall d.d \rightarrow <\overline{d}>d \leftarrow \mathit{STOP} \end{array}$$

Using CUT-1 twice we can reduce (4.21) to the following three subgoals:

$$\vdash P : \phi_1 \tag{4.22}$$

$$\vdash Q : \phi_2 \tag{4.23}$$

$$x : \phi_1, y : \phi_2 \vdash x \mid y : <\tau><\tau>[\tau]\bot \tag{4.24}$$

To prove (4.22) first apply ($<l>$-ACT) and (REFL) to reduce to

$$\vdash \nu b.\langle b\rangle b(c).0 : \nu b \leftarrow <b>\forall c.c \rightarrow \mathit{STOP} \tag{4.25}$$

and then using (NEW2) and ($\nu \leftarrow$-OUT) to

$$\vdash b(c).0 : <b>\forall c.c \rightarrow STOP \tag{4.26}$$

A few more steps in a similar vein suffices to complete the proof of (4.22), and the proof of (4.23) is similar. To prove (4.24) we first reduce to the following list of subgoals

$$x : \phi_1, y : \phi_2 \vdash x : <\bar{a}>\nu b \leftarrow <b>\forall c.c \rightarrow STOP \tag{4.27}$$

$$x : \phi_1, y : \phi_2 \vdash y : <a>\forall d.d \rightarrow <\bar{d}>d \leftarrow STOP \tag{4.28}$$

$$x : \phi_1, y : \phi_2, x' : \nu b \leftarrow <b>\forall c.c \rightarrow STOP \vdash x' : isCb \tag{4.29}$$

$$x : \phi_1, y : \phi_2, y' : \forall d.d \rightarrow <\bar{d}>d \leftarrow STOP \vdash y' : isA \tag{4.30}$$

$$x : \phi_1, y : \phi_2, x' : \nu b \leftarrow <b>\forall c.c \rightarrow STOP, y' : \forall d.d \rightarrow <\bar{d}>d \leftarrow STOP \\ \vdash x' \mid y' : <\tau>[\tau]\bot \tag{4.31}$$

using ($<\tau>$-PAR). The first two subgoals are just instances of (I). For (4.29) we just need to use the monotonicity rule ($\nu \leftarrow$-MON), and for (4.30) the rules ($\vee$-R), (W-R), ($\forall$-L), ($\forall$-R), and ($\rightarrow$-MON). Finally, for (4.31), we first use ($\forall$-L) to get

$$x : \cdots, y : \cdots, x' : \nu b \leftarrow <b>\forall c.c \rightarrow STOP, y' : d \rightarrow <\bar{d}>d \leftarrow STOP \\ \vdash x' \mid y' : <\tau>[\tau]\bot \tag{4.32}$$

where d is fresh, and then, using ($\nu \rightarrow$-$\rightarrow$-CONV) along with (CUT-1), to

$$x : \cdots, y : \cdots, x' : \nu b \leftarrow <b>\forall c.c \rightarrow STOP, y' : \cdots, y'' : \nu d \rightarrow <\bar{d}>d \leftarrow STOP \\ \vdash x' \mid y'' : <\tau>[\tau]\bot \tag{4.33}$$

which is reduced, by ($\nu \rightarrow$-$\nu \leftarrow$-COM), to

$$x : \cdots, y : \cdots, x' : <b>\forall c.c \rightarrow STOP, y' : \cdots, y'' : <\bar{b}>b \leftarrow STOP \\ \vdash x' \mid y'' : <\tau>[\tau]\bot \tag{4.34}$$

along with two goals resolved immediately by (I). Proceeding, we use ($<\tau>$-PAR) to reduce to

$$\cdots, x'' : \forall c.c \rightarrow STOP, \cdots, y''' : b \leftarrow STOP \vdash x'' \mid y''' : [\tau]\bot. \tag{4.35}$$

We now use ($\forall$-L) and ($\rightarrow$-$\leftarrow$-COM) to reduce to

$$\cdots, x''' : STOP, \cdots, y'''' : \ STOP \vdash x''' \mid y'''' : [\tau]\bot. \tag{4.36}$$

which is resolved very easily using ($[\tau]$-PAR) and some elementary reasoning.

Variable naming. Observe that, because of the shape of the rules, proofs tend to introduce long sequences of variables like x, x', x'', x''' in the example above. It is very often the case that, once transitions or input-output actions are taken (in the form of an application of a modal or an input-output rule), old variables can immediately be forgotten using weakening. In the examples that follow, for this reason we often identify variables like x and x', assuming implicitly that when x' is introduced, assumptions concerning x are immediately forgotten about, whence x' can be renamed to x.

6. Soundness and completeness for the modal fragment

Before pushing on to add rules of discharge that allow interesting recursive properties to be proved we pause to establish soundness and completeness for the modal fragment. Proofs in this section have been deferred to Appendix A. Soundness of the most delicate rules was shown as the proof system was presented, so here we can just state soundness as a fact.

Proposition 18 (Soundness, modal fragment) *If* $\Gamma \vdash^s E : \Delta$ *in the proof system of Section 5 then* $\Gamma \models^s E : \Delta$. ∎

Concerning completeness we consider this only in a rather weak form, namely that if $\Gamma \models^s E : \phi$ where Γ is elementary (i.e. all formulas in Γ are elementary) then $\Gamma \vdash^s E : \phi$ is provable. We first introduce some basic tools.

Definition 19 Let η be a partition and N a finite set of names.

1 *hyp*(η, N) denotes a sequence of the form $x_1 : \phi_1, \ldots, x_n : \phi_n$ where each ϕ_i has the form $a = b$ or $a \neq b$ with $a, b \in N$, and where $a = b$ $(a \neq b)$ is in *hyp*(η, N) if and only if $\eta \models a = b$ $(\eta \models a \neq b)$.

2 Let $\Gamma = x_1 : \psi_1, \ldots, x_n : \psi_n$ where all ψ_i are elementary. Then $\eta \models \Gamma$ if and only if $\eta \models \psi_i$ for all $i : 1 \leq i \leq n$.

We can now show the following property that deals with the first-order part.

Proposition 20 *Let* N *include the set of non-fresh names of* $\Gamma \vdash^s E : \phi$.

1 If ϕ *is boolean then* *hyp*$(\eta, N) \models^s E : \phi$ *if and only if* *hyp*$(\eta, N) \vdash^s E : \phi$.

2 $\Gamma \vdash^s E : \phi$ *if and only if* *hyp*$(\nu s.\eta, N) \vdash^s E : \phi$ *for all* η *such that* $\nu s.\eta \models \Gamma$.

From this point onwards we shall not be very explicit about the handling of elementary connectives. We prove completeness via a kind of "Decomposition Lemma" allowing us to decompose a goal into subgoals for subagents.

Lemma 21 (Decomposition) *Let N include fn$(E\{E_1/x_1,\ldots,E_n/x_n\})$, fn$(s)$, and fn$(\phi)$. Suppose $E\{E_1/x_1,\ldots,E_n/x_n\}$ is closed, and suppose ϕ is non-recursive. Let $\Gamma = hyp(\nu s.\eta, N)$. If $\Gamma \models^s E\{E_1/x_1,\ldots,E_n/x_n\} : \phi$ then there are $\phi_1,\ldots,\phi_n$ of modal depth not exceeding that of ϕ such that*

1. *for all $i : 1 \leq i \leq n$, $\Gamma, s\ fresh \models^{\emptyset} E_i : \phi_i$, and*

2. $\Gamma, x_1 : \phi_1, \ldots, x_n : \phi_n \vdash^s E : \phi$.

Theorem 22 (Completeness, modal fragment) *Suppose that $\Gamma \models^s E : \Delta$, Γ is boolean, and all formulas in Δ are non-recursive. Then $\Gamma \vdash^s E : \Delta$.* ∎

7. Proof rules for recursive formulas

We now proceed to address fixed points. All approaches to analysis or verification of μ-calculus relies at some level on approximation ordinals and well-founded induction, using the Knaster-Tarski Fixed Point Theorem. So indeed does ours. In some cases when fixed point formulas are unfolded it is possible to determine suitable approximation ordinals which provide progress measures towards satisfaction. This applies, in particular, when unfolding least fixed point formulas to the left of the turnstile, and when unfolding greatest fixed point formulas to the right. Approximation ordinals are reflected explicitly in the proof system, by specific ordinal variables. This provides a simple framework for dealing with a variety of complications including alternation of fixed points and, more importantly in fact, a number of complications related to fixed point interference which we explain below.

The material in this section is based on corresponding material in the paper [6]. For this reason, proofs of some theorems have been left out of this presentation.

Ordinal approximations. Soundness of fixed point induction relies on the well-known iterative characterisation where least and greatest fixed points are "computed" as iterative limits of their ordinal approximations. Let κ range over ordinal variables. Name interpretations are extended to map ordinal variables to ordinals. Let U, V range over fixed point formula abstractions of the form $\sigma X(\vec{a}).\phi$. New formulas are introduced of the shape U^κ and $\kappa < \kappa'$. Ordinal inequalities have their obvious semantics, and $\kappa \leq \kappa'$ abbreviates $\kappa < \kappa' \vee \kappa = \kappa'$ as usual. For approximated fixed point abstractions suppose first that $U = \sigma X(\vec{a}).\phi$ and $\sigma = \nu$. Then

$$\|U^\kappa(\vec{b})\|(\rho,\eta) = \begin{cases} \mathcal{P} \cup \mathcal{A} \cup \mathcal{C}, & (\eta(\kappa) = 0) \\ \|\phi\|(\rho\{\|U^\kappa\|/X\}, \eta\{\eta(\kappa) - 1/\kappa, \vec{b}/\vec{a}\}), & (\eta(\kappa)\ suc.o.) \\ \bigcap\{\|U^\kappa(\vec{b})\|(\rho,\eta') \mid (*)\}, & (\eta(\kappa)\ lim.o.) \end{cases}$$

where the condition $(*)$ is that $\eta'(\kappa) < \eta(\kappa)$ and whenever $x \neq \kappa$ then $\eta'(x) = \eta(x)$. Dually, if $\sigma = \mu$:

$$\|U^\kappa(\vec{b})\|(\rho,\eta) = \begin{cases} \emptyset, & (\eta(\kappa) = 0) \\ \|\phi\|(\rho\{\|U^\kappa\|/X\}, \eta\{\eta(\kappa) - 1/\kappa, \vec{b}/\vec{a}\}), & (\eta(\kappa)\ \mathit{suc.o.}) \\ \bigcup\{\|U^\kappa(\vec{b})\|(\rho,\eta') \mid (*)\}, & (\eta(\kappa)\ \mathit{lim.o.}) \end{cases}$$

with the same side condition $(*)$ as above. We get the following basic monotonicity properties of ordinal approximations:

Proposition 23 *Suppose that $\eta(\kappa) \leq \eta'(\kappa)$ and whenever $x \neq \kappa$ then $\eta'(x) = \eta(x)$.*

1 If U is a greatest fixed point abstraction then

$$\|U^\kappa(\vec{b})\|(\rho,\eta') \subseteq \|U^\kappa(\vec{b})\|(\rho,\eta)$$

2 If U is a least fixed point abstraction then

$$\|U^\kappa(\vec{b})\|(\rho,\eta) \subseteq \|U^\kappa(\vec{b})\|(\rho,\eta')$$

PROOF. By wellfounded induction. ∎

Moreover, and most importantly, we get the following straightforward application of the well-known Knaster-Tarski fixed point theorem.

Theorem 24 (Knaster–Tarski) *Suppose that $U = \sigma X(\vec{a}).\phi$. Then*

$$\|U(\vec{b})\|(\rho,\eta) = \begin{cases} \bigcap\{\|U^\kappa(\vec{b})\|(\rho,\eta\{\alpha/\kappa\}) \mid \alpha \text{ an ordinal}\}, & \text{if } \sigma = \nu \\ \bigcup\{\|U^\kappa(\vec{b})\|(\rho,\eta\{\alpha/\kappa\}) \mid \alpha \text{ an ordinal}\}, & \text{if } \sigma = \mu \end{cases}$$

As the intended model is countable the quantification in Theorem 24 can be restricted to countable ordinals.

7.1 Rules for fixed point unfolding and approximation

The main rules to reason locally about fixed point formulas are the unfolding rules. These come in four flavours, according to whether the fixed point abstraction concerned has already been approximated or not, and to the nature and position of the fixed point relative to the turnstile.

$$(\text{APPRX-L})\quad \frac{\Gamma, x : U^\kappa(\vec{b}) \vdash^s E : \Delta}{\Gamma, x : U(\vec{b}) \vdash^s E : \Delta}\ U \text{ lfp},\ \kappa \text{ fresh}$$

$$(\text{APPRX-R})\quad \frac{\Gamma \vdash^s E : U^\kappa(\vec{b}), \Delta}{\Gamma \vdash^s E : U(\vec{b}), \Delta}\ U \text{ gfp},\ \kappa \text{ fresh}$$

$$\text{(UNF-L-1)}\quad \frac{\Gamma, x : \phi\{U/X, \vec{b}/\vec{a}\} \vdash^s E : \Delta}{\Gamma, x : U(\vec{b}) \vdash^s E : \Delta}\ U = \sigma X(\vec{a}).\phi$$

$$\text{(UNF-R-1)}\quad \frac{\Gamma \vdash^s E : \phi\{U/X, \vec{b}/\vec{a}\}, \Delta}{\Gamma \vdash^s E : U(\vec{b}), \Delta}\ U = \sigma X(\vec{a}).\phi$$

$$\text{(UNF-L-2)}\quad \frac{\Gamma, x : \phi\{U^{\kappa_1}/X, \vec{b}/\vec{a}\}, \kappa_1 < \kappa \vdash^s E : \Delta}{\Gamma, x : U^{\kappa}(\vec{b}) \vdash^s E : \Delta}\ \begin{array}{c} U = \mu X(\vec{a}).\phi, \\ \kappa_1 \text{fresh} \end{array}$$

$$\text{(UNF-R-2)}\quad \frac{\Gamma, \kappa_1 < \kappa \vdash^s E : \phi\{U^{\kappa_1}/X, \vec{b}/\vec{a}\}, \Delta}{\Gamma \vdash^s E : U^{\kappa}(\vec{b}), \Delta}\ U = \nu X(\vec{a}).\phi, \kappa_1 \text{ fresh}$$

$$\text{(UNF-L-3)}\quad \frac{\Gamma, x : \kappa_1 < \kappa \supset \phi\{U^{\kappa_1}/X, \vec{b}/\vec{a}\} \vdash^s E : \Delta}{\Gamma, x : U^{\kappa}(\vec{b}) \vdash^s E : \Delta}\ \begin{array}{c} U = \nu X(\vec{a}).\phi, \\ \kappa_1 \text{ fresh} \end{array}$$

$$\text{(UNF-R-3)}\quad \frac{\Gamma \vdash^s E : \kappa_1 < \kappa \wedge \phi\{U^{\kappa_1}/X, \vec{b}/\vec{a}\}, \Delta}{\Gamma \vdash^s E : U^{\kappa}(\vec{b}), \Delta}\ \begin{array}{c} U = \mu X(\vec{a}).\phi, \\ \kappa_1 \text{ fresh} \end{array}$$

The first unfolding rules, (UNF-L-1) and (UNF-R-1), are the expected unfolding rules. These rules are always used when unfolding least fixed point formulas occurring to the right of the turnstile, or, dually, greatest fixed point formulas occurring to the left. In these cases the proof task is an existential one, to identify some approximation ordinal making the statement true, which the first unfolding rules merely serve to delay. On the other hand, in the case of least fixed point formulas occurring to the left of the turnstile, or greatest fixed point formulas occurring to the right, the task is a universal one, suggesting well-founded induction as a suitable proof strategy. The approximation rules, (APPRX-L) and (APPRX-R), serve to introduce ordinal variables for this purpose. Having introduced ordinal variables they need to decremented as approximated formulas are unfolded. This is the purpose of the second pair of unfolding rules, (UNF-L-2) and (UNF-R-2).

Now, since ordinal approximations are introduced at only certain positions in a judgment (to the left for least fixed points and to the right for greatest ones), if the positions of approximated formulas would be unaffected by the local proof rules, the six rules so far discussed would have been sufficient. Unfortunately, due to the cut rules, this is not so. Consider for instance the following (quite typical) application of the process cut rule:

$$\frac{\Gamma \vdash^{()} Q : U^{\kappa} \quad \Gamma, x : U^{\kappa} \vdash^{()} P : U^{\kappa}}{\Gamma \vdash^{()} P\{Q/x\} : U^{\kappa}}$$

In this example U may be a greatest fixed point formula which, through some earlier application of (APPRX-R) has been assigned the ordinal variable κ. The second antecedent has U^{κ} occurring to the left of the turnstile. The third pair of unfolding rules are needed to handle this situation.

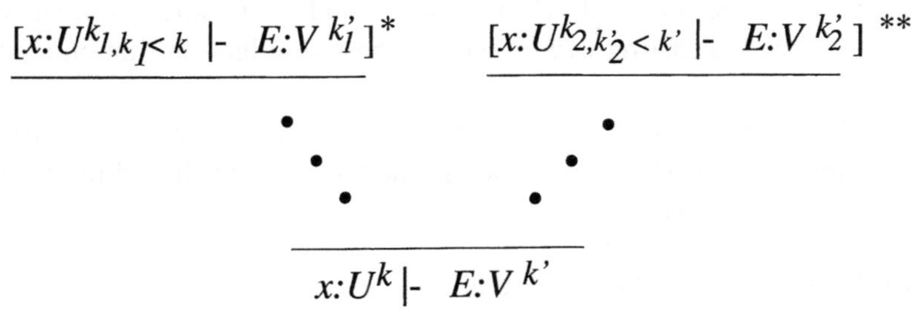

Figure 4.2. Fixed point interference

In addition to the above 8 rules it is useful also to add versions of the identity rules reflecting the monotonicity properties of ordinal approximations, Proposition 23:

$$\mathsf{IdMon1} \quad \frac{\Gamma \vdash \kappa < \kappa'}{\Gamma, x : U^{\kappa}(\vec{b}) \vdash^{()} x : U^{\kappa'}(\vec{b}), \Delta} \quad U \text{ lfp}$$

$$\mathsf{IdMon2} \quad \frac{\Gamma \vdash \kappa > \kappa'}{\Gamma, x : U^{\kappa}(\vec{b}) \vdash^{()} x : U^{\kappa'}(\vec{b}), \Delta} \quad U \text{ gfp}$$

Additionally a set of elementary rules are needed to support reasoning about well-orderings, including transitivity and irreflexivity of $<$. These rules are left out of the presentation. For the soundness proof (which is uncontroversial) we refer to [6].

Theorem 25 *The rules* (APPRX-L), (APPRX-R), (UNF-L-1), (UNF-R-1), (UNF-L-2), (UNF-R-2), (UNF-L-3), (UNF-R-3), IdMon1, *and* IdMon2 *are sound.* ∎

7.2 Rule of discharge

In addition to the local rules for unfolding and approximating fixed point formulas, a rule is needed for discharging valid induction hypothesis instances. The fundamental problem in devising such a rule is that fixed points may interfere as proofs are elaborated. The problem is illustrated in Figure 4.2. The formula U is assumed to be a least fixed point formula, and the formula V is a greatest fixed point formula. The node labelled $*$ can be interpreted locally as an instance of an induction hypothesis, using induction on κ, and the node labelled $**$ similarly uses induction on κ'. However, for the node $*$ there is no information as to the relationship between κ_1' and κ', and similarly for the node $**$ there is nothing relating κ_2' and κ'. This easily happens in practice, viz. the examples below. The problem is that in this case the unfolding of fixed points interfere: With the information provided we are unable to cast the proof as a proof by well-founded induction with the nodes $*$ and $**$ corresponding to applications of an inductive hypothesis for the simple reason that such an argument would be unsound. On the other hand, if we knew at node $**$, say, that $\kappa_2 \leq \kappa$, such a casting *would* exist, as nested induction first on κ and then in κ'.

That this problem indeed arises in practice is illustrated by the following two examples. The first example shows where discharge should fail because of fixed point interference.

Example 26 Consider the proof goal

$$x : \nu Z_1.\mu Z_2.[\tau]Z_1 \wedge [a]Z_2 \vdash^{()} x : \mu Z_3.\nu Z_4.[\tau]Z_4 \wedge \forall a.[a]Z_3 \tag{4.37}$$

The assumption states (in the absence of name passing which is not needed to illustrate the problems) that any infinite sequence of transitions labelled τ or a can only contain a finite number of consecutive transitions labelled a, while the conclusion states that any infinite sequence of transitions labelled τ or a can only contain a finite number of a transitions, never mind if consecutive or not. Thus (4.37) is false. We attempt to build a proof for (4.37) to see where the construction breaks down.

Let us introduce the following abbreviations:

$$\begin{aligned} U_1 &= \nu Z_1.\mu Z_2.[\tau]Z_1 \wedge [a]Z_2 \\ U_2 &= \mu Z_2.[\tau]U_1 \wedge [a]Z_2 \\ U_3 &= \mu Z_3.\nu Z_4.[\tau]Z_4 \wedge [a]Z_3 \\ U_4 &= \nu Z_4.[\tau]Z_4 \wedge [a]U_3 \end{aligned}$$

We start by refining (4.37) to the subgoal

$$x : U_2^{\kappa_2} \vdash^{()} x : U_4^{\kappa_4} \tag{4.38}$$

using the rules (UNF-L-1), (UNF-R-1), (APPRX-L) and (APPRX-R). Continuing a few steps further (by unfolding the fixed point formulas and treating the conjunctions on the left and on the right) we obtain the two subgoals

$$x : [\tau]U_1, x : [a]U_2^{\kappa_2'}, \kappa_2' < \kappa_2, \kappa_4' < \kappa_4 \vdash^{()} x : [\tau]U_4^{\kappa_4'} \tag{4.39}$$

$$x : [\tau]U_1, x : [a]U_2^{\kappa_2'}, \kappa_2' < \kappa_2, \kappa_4' < \kappa_4 \vdash^{()} x : [a]U_3 \tag{4.40}$$

Subgoal (4.39) is refined via rule Mon2 to

$$x' : U_1, x : [a]U_2^{\kappa_2'}, \kappa_2' < \kappa_2, \kappa_4' < \kappa_4 \vdash^{()} x' : U_4^{\kappa_4'} \tag{4.41}$$

and after unfolding U_1 using (UNF-L-1) and then approximating we arrive at

$$x' : U_2^{\kappa_2''}, x : [a]U_2^{\kappa_2'}, \kappa_2' < \kappa_2, \kappa_4' < \kappa_4 \vdash^{()} x' : U_4^{\kappa_4'}. \tag{4.42}$$

This judgment we might hope to be able to discharge against (4.38) by induction on κ_4. By the same token when we refine (4.40) to

$$x : [\tau]U_1, x' : U_2^{\kappa_2'}, \kappa_2' < \kappa_2, \kappa_4' < \kappa_4 \vdash^{()} x' : U_4^{\kappa_4''} \tag{4.43}$$

we would expect to be able to discharge against (4.38) inductively on κ_2. This does not work, however, since derivation of (4.42) from (4.38) fails to preserve the induction variable κ_2 needed for (4.43), and vice versa, κ_4 is not preserved along the path from (4.38) to (4.43).

Secondly we give an example showing where discharge should succeed.

Example 27 Consider the (reversed) proof goal

$$x : \mu Z_1.\nu Z_2.[\tau]Z_2 \wedge [a]Z_1 \vdash^{()} x : \nu Z_3.\mu Z_4.[\tau]Z_3 \wedge [a]Z_4 \tag{4.44}$$

stating that if all infinite sequences of transitions labelled τ or a can only contain a finite number of a transitions, then these infinite sequences of τ or a transitions can only contain finite sequences of *consecutive* a transitions. This goal is clearly valid.
The abbreviations we shall use are:

$$\begin{aligned} U_1 &= \mu Z_1.\nu Z_2.[\tau]Z_2 \wedge [a]Z_1 \\ U_2 &= \nu Z_2.[\tau]Z_2 \wedge [a]U_1^{\kappa_1'} \\ U_3 &= \nu Z_3.\mu Z_4.[\tau]Z_3 \wedge [a]Z_4 \\ U_4 &= \mu Z_4.[\tau]U_3^{\kappa_3'} \wedge [a]Z_4 \end{aligned}$$

First we apply rules (APPRX-L), (APPRX-R), (UNF-L-2) and (UNF-R-2) to reduce (4.44) to the subgoal

$$x : U_2, \kappa_1' < \kappa_1, \kappa_3' < \kappa_3 \vdash^{()} x : U_4 \tag{4.45}$$

Continuing in much the same way as in the preceding example we arrive at the two subgoals

$$x' : U_2, x : [a]U_1^{\kappa_1'}, \kappa_1' < \kappa_1, \kappa_3' < \kappa_3 \vdash^{()} x' : U_3^{\kappa_3'} \tag{4.46}$$

$$x : [\tau]U_2, x' : U_1^{\kappa_1'}, \kappa_1' < \kappa_1, \kappa_3' < \kappa_3 \vdash^{()} x' : U_4 \tag{4.47}$$

These subgoals are refined, using (UNF-R-2) and (UNF-L-2) respectively, to

$$x' : U_2, x : [a]U_1^{\kappa_1'}, \kappa_1' < \kappa_1, \kappa_3' < \kappa_3, \kappa_3'' < \kappa_3'$$
$$\vdash^{()} x' : \mu Z_4.[\tau]U_3^{\kappa_3''} \wedge [a]Z_4 \quad (4.48)$$
$$x : [\tau]U_2, x' : \nu Z_2.[\tau]Z_2 \wedge [a]U_1^{\kappa_1''}, \kappa_1' < \kappa_1, \kappa_3' < \kappa_3, \kappa_1'' < \kappa_1'$$
$$\vdash^{()} x' : U_4 \quad (4.49)$$

In this case it is safe to discharge both judgments against (4.45), since the unfolding of U_2 does not interfere with that of U_4.

The rule of discharge. There are two key task in providing a sound and generally applicable rule of discharge:

- The first task is to ensure that each discharged node determines a "progressing cycle" in the proof structure: That is, that the node determines a cycle, and that along that cycle some ordinal variable is decreased in a recursive manner.
- The second task is to ensure that progress required for the safe discharge of one node can not be undone by cycles induced by the discharge of some other node. Or in other words, in traversing cycles induced by node discharge critical progress ordinals belonging to other cycles must be *regenerated.*

It would be possible to use games to completely characterize the conditions under which discharge is safe. Such a game-based characterisation, however, would be too global a condition to be very useful for proof construction (which is the main aim of the work reported here), and it is often helpful to trade a complete, but global, rule for an incomplete one which is more local. Other versions of a rule of discharge than the one presented here can be devised, cf. [7] for an example applied to CCS. The present rule of discharge is based on [6].

The rule of discharge relies on some fixed, but arbitrary linear ordering $<$ on fixed point formula abstractions U. Assuming such a single fixed linear ordering can be too restrictive when recursive proof structures are independent. For the purpose of the examples and theorems of the rest of the paper this is, however, not a problem. Below we briefly discuss ways of relaxing the construction to allow the linear ordering to be built incrementally.

Below we define the critical notions of regeneration, progress, and discharge. We use v to range over proof structure nodes. Discharge is applied when facing a proof goal v_n which is unelaborated (no rule has been applied to that node), such that, below v_n we find some already elaborated node v_1 such that v_n is in a sense an instance of v_1. This requires names and term variables present in v_1 to be interpreted as names in v_n. This is what the substitution η of the following definition serves to achieve.

Definition 28 (Regeneration, progress, discharge) Let $\Pi = v_1, \ldots, v_n$ be a path such that v_n is not elaborated. Suppose that v_i is labelled by $\Gamma_i \vdash^{s_i} E_i : \Delta_i$ for all $i : 1 \leq i \leq n$.

1 The path Π is *regenerative* for U and the name interpretation η, if whenever there is a κ_i such that U^{κ_i} is a subformula of Γ_i (Δ_i) then there are also $\kappa_1, \ldots, \kappa_{i-1}, \kappa_{i+1}, \ldots, \kappa_n$ such that for all $j : 1 < j \leq n$, U^{κ_j} is a subformula of Γ_j (Δ_j), and $\Gamma_j \vdash^{()} \kappa_j \leq \kappa_{j-1}$. Moreover we require that $\eta(\kappa_1) = \kappa_n$.

2 The path Π is *progressive* for U and η if we can find $\kappa_1, \ldots, \kappa_n$ such that:

(a) For all $i : 1 < i \leq n$, U^{κ_i} is a subformula of Γ_i (Δ_i), and $\Gamma_i \vdash^{()} \kappa_i \leq \kappa_{i-1}$.

(b) $\eta(\kappa_1) = \kappa_n$.

(c) For some $i : 1 < i \leq n$, $\Gamma_i \vdash^{()} \kappa_i < \kappa_{i-1}$.

3 The node v_n can be *discharged* against the node v_1 if we can find some U and substitution η such that:

(a) Π is regenerative for all $U' < U$ and η.

(b) Π is progressive for U and η.

(c) $E_n = E_1\eta$ and $s_n = s_1\eta$.

(d) For all assumptions $x : \phi$ in Γ_1, $\Gamma_n \vdash^{()} x\eta : \phi\eta$, and all assertions ϕ in Δ_1 then $y : \phi\eta \vdash^{()} y : \Delta_n$.

In this case we term v_n a *discharge node* and v_1 its *companion node*.

In this definition we are being slightly sloppy with our use of U's: Really we are identifying fixed point formula abstractions up to ordinal approximations except where they are explicitly stated.

Condition 28 first states that discharge can take place on an unelaborated node if we find some ancestral node which is "more general" (Condition 28.3.c and d) than the node v_n being discharged, provided that the cycle thus induced satisfies the regeneration and progress constraints given. Observe that an efficient approximation of Condition 28.3.d would be to test for membership, ie. $s\eta : \phi\eta \in \Gamma_n$ and $\phi\eta \in \Delta_n$ respectively. With this condition the name interpretation η becomes extendable to a substitution making $\Gamma_1 \vdash^{s_1} E_1 : \Delta_1$ identical to a weakened version of $\Gamma_n \vdash^{s_n} E_n : \Delta_n$.

There does not appear to be any obvious way in which the equality constraint 28.3.c on restriction sets can be eased. Instead the second cut rule (CUT-2) must be used to explicitly garbage-collect unused names as proof elaboration proceeds.

The progress condition, 28.2, requires the existence of a cycle on an ordinal variable such that, along the cycle, the value associated to that ordinal variable is somewhere strictly decreased.

Finally, the regeneration condition, 28.1, is the condition required to ensure that progress cycles on formulas V with "higher priority" (smaller in the ordering $<$) than the formula U currently being considered can not be undone when the cycles are nested.

Concerning the examples, it is quite easy to verify that for Example 26 no linearisation of the fixed point formulas can be devised such that the nodes (4.41) and (4.42) can be discharged. On the other hand, for Example 27, any linear ordering which (up to approximation ordinals) has $U_4 < U_2$ will do.

Observe that the linear ordering on fixed point formula abstractions can be chosen quite freely. One might expect some correlation between position in the linear ordering and depth of alternation, viz. Example 27 above. In practice this is in fact a good guide to choosing a suitable linear ordering. However, as we show, we do not need to require such a correlation a priori. Moreover one can construct examples, using cut's, of proofs for which the above rule of thumb does not work.

Now, the *compositional proof system* is obtained by adding the proof rules for fixed points, including the rule of discharge, to the local rules of Section 5. We write $\Gamma \vdash^s_C E : \Delta$ if the judgment $\Gamma \vdash^s E : \Delta$ is provable in the compositional proof system. For the proof of soundness of this proof system we refer the reader to [6],

Theorem 29 (Soundness, compositional proof system) *If* $\Gamma \vdash^s_C E : \Delta$ *then* $\Gamma \models^s E : \Delta$. ∎

8. Finite control completeness

We then turn to the issue of completeness and consider generalisations of the completeness result (Theorem 22) to recursive formulas. The generalisation of Theorem 22 which we seek is to general formulas and finite control processes.

Definition 30 (Finite control agent) Say that an agent term E *uses* the agent identifier D, if either D occurs in E or else D occurs in the body of an identifier D' such that E uses D'. A *finite control agent* is a closed agent term E such that the parallel operator $|$ does not occur in the body of any identifier used by E.

The notion of finite control process is a direct generalisation of the notion of finite state process in the case of CCS. Completeness for finite state (CCS) processes vs a compositional proof system related to the proof system considered here was proved in [5]. However, the details of that formalisation, in particular the rule of discharge, was quite different from the proof system presented here.

To prove completeness for finite control processes we formulate a *model checker*, an alternative, non-compositional proof system and show that whichever judgment is provable using the model checker is also provable in the proof system presented in the preceding sections. Model checkers based on non-compositional proof systems such as the one we go on to present have been considered and proved complete for finite control processes several times over [4, 1]. Similar techniques can be used to prove completeness of the present version.

Definition 31 (Model checking judgments) A *model checking judgment* is a judgment of the form $\Gamma \vdash E : \phi$ for which all formulas in Γ are elementary and for which E is of finite control.

We define a proof system to apply to model checking judgments. The proof system consists of all rules defining the compositional proof system (restricted to model checking judgments) excluding the transition and io modality rules. That is, the proof system includes the structural rules (which due to the restriction to model checking judgments become rather standard, for instance the first two cut rules become superfluous), the logical rules, and the rules equality/inequality.

The transition rules are replaced by the following two schemes:

(POSS) If for all η such that $\eta \models \Gamma$ there is an E such that $P\eta \overset{l_\tau \eta}{\to} E$ and $\Gamma \vdash E : \phi$ has been inferred, infer then $\Gamma \vdash P : <l_\tau>\phi$

(NEC) If $\Gamma \vdash E : \phi$ has been inferred for all E such that $P\eta \overset{l_\tau \eta}{\to} E$ and $\eta \models \Gamma$, infer then $\Gamma \vdash P : [l_\tau]\phi$

together with the rules ($\to$-IN), ($\nu \to$-IN), ($\leftarrow$-OUT), and

$$(\nu \leftarrow\text{-OUT'}) \quad \frac{\Gamma, a\, \mathit{fresh} \vdash C : \phi\{a/b\}}{\Gamma \vdash \nu a.\langle a\rangle C : \nu b \leftarrow \phi} \quad (a \text{ fresh})$$

The rule of discharge applies to the model checker without modification. In effect the conditions in this case degenerate to those of the model checker presented in [4]. Observe that the rules (POSS) and (NEC) are infinitary due to the quantification over name interpretations. Name interpretations are only significant, however, up to the names that are not fresh (of which there is only a finite supply). Write $\Gamma \vdash_{MC} E : \phi$ if there is a model checker proof of $\Gamma \vdash E : \phi$.

We state soundness and completeness of the model checker proof system without proof:

Theorem 32 (Soundness and completeness of model checker) *For all model checking judgments $\Gamma \vdash E : \phi$, $\Gamma \vdash_{MC} E : \phi$ if and only if $\Gamma \models E : \phi$.* ∎

One main obstacle in proving the completeness result we seek is that we need to devise a strategy for choosing cut-formulas in order to apply the dynamic rules in the compositional proof system. There are many ways of doing this. In practice one normally takes a lazy approach and just introduces a logical variable to stand for the required cut formula, and then gradually instantiate this variable as the need arises. In fact such a strategy may be quite efficient and it is moreover well suited for parallel or distributed implementations. Here, however, we choose a non-lazy approach instead, mainly because it makes the proof easier. Because processes are assumed not to contain $|$ in recursively defined contexts in turns out that such processes can be characterised completely up to (in this case strong late) bisimulation equivalence by a so-called characteristic formula. We have already shown this to be the case for a related π-calculus logic in earlier work [1]. Assume that P is a finite control process. Assume for each well-formed process term Q and name interpretation η a unique formula variable $X_{Q,\eta}$. Let vis range over sequences of pairs of process terms and name interpretations (those pairs that have already been visited once in constructing the characteristic formula). The formula $\mathit{Char}(P, \eta, vis)$ is the characteristic formula for P, given name interpretation η and visited list vis:

$$\mathit{Char}(E, \eta, vis) = \begin{cases} X_{E,\eta} & \text{if } (E,\eta) \in vis \\ \nu X_{E,\eta}. \bigwedge_{l_\tau} \phi(<l_\tau>) \wedge \bigwedge_{l_\tau} \phi([l_\tau]) & \text{otherwise} \end{cases}$$

Here, $\phi(<l_\tau>)$ describes the potential transitions possible from P assuming η, and $\phi([l_\tau])$ describes the property "necessarily" holding in the continuation:

$$\begin{aligned} \phi(<l_\tau>) &= \bigwedge\{<l_\tau>\mathit{Char}(E, \eta, (P, \eta) :: vis) \mid P\eta \overset{l_\tau\eta}{\to} E\} \\ \phi([l_\tau]) &= [l_\tau]\bigvee\{\mathit{Char}(E, \eta, (P, \eta) :: vis) \mid P\eta \overset{l_\tau\eta}{\to} E\} \end{aligned}$$

For abstraction and concretions we define:

$$\begin{aligned} \mathit{Char}((a)A, \eta, vis) &= \forall a.a \to \bigvee\{a=b \wedge \mathit{Char}(A\{b/a\}, \eta, vis) \mid b \in fn((a)A)\} \vee \\ &\quad (\bigwedge\{a \neq b \mid b \in fn((a)A)\} \wedge \mathit{Char}(A, \nu a.\eta, vis)) \\ \mathit{Char}(\langle a\rangle C, \eta, vis) &= a \leftarrow \mathit{Char}(C, \eta, vis) \\ \mathit{Char}(\nu a.\langle a\rangle C, \eta, vis) &= \nu a \leftarrow \mathit{Char}(C, \nu a.\eta, vis) \end{aligned}$$

Abbreviate $\mathit{Char}(E, \eta, \varepsilon)$ by $\mathit{Char}(E, \eta)$.

Observe that only finite conjunctions and disjunctions are used in the definition of *Char*. Thus, the only reasons why *Char*(E, η, vis) could be ill-defined would be if the computation of *Char*(E, η, vis) failed to terminate, and this could easily happen if care is not taken when choosing names of bound variables. However, if we adopt the *conventions* that

1 names are linearly ordered, and every time a variable is bound it is chosen to be minimal in the ordering, and

2 $\nu a.E$ is identified with E whenever a does not appear freely in E,

then it can be shown that *Char*(E, η, vis) is in fact well-defined using the techniques of e.g. [4] or [1]. Here we just state this as a fact.

Proposition 33 *Suppose that* P *is a finite control process. Then* Char(P, η) *is well-defined.* ∎

For the completeness proof we need to resort to the following correctness property for characteristic formulas. We leave out the proof of this quite easy lemma. A similar result was proved in [1]. In the statement of this lemma and for the remainder of this section we abbreviate $hyp(\eta, N)$ by $hyp(\eta)$ where the set N is any set of names including those free in the judgment under consideration.

Lemma 34 *For all finite control processes* P *and partitions* η,
$$hyp(\eta) \vdash_{MC} P : \mathrm{Char}(P, \eta).$$ ∎

Before embarking on the main completeness result we need one further lemma, used to pass from results concerning terms with substitutions to results concerning open terms governed by assumptions.

Lemma 35 *For all finite control processes* P *and partitions* η, *if* $hyp(\eta) \vdash_{MC} P : \phi$ *then* $hyp(\eta), x : \mathrm{Char}(P, \eta) \vdash_C x : \phi$.

PROOF. (Hint) The compositional proof mimicks the model checker proof by appealing to the monotonicity rules. ∎

The proof of the finite control completeness theorem is deferred to Appendix B.

Theorem 36 *If* $\Gamma \vdash_{MC} E : \phi$ *then* $\Gamma \vdash_C^{()} E : \phi$. ∎

In view of Theorem 32, completeness for finite control processes is then a direct corollary:

Corollary 37 (Finite control completeness) *For all model checking judgments* $\Gamma \vdash E : \Delta$, *if* $\Gamma \models E : \Delta$ *then* $\Gamma \vdash_C^{()} E : \Delta$. ∎

9. Natural numbers

In this section we consider the specification of Nat given in Section 4, and show how one can use the proof systems to formally demonstrate that the operations (of $ZERO$, $SUCC$, $COPY$, and ADD) satisfy the desired properties. Observe that while $ZERO$ and $SUCC$ are very simple static processes, $COPY$ and ADD are not.

Proposition 38 *The following judgments are derivable:*

1 $\vdash_C^{()} \mathit{ZERO}(n) : \mathit{Nat}(n)$

2 $x : \mathit{Nat}(n) \vdash_C^{n} \mathit{SUCC}(n, m) \mid x : \mathit{Nat}(m)$

3 $x : \mathit{Nat}(n) \vdash_C^{n} \mathit{COPY}(n, m) \mid x : \mathit{Nat}(m)$

4 $x : \mathit{Nat}(n_1), y : \mathit{Nat}(n_2) \vdash_C^{n_1, n_2} \mathit{ADD}(n_1, n_2, m) \mid x \mid y : \mathit{Nat}(m)$ ∎

We concentrate in this section on giving an outline proof of (3), proving (2) along the way. The proof of (1) is quite straightforward, and the proof of (4) is a variation on the theme.

The main problem which the proof (of (3)) has to face is process creation in the definition of $COPY$. That is, the continuation of $\mathit{COPY}(n, m)$ contains a term of the shape $\nu m_1.(\mathit{SUCC}(m_1, m) \mid \mathit{COPY}(n_1, m_1))$. To deal with this we use a process cut, replacing each of the subprocesses $\mathit{SUCC}(m_1, m)$ and $\mathit{COPY}(n_1, m_1)$ by abstract state-oriented descriptions, and continue proving correctness based on those.

Finding an appropriate cut formula for *SUCC* is easy:

$$
\begin{aligned}
\mathit{Succ}_0(n, m, \phi) &= \langle m\rangle\top \wedge \\
&\quad [\tau]\bot \wedge \\
&\quad \forall b.[b](b = m \wedge \nu s \to \nu z \to \phi(n, s)) \wedge \\
&\quad \forall b.[\bar{b}]\bot \\
\mathit{Succ}_1(n, s) &= \langle \bar{s}\rangle\top \wedge \\
&\quad [\tau]\bot \wedge \\
&\quad \forall b.[b]\bot \wedge \\
&\quad \forall b.[\bar{b}](b = s \wedge n \leftarrow \mathit{STOP} \\
\mathit{Succ}(n, m) &= \mathit{Succ}_0(n, m, \mathit{Succ}_1)
\end{aligned}
$$

The correctness of *Succ* is easily proved:

Proposition 39

1 $\vdash_C^{()} \mathit{SUCC}(n, m) : \mathit{Succ}(n, m)$

2 $x : Nat(n), y : Succ(n,m) \vdash_C^n y \mid x : Nat(m)$ ∎

Observe that this establishes Proposition 38.2 by means of a process cut.

The next step is to find a cut formula for *COPY*. Our intention is to define a property characterising, to a sufficiently precise degree, the behaviour of a process term of the shape

$$\nu m_1.(SUCC(m_1,m) \mid \nu m_2.SUCC(m_2,m_1) \mid \cdots \mid COPY(n,m_k)\cdots) \tag{4.50}$$

Again we use a state machine-oriented style of specification. The machine being predicated will, as (4.50) have two components, each with a two-state behaviour, namely a component corresponding to the successor processes, and a component corresponding to the copy process. Each of these component can to a large extent execute concurrently. We thus propose the following definition of the cut formula Gcopy (generalized copy) as a greatest fixed point formula:

$$\begin{aligned}
Gcopy_{0,0}(n,m) &= \langle\overline{n}\rangle\top \wedge [\tau]\bot \wedge \\
&\quad \forall b.[b](b = m \wedge \nu s \rightarrow \nu z \rightarrow Gcopy_{1,0}(n,s,z)) \wedge \\
&\quad \forall b.[\overline{b}](b = n \wedge \nu s_1 \leftarrow \nu z_1 \leftarrow Gcopy_{0,1}(m,s_1,z_1)) \\
Gcopy_{1,0}(n,s,z) &= \langle\overline{s}\rangle\top \wedge \langle\overline{n}\rangle\top \wedge [\tau]\bot \wedge \\
&\quad \forall b.[b]\bot \wedge \\
&\quad \forall b.[\overline{b}]((b = s \wedge \nu m_1 \leftarrow Gcopy_{0,0}(n,m_1)) \vee \\
&\quad (b = n \wedge \nu s_1 \leftarrow \nu z_1 \leftarrow Gcopy_{1,1}(s,z,s_1,z_1))) \\
Gcopy_{0,1}(m,s_1,z_1) &= \langle z_1\rangle\top \wedge \langle s_1\rangle\top \wedge [\tau]\bot \wedge \\
&\quad \forall b.[b]((b = z_1 \wedge Nat(m)) \vee \\
&\quad (b = s_1 \wedge \nu n_1 \rightarrow Gcopy_{0,0}(n_1,m)) \vee \\
&\quad (b = m \wedge \nu s \rightarrow \nu z \rightarrow Gcopy_{1,1}(s,z,s_1,z_1))) \\
Gcopy_{1,1}(s,z,s_1,z_1) &= \langle\overline{s}\rangle\top \wedge \langle z_1\rangle\top \wedge \langle s_1\rangle\top \wedge [\tau]\bot \wedge \\
&\quad \forall b.[b]((b = z_1 \wedge Nat_1(Nat)(s,z)) \vee \\
&\qquad\qquad (b = s_1 \wedge \nu n_1 \rightarrow Gcopy_{1,0}(n_1,s,z))) \wedge \\
&\quad \forall b.[\overline{b}](b = s \wedge \nu m_1 \leftarrow Gcopy_{0,1}(m_1,s_1,z_1))
\end{aligned}$$

The definition of *Gcopy* is given in equational terms. However, the definition is easily rewritten as a proper greatest fixed point formula. The correctness of *Gcopy* is reflected by the following proposition:

Proposition 40

1 $x : Succ(m_1,m), y : Gcopy_{0,0}(n,m_1) \vdash_C^{m_1} x \mid y : Gcopy_{0,0}(n,m)$

2 $\vdash_C^{()} COPY(n,m) : Gcopy_{0,0}(n,m)$

3 $x : Nat(n), y : Gcopy_{0,0}(n,m) \vdash^n y \mid x : Nat(m)$ ∎

The formal proof of Proposition 40 is quite sizable though (since we have identified the cut formulas) routine and likely to be mechanizable. Observe that the proof of 40.2 uses 40.1 through two process cuts. Finally Proposition 38.3 is a direct corollary of Proposition 40.2 and 40.3 using a process cut.

It may be worthwhile to make clear that we do not intend to advocate the use π-calculus or π-μ-calculus representations of data types in actual practice. There are much simpler accounts of data types around useful for practical programming and specification. What the example serves to illustrate, however, is that while fairly trivial as a data type, *as processes executing in parallel*, π-calculus representations of data type embody a simple sort of mobility protocol the behaviour and correctness of which is not trivial. Observe in this connection also that we achieve genuinely more than the corresponding type correctness results of [10] which are, in effect, results in the meta theory of the π-calculus whereas here the reasoning has been "internalised" using the π-μ-calculus proof system.

10. Buffers

In this section we consider buffer properties in the style of Example 10. Consider for instance the formula $NoSpuOut(i,o)$ describing absence of spurious output. Our task in this section is to show that $NoSpuOut(i,o)$ holds of the unbounded garbage-collecting buffer $GCBuf(i,o)$. As usual, since $GCBuf(i,o)$ creates processes dynamically, we use process cut's using cut-formulas which reflect a state machine-like behaviour. First, for start cells define:

$$\begin{aligned} Sc(i,o,n,a) \;=\;& [\tau]Sc(i,o,n,a) \wedge \\ & \forall b.[b]((b = i \wedge \forall d.d \to d = a \vee Sc(i,o,n,a)) \vee \\ & \quad (b = n \wedge \forall o',n'.i = o' \vee i = n' \vee \\ & \qquad o' \to n' \to Sc(i,o',n',a))) \wedge \\ & \forall b.[\bar{b}](b = o \wedge \exists d.d \leftarrow d \neq a \wedge Sc(i,o,n,a)) \end{aligned}$$

and then for buffer cells define

$$\begin{aligned} Bc(o,d,n_l,n_r) \;=\;& [\tau]\bot \wedge \\ & \forall b.[b](b = n_r \wedge \forall o',n'.i = o' \vee i = n' \vee \\ & \qquad o' \to n' \to Bc(o',d,n_l,n')) \wedge \\ & \forall b.[\bar{b}](b = o \wedge d \leftarrow Bc_1(o,n_l,n_r)) \end{aligned}$$

$$\begin{aligned} Bc_1(o,n_l,n_r) \;=\;& [\tau]\bot \wedge \\ & \forall b.[b]\bot \wedge \\ & \forall b.[\bar{b}](b = n_l \wedge o \leftarrow n_r \leftarrow STOP \end{aligned}$$

It is a straightforward task to translate the above equational property descriptions into proper greatest fixed point formulas. This task is left to the reader. The main lemmas and the final correctness property (item 5) is stated in the following proposition:

Proposition 41

1 $\vdash_C^{()} \mathit{BufCell}(o, d, n_l, n_r) : \mathit{Bc}(o, d, n_l, n_r)$

2 $i{\neq}o, i{\neq}n, d{\neq}a, x : \mathit{Sc}(i, o', n', a), y : \mathit{Bc}(o, d, n', n) \vdash_C^{o',n'} x \mid y : \mathit{Sc}(i, o, n, a)$

3 $i \neq o, i \neq n \vdash_C^{()} \mathit{StartCell}(i, o, n) : \mathit{Sc}(i, o, n, a)$

4 $x : \mathit{Sc}(i, o, n, a) \vdash_C^{()} x : \mathit{NoSpuOut}'(i, o, a)$

5 $i \neq o \vdash_C^{()} \mathit{GCBuf}(i, o) : \mathit{NoSpuOut}(i, o)$

Most of these items are proved in a straightforward manner. The exception, if any, is 41.2. The goal-directed proof starts with the desired judgment:

$$i{\neq}o, i{\neq}n, d{\neq}a, x{:}\mathit{Sc}(i, o', n', a), y{:}\mathit{Bc}(o, d, n', n) \vdash_C^{o',n'} x \mid y : \mathit{Sc}(i, o, n, a). \tag{4.51}$$

We start by approximating and unfolding the right hand side fixed point, then introducing the conjunctions to the right. The result is the following three subgoals:

$$i{\neq}o, \cdots, x : \cdots, y : \cdots \vdash_C^{o',n'} x \mid y : [\tau]\mathit{Sc}(i, o, n, a) \tag{4.52}$$

$$i{\neq}o, \cdots, x : \cdots, y : \cdots \vdash_C^{o',n'} x \mid y : \forall b.[b]((b = i \wedge \cdots) \vee (b = n \wedge \cdots)) \tag{4.53}$$

$$i{\neq}o, \cdots, x : \cdots, y : \cdots \vdash_C^{o',n'} x \mid y : \forall b.[\overline{b}](b = o \wedge \cdots) \tag{4.54}$$

First, for (4.52), we observe that τ-steps of each of x or y give immediately cause for discharge. Furthermore, no communication between x and y is enabled: x may input along i or n', but y may only output along o – neither of these can be identified given the assumption and the restriction set. Also x may only output along o' but y may only input along n – again these can not be identified. Thus we can dispense with subgoal (4.52).

Then, for subgoal (4.53), there are two possibilities for input, along i and along n (and these are known to be distinct channels). First, in the case of input along i the subgoal is (after some initial reasoning steps) reduced to

$$\begin{aligned} &i{\neq}o, \cdots, x : d = a \vee \mathit{Sc}(i, o', n', a), y : \mathit{Bc}(o, d, n', n) \\ &\vdash_C^{o',n'} \quad x \mid y : d = a \vee \mathit{Sc}(i, o, n, a) \end{aligned} \tag{4.55}$$

which is resolved almost immediately. Second, in the case of input along n the subgoal is reduced to

$$\begin{aligned} & i{\neq}o'', i \neq n'', \cdots, x : Sc(i, o', n', a), y : Bc(o'', d, n', n'') \\ & \vdash_C^{o',n'} \quad x \mid y : Sc(i, o'', n'', a) \end{aligned} \tag{4.56}$$

which can be discharged against the top-level goal (Proposition 41.2).

For subgoal (4.54), then, we need to consider output along o. We obtain the following reduced subgoal:

$$i{\neq}o, \cdots, x : Sc(i, o', n', a), y : Bc_1(o, n', n) \tag{4.57}$$

$$\vdash_C^{o',n'} \quad x \mid y : Sc(i, o, n, a). \tag{4.58}$$

Again we need to approximate and unfold the right hand-side formula, and consider each case of action type in turn. This analysis is simpler than the one we've already done except for the situation characteristic of Bc_1 of outputting along (in this case) n' to x. The result (after enacting the reductions corresponding to this scenario) in this case is a subgoal of the shape

$$i \neq o, \cdots, x : Sc(i, o, n, a), y : STOP \vdash x \mid y : Sc(i, o, n, a) \tag{4.59}$$

which is an instance of a generally (and easily) provable fact, that parallel composition with the *STOP* process does not affect behaviour. We can thus regard absence of spurious output for our garbage-collecting unbounded buffer as proved.

11. Conclusion

Earlier work on modal and temporal logic and the π-calculus includes [13], [4] and [1]. The work by Milner, Parrow and Walker did not consider temporal connectives. In [4] we attempted an automated, model checking approach restricted to finite control processes. In [1] we reconsidered the model checking problem and also gave a proof system, however for non-recursive formulas only. Also in the present paper the details are very different: the temporal logic is somewhat different, the proof system is cleaner, and we use a symbolic approach which is essential for efficiency in practical applications.

There are several important lines of enquiry for future work. The first concerns the practical applicability of the proof system. The sheer number of rules involved in the proof system may seem disheartening. On the other hand most rules are actually very intuitive – given the number and nature of the basic process and formula connectives there are just a lot of cases to be considered. Thus, for any *given* judgment, only a small and easily comprehensible collection of rules are actually applicable. Moreover by far the large majority of proof steps are entirely mechanical, and only at very specific places (choice points, cut's, applications of the (DILEMMA) rule) is intelligence required. Thus the proof system is much better geared to computer aided tools than to pen-and-paper. At SICS we are currently extending the work reported here to a core fragment of Ericsson's Erlang programming language [3] including features such as asynchronous buffered communication, data types such as natural numbers, lists, atoms, pairing and process identifiers (pid's), dynamic pid creation, process spawning, sequential composition, pattern matching.

An equally important and related line of enquiry concerns the source of incompleteness of the proof system. Intuitively the key problem is that in proving properties of a parallel composition one must guess properties of the components. But it is not always possible to find such properties as it may very well be the case that $P \mid Q$ has a property like divergence (the capability of performing an infinite sequence of internal computation steps) because P and Q have properties (context-free, context-sensitive, or beyond) that are inexpressible in our logic. It has sometimes been argued that this problem makes the entire problem of devising compositional proof systems a futile one. We do not at all subscribe to this view, however. Compositionality should not be viewed as an all-or-nothing matter, rather, compositionality is a useful tool, to be brought to bear when warranted by the specific situation. But: is expressiveness the only source of incompleteness? If this is the case we ought to be able to prove completeness, or maybe even decidability, for judgments like

$$\Gamma, x : \phi, y : \phi \vdash^{s} x \mid y : \gamma. \tag{4.60}$$

If it is not we would still like to ask whether derivability of judgments like (4.60) is decidable as this would have obvious implications for the utility of our approach. Results like these would be particularly important as we as yet only have examples such as the ones given to bring out our intuition that the principles we are exploring are essential improvements upon those basing themselves solely on global state exploration (including, for instance, the approach of [2]). For observe that the completeness results for finite control agents established here only suffice to show that the power of our compositional proof system is not worse than what one can achieve using much simpler global approaches such as [4].

Other issues concern the fundamentals of the proof system. The rule of discharge needs better motivation than the one we are giving here. Really one should view the discharge conditions given here as finitary approximations of conditions applying to infinite proof structures, leading to automata based characterisations. Even the shape of the local rules themselves, as well as the choice of logical connectives is open to debate. Why do we choose some connectives over others? Certainly the private input modality considered in [1] is potentially very valuable in applications. Does it make sense to devise a separate logical connective for restriction? Concerning the shape of basic judgments, how important is the use of the relativized turnstile? We use restriction sets in the style of [1] (and [18]) in order not to violate alpha-convertibility, even in the presence of free process variables. On the other hand, expecting alpha-convertibility of open terms may be unreasonable and counterintuitive, and by abandoning this requirement judgments, and hence the entire proof system, may be simplified rather considerably.

Further, we need to obtain characterisations of expressiveness along the lines explored for the modal μ-calculus itself by a number of authors, e.g. [8]. Concerning the choice of proof rules better completeness and decidability results will give much sharper handles on the kind of rules that should be admitted. A promising approach is to embed directly into the proof system the operational semantics proof rules in the style suggested by Simpson [17] for Hennessy-Milner logic (without fixed points). First investigations in this direction for CCS and the modal μ-calculus are reported in [7]. Progress in this direction would be useful to remove some apparent arbitrariness in our choice of process calculus. Really we should expect to be able to construct similar logics and proof systems for whichever versions of π-calculus, or other calculus for that matter, one might want to come up with. Not all logics or proof systems would be equally attractive, and indeed some effort has been put into the choice of a version of the π-calculus which remains both faithful to the original π-calculus, while at the same time permitting as orthogonal treatments of the different modalities as possible. In particular we have chosen to avoid the issue of sorting. In our calculus local deadlocks may arise in case the number of arguments in sending and receiving actions does not match. What the deeper relations should be between sorting (and more generally types and static analysis techniques), and semantically based proof systems as the ones we consider here, however, is outside the scope of this paper.

Acknowledgements

Thanks are due to Roberto Amadio, Lars-åke Fredlund, and Dilian Gurov for many discussions on these and related topics. Early parts of the work were developed in collaboration with Roberto Amadio and reported in [1]. Dilian Gurov helped, among other matters, with formulating the rule of discharge. Finally, thanks are due also to Joachim Parrow for his examples of buffer specifications.

References

[1] R. Amadio and M. Dam. A modal theory of types for the π-calculus. In *Proc. FTRTFT'96, Lecture Notes in Computer Science*, 1135:347–365, 1996.

[2] Henrik Reif Andersen, Colin Stirling, and Glynn Winskel. A compositional proof system for the modal μ-calculus. In *Proceedings, Ninth Annual IEEE Symposium on Logic in Computer Science*, pages 144–153, Paris, France, 4–7 July 1994. IEEE Computer Society Press.

[3] J. Armstrong, R. Virding, C. Wikström, and M. Williams. *Concurrent Programming in Erlang (Second Edition).* Prentice-Hall International (UK) Ltd., 1996.

[4] M. Dam. Model checking mobile processes. *Information and Computation*, 129:35–51, 1996.

[5] M. Dam. Proving properties of dynamic process networks. *Information and Computation*, 140:95–114, 1998.

[6] M. Dam, L.-å. Fredlund, and D. Gurov. Toward parametric verification of open distributed systems. In *Compositionality: the Significant Difference,* H. Langmaack, A. Pnueli and W.-P. de Roever (eds.), Springer, 1536:150–185, 1998.

[7] M. Dam and D. Gurov. Compositional verification of CCS processes. In *Proc. PSI'99*, 1999.

[8] D. Janin and I. Walukiewicz. On the expressive completeness of the propositional mu-calculus with respect to the monadic second order logic. In *Proc. CONCUR'94,* Lecture Notes in Computer Science, 1119:263–277, 1996.

[9] R. Milner. *Communication and Concurrency*. Prentice Hall International, 1989.

[10] R. Milner. The polyadic π-calculus: A tutorial. Technical Report ECS-LFCS-91-180, Laboratory for the Foundations of Computer Science, Department of Computer Science, University of Edinburgh, 1991.

[11] R. Milner. Functions as processes. *Mathematical Structures in Computer Science*, 2:119–141, 1992.

[12] R. Milner, J. Parrow, and D. Walker. A calculus of mobile processes, I and II. *Information and Computation*, 100(1):1–40 and 41–77, 1992.

[13] R. Milner, J. Parrow, and D. Walker. Modal logics for mobile processes. *Theoretical Computer Science*, 114:149–171, 1993.

[14] D. Park. Finiteness is mu-Ineffable. *Theoretical Computer Science*, **3**:173–181, 1976.

[15] D. Sangiorgi. A theory of bisimulation for the π-calculus. *Acta Informatica*, 33:69–97, 1996.

[16] Davide Sangiorgi. From π-calculus to Higher-Order π-calculus – and back. in *Proc. TAPSOFT'93* Lecture Notes in Computer Science, 668:151–166, 1993.

[17] A. Simpson. Compositionality via cut-elimination: Hennessy-Milner logic for an arbitrary GSOS. In *Proc. LICS*, pages 420–430, 26–29 1995.

[18] C. Stirling. Modal logics for communicating systems. *Theoretical Computer Science*, 49:311–347, 1987.

[19] D. Walker. Objects in the π-calculus. *Information and Computation*, 116:253–271, 1995.

Appendix A. Proofs for Section 6

Proposition 20. *Let N include the set of non-fresh names of $\Gamma \vdash^s E : \phi$.*

1 *If ϕ is elementary then $hyp(\eta, N) \models^s E : \phi$ if and only if $hyp(\eta, N) \vdash^s E : \phi$.*

2 *$\Gamma \vdash^s E : \phi$ if and only if $hyp(\nu s.\eta, N) \vdash^s E : \phi$ for all η such that $\nu s.\eta \models \Gamma$.*

PROOF. (1) If: Use soundness. Only-if: By induction on the structure of ϕ. Let $\phi = \forall a.\psi$. Reduce $hyp(\eta, N) \vdash^s E : \forall a.\psi$ to $hyp(\eta, N) \vdash^s E : \psi\{b/a\}$ where $b \notin N$ by ($\forall$-R). Using (D-L) this is reduced, using elementary reasoning, to the set of judgments $hyp(\eta', N \cup \{b\}) \vdash^s E : \psi\{b/a\}$ where η' is required to agree with η on N. But each element in this set is provable by the induction hypothesis. Let then $\phi = \exists a.\psi$. Since $hyp(\eta, N) \models^s E : \phi$ we find a $b \in \mathcal{G}$ such that $hyp(\eta, N) \models^s E : \psi\{b/a\}$. Either b can be chosen in N in which case $hyp(\eta, N) \vdash^s E : \psi\{b/a\}$ by the ind. hyp. so that $hyp(\eta, N) \vdash^s E : \phi$ by ($\exists$-R), or else $hyp(\nu b.\eta, N \cup \{b\}) \models^s E : \psi\{b/a\}$ in which case $hyp(\nu b.\eta, N \cup \{b\}) \vdash^s E : \psi\{b/a\}$, thus also $hyp(\nu b.\eta, N \cup \{b\}) \vdash^s E : \phi$ by ($\exists$-R). But then

$$hyp(\eta, N), x : (b \neq a_1) \wedge \cdots \wedge (b \neq a_n) \vdash^s E : \phi$$

too (by definition (19)), and then

$$hyp(\eta, N), x : \exists b.(b \neq a_1) \wedge \cdots \wedge (b \neq a_n) \vdash^s E : \phi.$$

But then it suffices to use (INFTY) and (CUT-1) to prove $hyp(\eta, N) \vdash^s E : \phi$ as desired. The remaining cases are quite easy.

(2) By induction on the complexity of Γ.

$\Gamma = ()$. If $\vdash^s E : \phi$ then $hyp(\nu s.\eta, N) \vdash^s E : \phi$ by (W-L). In the other direction use (D-L).

$\Gamma = \Gamma', x : \forall a.\psi$. Suppose first that $\Gamma \vdash^s E : \phi$ and $\nu s.\eta \models \Gamma$. Reduce the goal $hyp(\nu s.\eta, N) \vdash^s E : \phi$ to the two subgoals

$$hyp(\nu s.\eta, N) \vdash^{()} x : \forall a.\psi \tag{4.61}$$

$$hyp(\nu s.\eta, N), x : \forall a.\psi \vdash^s E : \phi \tag{4.62}$$

The second subgoal is easily dealt with using (CUT-1) and the induction hypothesis by showing that $hyp(\nu s.\eta, N) \vdash^{()} y : \gamma$ whenever $y : \gamma$ is a hypothesis in Γ'. The first subgoal is proved first using ($\forall$-R) to reduce to $hyp(\nu s.\eta, N) \vdash^{()} x : \psi\{b/a\}$ and then by (D-L) to all goals of the form $hyp(\nu s.\eta', N \cup \{b\}) \vdash^{()} x : \psi\{b/a\}$ where η' agrees with η on N. Using the induction hypothesis this can be reduced to the single goal $\Gamma', x_1 : \psi\{a_1/a\}, \ldots, x_m : \psi\{a_m/a\}, x_{m+1} : \psi\{b/a\} \vdash^{()} x : \psi\{b/a\}$ which is directly provable by (I) and (E-L). Conversely we need to show $\Gamma \vdash^s E : \phi$ from the assumption that $hyp(\nu s.\eta, N) \vdash^s E : \phi$ whenever $\nu s.\eta \models \Gamma$. First reduce $\Gamma \vdash^s E : \phi$ to

$$\Gamma', x_1 : \forall a.\psi, \ldots, x_{n+1} : \forall a.\psi \vdash^s E : \phi$$

where $N = \{a_1, \ldots, a_n\}$. Using (CUT-1) and (INFTY) this is reduced to

$$\Gamma', x_1 : \forall a.\psi, \ldots, x_n : \forall a.\psi, x_{n+1} : \forall a.\psi \wedge \exists b.b \neq a_1 \wedge \cdots \wedge b \neq a_n \vdash^s E : \phi.$$

Now, by ($\exists$-L) and ($\forall$-L) this is reduced to

$$\Gamma', x_1 : \forall a.\psi, \ldots, x_n : \forall a.\psi, x_{n+1} : \psi\{b/a\} \wedge b \neq a_1 \wedge \cdots \wedge b \neq a_n \vdash^s E : \phi$$

where b is fresh, and then using (CUT-1) and ($\exists$-R) finally to

$$\Gamma', x_1 : \forall a.\psi, \ldots, x_n : \forall a.\psi, x_{n+1} : \exists b.\psi\{b/a\} \wedge b \neq a_1 \wedge \cdots \wedge b \neq a_n \vdash^s E : \phi. \tag{4.63}$$

But observe that $\nu s.\eta \models \Gamma$ if and only if $\nu s.\eta \models \Gamma', x_1 : \forall a.\psi, \ldots, x_n : \forall a.\psi, x_{n+1} : \exists b.\psi\{b/a\} \wedge b \neq a_1 \wedge \cdots \wedge b \neq a_n$ whence the result follows by the induction hypothesis.
The remaining cases are left for the reader. ∎

Lemma 21 (Decomposition). *Let N include $fn(E\{E_1/x_1, \ldots, E_n/x_n\})$, $fn(s)$, and $fn(\phi)$, suppose $E\{E_1/x_1, \ldots, E_n/x_n\}$ is closed, and suppose ϕ is non-recursive. Let $\Gamma = hyp(\nu s.\eta, N)$. If $\Gamma \models^s E\{E_1/x_1, \ldots, E_n/x_n\} : \phi$ then there are $\phi_1, \ldots, \phi_n$ of modal depth not exceeding that of ϕ such that*

1 for all $i : 1 \leq i \leq n$, Γ, s fresh $\models^{()} E_i : \phi_i$, and

2 $\Gamma, x_1 : \phi_1, \ldots, x_n : \phi_n \vdash^s E : \phi$

PROOF. Assume the preconditions of the Lemma hold. The proof is by induction on the modal depth of ϕ, then structure of E, and then structure of ϕ. The first-order connectives can be dealt with in a generic manner.

$\phi = \bot$. Contradiction.

$\phi = a = b$. Let $\phi_i = \top$ for all $i : 1 \leq i \leq n$. We obtain Γ, s *fresh* $\models^{()} E_i : \phi_i$, and $\Gamma, x_1 : \phi_1, \ldots, x_n : \phi_n \vdash^s E : \phi$ by elementary reasoning and Proposition 20.1. The cases for ϕ an inequation $a \neq b$ or $\phi = \top$ are similar.

$\phi = \psi_1 \vee \psi_2$. For $j = 1$ or $j = 2$, $\Gamma \models^s E\{E_1/x_1, \ldots, E_n/x_n\} : \psi_j$. By the induction hypothesis we find $\phi_1, \ldots, \phi_n$ (of sufficiently small modal depth) such that $\Gamma, s\ \mathit{fresh} \models^{()} E_i : \phi_i$ for all $i : 1 \leq i \leq n$, and $\Gamma, x_1 : \phi_1, \ldots, x_n : \phi_n \vdash^s E : \psi_j$. Then we are done by $\vee$-R.

$\phi = \forall a.\psi$. Suppose that a is not in N. We obtain $\Gamma \models^s E\{E_1/x_1, \ldots, E_n/x_n\} : \psi$. By the induction hypothesis we find appropriate $\psi_1, \ldots, \psi_n$ such that

$$\Gamma, s\ \mathit{fresh} \vdash^{()} E_i : \psi_i \quad (\forall i : 1 \leq i \leq n)$$
$$\Gamma, x_1 : \psi_1, \ldots, x_n : \psi_n \vdash^s E : \psi$$

Let $\phi_i = \forall a.\psi_i$. Since a was chosen not in N, $\Gamma \vdash^{()} E_i : \phi_i$, and by the $\forall$-rules, $\Gamma, x_1 : \phi_1, \ldots, x_n : \phi_n \vdash^s E : \phi$ too. The case for $\wedge$ is similar.

$\phi = \exists a.\psi$. We find a b such that $\Gamma \models^s E\{E_1/x_1, \ldots, E_n/x_n\} : \psi\{b/a\}$, thus by the induction hypothesis we find (appropriate) $\phi_1, \ldots, \phi_n$ such that $\Gamma \vdash^{()} E_i : \phi_i$ for all i, and $\Gamma, x_1 : \phi_1, \ldots, x_n : \phi_n \vdash^s E : \psi$. Then, by ($\exists$-R), $\Gamma, x_1 : \phi_1, \ldots, x_n : \phi_n \vdash^s E : \phi$ and we are done.

Secondly we can deal with some process connectives in a generic manner too:

$E = \textbf{if } a = b \textbf{ then } F_1 \textbf{ else } F_2$. Assume $a, b \notin s$. Assume $\Gamma \models^s E : a = b$ (other case is symmetric). Then $\Gamma \models^s F_1 : \phi$. By the 2nd level induction hypothesis we find $\phi_1, \ldots, \phi_n$ such that $\Gamma, s\ \mathit{fresh} \models^{()} E_i : \phi_i$ for all i, and $\Gamma, x_1 : \phi_1, \ldots, x_n : \phi_n \vdash^s F_1 : \phi$. Since $\Gamma \vdash^s F_1 : a = b$ by Proposition 20.1, by (W-L) also $\Gamma, x_1 : \phi_1, \ldots, x_n : \phi_n \vdash^s F_1 : a = b$. Then we use (COND) along with elementary reasoning to conclude $\Gamma, x_1 : \phi_1, \ldots, x_n : \phi_n \vdash^s E : \phi$ as desired. Let then (e.g.) $a \in s$ and $a \neq b$. We obtain that $\Gamma \models^s F_2 : \phi$ since in the context s, a and b can not be identified. We then use the 2nd level induction hypothesis to find $\phi_1, \ldots, \phi_n$ such that $\Gamma, s\ \mathit{fresh} \models^{()} E_i : \phi_i$ and $\Gamma, x_1 : \phi_1, \ldots, x_n : \phi_n \vdash^s F_2 : \phi$. By (W-L), $\Gamma, x_1 : \phi_1, \ldots, x_n : \phi_n, y : a \neq b \vdash^s F_2 : \phi$, and by (NEW1), $\Gamma, x_1 : \phi_1, \ldots, x_n : \phi_n, y : a = b \vdash^s F_1 : \phi$, so by (COND), $\Gamma, x_1 : \phi_1, \ldots, x_n : \phi_n \vdash^s E : \phi$ as desired. Remaining is the case for $a = b$ and $a \in s$. In that case we obtain $\Gamma \models^s F_1 : \phi$ so by the 2nd level induction hypothesis we find $\phi_1, \ldots, \phi_n$ such that $\Gamma, s\ \mathit{fresh} \models^{()} E_i : \phi_i$ and $\Gamma, x_1 : \phi_1, \ldots, x_n : \phi_n \vdash^s F_1 : \phi$. By (W-L), $\Gamma, x_1 : \phi_1, \ldots, x_n : \phi_n, y : a = b \vdash^s F_1 : \phi$, and by (IRR), $\Gamma, x_1 : \phi_1, \ldots, x_n : \phi_n, y : a \neq b \vdash^s F_2 : \phi$. Thus $\Gamma, x_1 : \phi_1, \ldots, x_n : \phi_n \vdash^s E : \phi$ as desired.

$E = \nu a.F$. We obtain $\Gamma \models^{s,a} F : \phi$, so by the 2nd level induction hypothesis we find $\phi_1, \ldots, \phi_n$ such that $\Gamma, (s, a)\ \mathit{fresh} \models^{()} E_i : \phi_i$ for all i, and $\Gamma, x_1 : \phi_1, \ldots, x_n : \phi_n \vdash^{s,a} F : \phi$, thus $\Gamma, x_1 : \phi_1, \ldots, x_n : \phi_n \vdash^s \nu a.F : \phi$ by (NEW2). Moreover, $\Gamma, s\ \mathit{fresh} \models^{()} E_i : \phi_i$ too since a is fresh.

For the remaining connectives we proceed by induction first on modal depth and then on agent structure.

$\phi = <l>\psi$. We continue by induction on the structure of E.

- $E = 0$. Contradiction.

- $E = F_1 + F_2$. $\Gamma \models^s F_j : <a>\psi$ for $j = 1$ or $j = 2$. By the 2nd level induction hypothesis find $\phi_1, \ldots, \phi_n$ of desired properties. Then $\Gamma, s\ \mathit{fresh} \models^{()} E_i : \phi_i$ for all i, and $\Gamma, x_1 : \phi_1, \ldots, x_n : \phi_n \vdash^s E : <a>\psi$ by ($<l_\tau>$-PLUS).

- $E = \tau.F$. Contradiction.

- $E = l'.F$. In this case $\Gamma \models^s F : l = l'$ and $\Gamma \models^s F : \psi$. By the outer level induction hypothesis and ($<l>$-ACT) we are then done.

- $E = F_1 \mid F_2$. Assume $F_1 \xrightarrow{l} F_1'$ and $\Gamma \models^s F_1' \mid F_2 : \psi$ (the other case is symmetrical). By the outer level induction hypothesis we find ψ_1', ψ_2' such that $\Gamma, s\ \mathit{fresh} \models^{()} F_1' : \psi_1'$, $\Gamma, s\ \mathit{fresh} \models^{()} F_2 : \psi_2'$, and $\Gamma, y_1 : \psi_1', y_2 : \psi_2' \vdash^s y_1 \mid y_2 : \psi$. Let $\psi_1 = <l>\psi_1'$ and $\psi_2 = \psi_2'$. We obtain $\Gamma, s\ \mathit{fresh} \models^{()} F_1 : \psi_1, \Gamma, s\ \mathit{fresh} \models^{()} F_2 : \psi_2$, and $\Gamma, y_1 : \psi_1, y_2 : \psi_2 \vdash^s y_1 \mid y_2 : <l>\psi$ by ($<l>$-PAR) and some elementary reasoning. Now use the 2nd level induction hypothesis to find $\phi_{j,1}, \ldots, \phi_{j,n}, j \in \{1, 2\}$, such that $\Gamma, s\ \mathit{fresh} \models^{()} E_i : \phi_{j,i}$ for all i and j, and $\Gamma, x_1 : \phi_{j,1}, \ldots, x_n : \phi_{j,n}, s\ \mathit{fresh} \vdash^{()} F_j : \psi_j$. Let then $\phi_i = \phi_{1,i} \wedge \phi_{2,i}$, and the result is obtained using elementary reasoning and some cuts.

- $E = \lambda a.F$ and $E = [a]F$. Contradiction.

- $E = x$. Since $E\{a_1/x_1, \ldots, a_n/x_n\}$ is closed, $x = x_k$ for some $k : 1 \leq k \leq n$. We can then proceed by induction on the structure of E_k.

$\phi = <\tau>\psi$. Given the previous case, and observing that all rules relevant to the $<l>$ modality have correlates for $<\tau>$, the only subcase warranting attention is the case for $E = F_1 \mid F_2$. Assume first that $F_1 \xrightarrow{\overline{a}} \nu b.\langle b\rangle F_1'$ and $F_2 \xrightarrow{a} (c)F_2'$ such that $E \xrightarrow{\tau} \nu b.(F_1' \mid (F_2'\{b/c\}))$, and $\Gamma \models^s \nu b.(F_1'(F_2'\{b/c\})) : \psi$. Then $\Gamma \models^{s,b} F_1'(F_2'\{b/c\}) : \psi$. By the outer induction hypothesis we find ψ_1' and ψ_2' such that (since b is fresh)

$$\Gamma, (s, b)\ \mathit{fresh} \models^{()} F_1' : \psi_1', \quad \Gamma, (s, b)\ \mathit{fresh} \models^{()} F_2' : \psi_2', \tag{4.64}$$

$$\Gamma, y_1 : \psi_1', y_2 : \psi_2'\{b/c\} \vdash^{s,b} y_1 \mid y_2 : \psi. \tag{4.65}$$

Let $\psi_1 = <\overline{a}>(\nu b \leftarrow \psi_1')$ and $\psi_2 = <a>\nu b \rightarrow \psi_2'$. Then $\Gamma, s\ \mathit{fresh} \models^{()} F_j : \psi_j$, and the goal we need to show is

$$\Gamma, y_1 : \psi_1, y_2 : \psi_2 \vdash^s y_1 \mid y_2 : <\tau>\psi. \tag{4.66}$$

Use (<τ>-PAR) to reduce this to the subgoals

$$\Gamma, y_1 : \nu b \leftarrow \psi_1', y_2 : \psi_2, s\ fresh \vdash^{()} y_1 : isCb \tag{4.67}$$
$$\Gamma, y_1 : \psi_1, y_2 : \psi_2', s\ fresh \vdash^{()} y_1 : isA \tag{4.68}$$
$$\Gamma, y_1 : \nu b \leftarrow \psi_1', y_2 : \nu b \rightarrow \psi_2' \vdash^{s} y_1 \mid y_2 : \psi. \tag{4.69}$$

The first two subgoals are quite trivial, and 4.69 is resolved by ($\nu \rightarrow$-$\nu \leftarrow$-COM) to reduce to

$$\Gamma, y_1 : \psi_1', y_2 : \psi_2' \vdash^{s,b} y_1 \mid y_2 : \psi. \tag{4.70}$$

Use then ($\vee$-L) and (NEW1) to obtain the subgoal

$$\Gamma, y_1 : \psi_1', y_2 : \psi_2' \vdash^{s,b} y_1 \mid y_2 : \psi \tag{4.71}$$

which is proved (4.65). The other case, where the communication between F_1 and F_2 is free, is similar and left for the reader.

$\phi = [l]\psi$. Similar to the case for <a>.

$\phi = [\tau]\psi$. We consider only the case for $E = F_1 \mid F_2$. We identify formulas $\psi_1, \psi_2, \psi_3, \psi_4$ for F_1 such that $\Gamma, s\ fresh \models^{()} F_1 : \psi_1 \wedge [\tau]\psi_2 \wedge \forall a.[a]\psi_3 \wedge \forall a.[\overline{a}]\psi_4$, and similarly formulas $\gamma_1, \gamma_2, \gamma_3, \gamma_4$ for F_2 such that $\Gamma, s\ fresh \models^{()} F_2 : \gamma_1 \wedge [\tau]\gamma_2 \wedge \forall a.[a]\gamma_3 \wedge \forall a.[\overline{a}]\gamma_4$, such that the premises of ([τ]-PAR) become derivable. So, assume first that $E\{E_1/x_1, \ldots, E_n/x_n\}(\nu s.\eta) \xrightarrow{\tau} F$ so that $\Gamma \models^{s} F : \psi$. There are six cases of interest:

1 $F_1(\nu s.\eta) \xrightarrow{\tau} F_1'$ such that $F = F_1' \mid F_2(\nu s.\eta)$.

2 $F_1(\nu s.\eta) \xrightarrow{\overline{a}} \langle b\rangle F_1'$ and $F_2(\nu s.\eta) \xrightarrow{a} (c)F_2'$ such that $F = F_1' \mid (F_2'\{b/c\})$.

3 $F_1(\nu s.\eta) \xrightarrow{\overline{a}} \nu b.\langle b\rangle F_1'$ and $F_2(\nu s.\eta) \xrightarrow{a} (c)F_2'$ such that $F = \nu b.(F_1' \mid (F_2'\{b/c\}))$.

Cases (4)–(6) are symmetric to the above. In case (1) we find, using the outer level induction hypothesis, $\psi(F)$ and $\gamma(F)$ such that $\Gamma, s\,fresh \models^{()} F_1' : \psi(F)$, $\Gamma, s\,fresh \models^{()} F_2 : \gamma(F)$, and $\Gamma, y_1 : \psi(F), y_2 : \gamma(F) \vdash^{s} y_1 \mid y_2 : \psi$. In case (2) we find $\psi(F)$ and $\gamma(F)$ such that $\Gamma, s\,fresh \models^{()} F_1' : \psi(F)$, $\Gamma, s\,fresh \models^{()} F_2'\{b/c\} : \gamma(F)$, and $\Gamma, y_1 : \psi(F), y_2 : \gamma(F) \vdash^{s} y_1 \mid y_2 : \psi$. In case (3) we find $\psi(F)$ and $\gamma(F)$ such that $\Gamma, s\,fresh \models^{()} F_1' : \psi(F)$, $\Gamma, s\,fresh \models^{()} F_2'\{b/c\} : \gamma(F)$, and $\Gamma, y_1 : \psi(F), y_2 : \gamma(F) \vdash^{s} \nu b.y_1 \mid y_2 : \psi$. In the symmetric cases $\psi(F)$ and $\gamma(F)$ are identified similarly. Define now

$$\psi_1 = \bigwedge\{\psi(F_1 \mid F_2') \mid F_2(\nu s.\eta) \xrightarrow{\tau} F_2'\}$$
$$\psi_2 = \bigvee\{\psi(F_1' \mid F_2) \mid F_1(\nu s.\eta) \xrightarrow{\tau} F_1'\}$$

$$\begin{aligned}\psi_3 &= (\mathit{isA} \supset \bigvee\{\bigwedge\{b \rightarrow \psi(F_1'\{b/c\} \mid F_2') \mid F_2(\nu s.\eta) \xrightarrow{\overline{a}} \langle b\rangle F_2'\} \mid \\ &\quad F_1(\nu s.\eta) \xrightarrow{a} (c)F_1'\}) \wedge (\mathit{isA} \supset \bigvee\{\bigwedge\{\nu b \rightarrow \psi(\nu b.F_1'\{b/c\} \mid F_2'\{b/d\}) \mid \\ &\qquad F_2(\nu s.\eta) \xrightarrow{\overline{a}} \nu d.\langle d\rangle F_2'\} \mid F_1(\nu s.\eta) \xrightarrow{a} (c)F_1'\}) \\ \psi_4 &= (\mathit{isCf} \supset \bigvee\{\bigwedge\{b \leftarrow \psi(F_1' \mid F_2'\{b/c\}) \mid F_2(\nu s.\eta) \xrightarrow{a} (c)F_2'\} \mid F_1(\nu s.\eta) \xrightarrow{\overline{a}} \langle b\rangle F_1'\}) \\ &\quad \wedge (\mathit{isCb} \supset \bigvee\{\bigwedge\{\nu b \leftarrow \psi(\nu b.(F_1'\{b/d\} \mid F_2'\{b/c\})) \mid F_2(\nu s.\eta) \xrightarrow{a} (c)F_2'\} \mid \\ &\qquad F_1(\nu s.\eta) \xrightarrow{\overline{a}} \nu d.\langle d\rangle F_1'\})\end{aligned}$$

where b *fresh* abbreviates the conjunction of inequations $b \neq c$ for all c that occurs freely in F. The formulas $\gamma_1, \ldots, \gamma_3$ are defined symmetrically. We need to show that $\Gamma, s\ \mathit{fresh} \models^{()} F_1 : \psi_1 \wedge [\tau]\psi_2 \wedge \forall a.[a]\psi_3 \wedge \forall a.[\overline{a}]\psi_4$. Each conjunct is considered separately. The first two are quite easy and left aside. For the third we need to show, e.g., that

$$\begin{aligned}\Gamma, s\ \mathit{fresh} \models^{()} (c)F_1' : \mathit{isA} \supset \bigvee\{\bigwedge\{\nu b \rightarrow \psi(\nu b.F_1'\{b/c\} \mid F_2'\{b/d\}) \mid \\ F_2(\nu s.\eta) \xrightarrow{\overline{a}} \nu d.\langle d\rangle F_2'\} \mid F_1(\nu s.\eta) \xrightarrow{a} (c)F_1'\}.\end{aligned}$$

This is reduced to

$$\Gamma, s\ \mathit{fresh} \models^{()} (c)F_1' : \forall b.b \rightarrow b\ \mathit{fresh} \supset \psi(\nu b.F_1'\{b/c\} \mid F_2'\{b/d\})$$

for all F_2' such that $F_2(\nu s.\eta) \xrightarrow{\overline{a}} \nu d.\langle d\rangle F_2'$. So assume $F_2(\nu s.\eta) \xrightarrow{\overline{a}} \nu d.\langle d\rangle F_2'$. Assuming b fresh we then need to show

$$\Gamma, s\ \mathit{fresh} \models^{()} F_1'\{b/c\} : \psi(\nu b.F_1'\{b/c\} \mid F_2'\{b/d\})$$

but this follows by the induction hypothesis. The remaining conjuncts of ψ_3 and $\psi_4$are verified similarly.

We also need to show

$$\Gamma, y_1 : \psi_1 \wedge [\tau]\psi_2 \wedge \forall a.[a]\psi_3 \wedge \forall a.[\overline{a}]\psi_4, y_2 : \gamma_1 \wedge [\tau]\gamma_2 \wedge \forall a.[a]\gamma_3 \wedge \forall a.[\overline{a}]\gamma_4 \vdash^s y_1 \mid y_2 : [\tau]\psi.$$

By ($[\tau]$-PAR) we need to show the following:

$$\Gamma, y_1 : \psi_1, y_2 : \gamma_2 \vdash^s y_1 \mid y_2 : \psi \tag{4.72}$$

$$\Gamma, y_1 : \psi_2, y_2 : \psi_1 \vdash^s y_1 \mid y_2 : \psi \tag{4.73}$$

$$\Gamma, y_1 : \psi_3, y_1 : \mathit{isA}, y_2 : \gamma_4, y_2 : \mathit{isCf} \vdash^s y_1 \mid y_2 : \psi \tag{4.74}$$

$$\Gamma, y_1 : \psi_3, y_1 : \mathit{isA}, y_2 : \gamma_4, y_2 : \mathit{isCb} \vdash^s y_1 \mid y_2 : \psi \tag{4.75}$$

$$\Gamma, y_1 : \psi_4, y_1 : \mathit{isCf}, y_2 : \gamma_3, y_2 : \mathit{isA} \vdash^s y_1 \mid y_2 : \psi \tag{4.76}$$

$$\Gamma, y_1 : \psi_4, y_1 : \mathit{isCb}, y_2 : \gamma_3, y_2 : \mathit{isA} \vdash^s y_1 \mid y_2 : \psi \tag{4.77}$$

along with a few other subgoals which are easily proven from the assumptions. To show (4.72) use first ($\vee$-L) and then ($\wedge$-L) to reduce to a goal of the form

$\Gamma, y_1 : \psi(F_1 \mid F_2'), y_2 : \gamma(F_1 \mid F_2') \vdash^s y_1 \mid y_2 : \psi$ which is provable by the induction hypothesis. (4.73) is proved similarly. To show (4.75) reduce the goal, using elementary reasoning, to a subgoal of the form

$$\begin{aligned}\Gamma, y_1 : \bigvee\{\bigwedge\{\nu b \to \psi(F_1'\{b/c\} \mid F_2') \mid F_2(\nu s.\eta) \xrightarrow{\bar{a}} \langle b\rangle F_2'\} \mid \\ F_1(\nu s.\eta) \xrightarrow{a} (c)F_1'\}, \\ y_2 : \bigvee\{\bigwedge\{\nu b \leftarrow \gamma(F_1' \mid F_2'\{b/c\}) \mid F_1(\nu s.\eta) \xrightarrow{a} (c)F_1'\} \mid \\ F_2(\nu s.\eta) \xrightarrow{\bar{a}} \nu d.\langle b\rangle F_2'\} \\ \vdash^s y_1 \mid y_2 : \psi.\end{aligned}$$

To show this let $F_1(\nu s.\eta) \xrightarrow{a} (c)F_1'$ and $F_2(\nu s.\eta) \xrightarrow{a} \nu d.\langle d\rangle F_2'$, and, using ($\bigvee$-L), we have to show

$$\begin{aligned}\Gamma, y_1 : \bigwedge\{\forall b.b \to b\,\textit{fresh} \supset \psi(F_1'\{b/c\} \mid F_2') \mid F_2(\nu s.\eta) \xrightarrow{\bar{a}} \langle b\rangle F_2'\}, \\ y_2 : \bigwedge\{\nu b \leftarrow \gamma(F_1' \mid F_2'\{b/c\}) \mid F_1(\nu s.\eta) \xrightarrow{a} (c)F_1'\} \\ \vdash^s y_1 \mid y_2 : \psi.\end{aligned}$$

This is then reduced, using ($\bigwedge$-L) to

$$\Gamma, y_1 : \nu b \to \psi(F_1'\{b/c\} \mid F_2'), y_2 : \nu b \leftarrow \gamma(F_1' \mid F_2'\{b/c\}) \vdash^s y_1 \mid y_2 : \psi$$

which is in turn reduced, using ($\nu \to$-$\nu \leftarrow$-COM), to

$$\Gamma, y_1 : \psi(F_1'\{b/c\} \mid F_2'), y_2 : \gamma(F_1' \mid F_2'\{b/c\}) \vdash^{s,b} y_1 \mid y_2 : \psi$$

which is provable by the induction hypothesis. Now it just remains, using the 2nd level induction hypothesis, and observing that modal depth has not increased, to identify the desired $\phi_1, \ldots, \phi_n$ and to put the desired proof together using (CUT-1). This is routine.

$\phi = a \to \psi$. Again we proceed by induction on agent structure.

- First we deal in one blow with any E which is not an abstraction since these cases are contradictory. The remaining cases are:
- $E = F_1 \mid F_2$. The only relevant case is when F_1 is an abstraction and F_2 is a process (or symmetrically). We can write F_1 as $(b)F_1'$. Then $\Gamma \models^s F_1'\{a/b\} \mid F_2 : \psi$. Since $F_1'\{a/b\}$ is of smaller size than E the 2nd level induction hypothesis applies to produce ψ_1, ψ_2 such that $\Gamma, s\,\textit{fresh} \models^{()} F_1'\{a/b\} : \psi_1$ and $\Gamma, s\,\textit{fresh} \models^{()} F_2 : \psi_2 \wedge \textit{isP}$, and $\Gamma, y_1 : \psi_1, y_2 : \psi_2 \wedge \textit{isP} \vdash^s y_1 \mid y_2 : \psi$. Now $\Gamma, s\,\textit{fresh} \models^{()} F_1 : a \to \psi_1$ and $\Gamma, y_1 : a \to \psi_1, y_2 : \psi_2 \wedge \textit{isP} \vdash^s y_1 \mid y_2 : a \to \psi$ as desired, by ($\to$-PAR) and a little elementary reasoning.

- $E = (b)F$. Then $\Gamma \models^s F\{E_1/x_1, \ldots, E_n/x_n, a/b\} : \psi$ so by the 2nd level induction hypothesis we find $\phi_1, \ldots, \phi_n$ such that Γ, s *fresh* $\models^{()} E_i : \phi_i$ for all i, and $\Gamma, x_1 : \phi_1, \ldots, x_n : \phi_n \vdash^s F\{a/b\} : \psi$. Then we are done by ($\rightarrow$-IN).

- $E = x$. In this case $x = x_k$ and we can proceed by induction in the structure of E_k.

$\phi = a \leftarrow \psi$, $\phi = \nu a \leftarrow \psi$, or $\phi = \nu a \rightarrow \psi$. Similar to previous case. ∎

Appendix B. Proof of Theorem 36

Theorem 36. *If* $\Gamma \vdash_{MC} E : \phi$ *then* $\Gamma \vdash_{C} E : \phi$.

PROOF. Observe first that it is safe to assume that, up to names that are not fresh, Γ determines a unique η such that $\eta \models \Gamma$. If this is not the case then the dilemma rules apply on both sides to reduce the problem to a number of problems that do possess this property. Now, since η is uniquely determined we might as well assume that η is the identity on names. Let then η be the substitution $\{E_1/x_1, \ldots, E_n/x_n\}$. Consider a model checker node v labelled

$$\Gamma \vdash \nu s.E\eta : \phi, \tag{4.78}$$

where the variables $x_1, \ldots, x_n$ occurs linearly in E. The aim is to produce a proof tree of $\Gamma, \Delta \vdash^{s} E : \phi$ where $\Delta = x_1 : C(E_1, \eta), \ldots, x_n : C(E_n, \eta)$, and show that, for the tree constructed from the root model checker node, all non-axiom leaves can be discharged. The proof is by induction on the number of occurrences of $|$ in $E\eta$, and then following the structure of the model checker proof. Most subcases of the base and inductive steps for the outer induction are common so we proceed directly to a case analysis on the structure of the model checker proof.

Suppose first that v is an axiom leaf, i.e. an instance of (I), (REFL), (IRR), (NEW1), (INFTY), or (NEC) (where, in the last case, there is no E' such that $E\eta \xrightarrow{l_\tau} E'$). In this case we obtain $\Gamma \vdash^{s}_{C} E\eta : \phi$ by Theorem 22.

Leaves that are discharged are not considered until later.

So, assume that v is an internal node. We consider each possible rule in turn. There is nothing to do until we get to the rules that are unique to the model checker proof system.

(POSS) Assume that $\phi = <l_\tau>\phi'$ and that the model checker rule application infers $\Gamma \vdash \nu s.E\eta : <l_\tau>\phi'$ by showing that there is an E' such that $\nu s.E\eta \xrightarrow{l_\tau} E'$, and $\Gamma \vdash E' : \phi'$. Observe that E' will have the form $\nu s.E''$. We proceed by induction on size of inference that $\nu s.E\eta \xrightarrow{l_\tau} E'$ and then by cases on E.

1 E is a variable x_i. In this case $\nu s.E_i \xrightarrow{l_\tau} E'$ and we obtain (by Lemma 35) that $\Gamma, \Delta \vdash^{s}_{C} E : \phi$.

2 (SUM). Let $E = F_1 + F_2$. If $\nu s.E\eta \xrightarrow{l_\tau} E'$ then $\nu s.F_j\eta \xrightarrow{l_\tau} E'$ for $j = 1$ or $j = 2$, and then by the innermost induction hypothesis, $\Gamma, \Delta \vdash^{s}_{C} F_j : \phi$, so by ($<l_\tau>$-PLUS), $\Gamma, \Delta \vdash^{s}_{C} E : \phi$.

3 (PRE). Let $E = l_\tau.F$. In this case $\Gamma \vdash \nu s.F\eta : \phi'$, so by the (2nd or 3rd) level induction hypothesis, $\Gamma, \Delta \vdash^{s}_{C} F : \phi'$, whence by ($(\tau)$-TAU) or ($<l>$-ACT), $\Gamma, \Delta \vdash^{s}_{C} E : \phi$.

4 (PAR). Let $E = F_1 \mid F_2$. Assume for simplicity that $s = ()$ and that the inference of $E\eta \xrightarrow{l_\tau} E'$ has the shape

$$\frac{F_1\eta \xrightarrow{\tau} F_1'}{(F_1 \mid F_2)\eta \xrightarrow{\tau} F_1' \mid (F_2\eta)}$$

(such that $E' = F_1' \mid (F_2\eta)$). To build up the proof tree from the node $\Gamma, \Delta \vdash^s E : \phi$ first apply (CUT-1) to reduce to the following two subgoals:

$$\Gamma, \Delta \vdash^{()} F_1 : \mathit{Char}(F_1\eta, \eta) \tag{4.79}$$
$$\Gamma, \Delta, x : \mathit{Char}(F_1\eta, \eta) \vdash^{()} x \mid F_2 : \phi \tag{4.80}$$

Now, by Lemma 34 we know that $\Gamma \vdash_{MC} F_1\eta : \mathit{Char}(F_1\eta, \eta)$ and so, by the outermost induction hypothesis,

$$\Gamma, \Delta \vdash_C F_1 : \mathit{Char}(F_1\eta, \eta) \tag{4.81}$$

too. Thus only (4.80) is left. By unfolding the lhs fixed point, and by introducing $\wedge$ to the left (4.80) is reduced to

$$\Gamma, \Delta, x : {<}\tau{>}\mathit{Char}(F_1', \eta) \vdash^{()} x \mid F_2 : \phi \tag{4.82}$$

which in turn is reduced to

$$\Gamma, \Delta, x : \mathit{Char}(F_1', \eta) \vdash^{()} x \mid F_2 : \phi' \tag{4.83}$$

along with the subgoal $\Gamma, \Delta \vdash^{()} F_2 : \mathit{isP}$ which is proved by Theorem 22. The goal (4.83) stands in the desired relation to the model checker node $\Gamma \vdash E' : \phi'$.

The remaining cases for (Par) are easily proved in a similar fashion.

5 (COM) – free communication. Let $E = F_1 \mid F_2$. Again we assume that $s = ()$ and that the inference of $E\eta \xrightarrow{l_\tau} E'$ has the shape

$$\frac{F_1\eta \xrightarrow{\overline{a}} \langle b\rangle F_1' \quad F_2\eta \xrightarrow{a} (d)F_2'}{E\eta \xrightarrow{\tau} F_1' \mid (F_2'\{b/d\})}$$

such that $l_\tau = \tau$. First use Lemma 34 to obtain

$$\Gamma \vdash_{MC} F_1\eta : \mathit{Char}(F_1\eta, \eta) \tag{4.84}$$
$$\Gamma \vdash_{MC} F_2\eta : \mathit{Char}(F_2\eta, \eta) \tag{4.85}$$

Let

$$\begin{aligned} \psi_1 &= \mathit{Char}(\langle b\rangle F_1', \eta) \\ \psi_2 &= \mathit{Char}((d)F_2', \eta). \end{aligned}$$

The goal is to prove

$$\Gamma, \Delta \vdash^{()} F_1 \mid F_2 : <\tau>\phi'. \tag{4.86}$$

Use two cuts to reduce to

$$\Gamma, \Delta \vdash^{()} F_1 : <\overline{a}>\psi_1 \tag{4.87}$$
$$\Gamma, \Delta \vdash^{()} F_2 : <a>\psi_2 \tag{4.88}$$
$$\Gamma, \Delta, y_1 : <\overline{a}>\psi_1, y_2 : <a>\psi_2 \vdash^{()} y_1 \mid y_2 : <\tau>\phi'. \tag{4.89}$$

Subgoals (4.87) and (4.88) are obtained, as before, by the outermost induction hypothesis. By ($<\tau>$-PAR), (4.89) is reduced to the subgoal

$$\Gamma, \Delta, y_1 : \psi_1, y_2 : \psi_2 \vdash^{()} y_1 \mid y_2 : \phi'. \tag{4.90}$$

We are now almost done except that the communication needs to be completed. Observe that

$$\psi_1 = b \leftarrow Char(F_1', \eta) \tag{4.91}$$
$$\psi_2 = \forall d.d \rightarrow \bigvee\{d = e \wedge Char(E\{e/d\}, \eta) \mid e \in fn((d)F_2')\} \vee (\bigwedge\{d \neq e \mid e \in fn((d)F_2')\} \wedge Char(F_2', \nu d.\eta)) \tag{4.92}$$

We now reduce (4.90), using ($\forall$-L), to

$$\Gamma, \Delta, y_1 : \psi_1, y_2 : b \rightarrow \bigvee\{b = e \cdots\} \vee (\cdots) \vdash^{()} y_1 \mid y_2 : \phi' \tag{4.93}$$

and then, by ($\rightarrow$-$\leftarrow$-COM) and boolean reasoning, to

$$\Gamma, \Delta, y_1 : Char(F_1', \eta), y_2 : Char(F_2'\{b/d\}, (\nu b.)\eta) \vdash^{()} y_1 \mid y_2 : \phi' \tag{4.94}$$

where the νb is present in the case b is provably distinct from all free names in $(d)F_2'$. If the νb is absent we are done immediately, and if not we only have to observe that $Char(F_2', \nu b.\eta) = Char(F_2', \eta)$, and we are done.

6 (COM) – bound communication. Let $E = F_1 \mid F_2$. Assume again that $s = ()$ and that the inference of $E\eta \xrightarrow{\tau} E'$ has the shape

$$\frac{F_1\eta \xrightarrow{\overline{a}} \nu b.\langle b\rangle F_1' \quad F_2\eta \xrightarrow{a} (d)F_2'}{E\eta \xrightarrow{\tau} \nu b.(E_1' \mid (E_2'\{b/d\}))}$$

The proof follows the previous subcase quite closely. Let

$$\psi_1 = Char(\nu b.\langle b\rangle F_1', \eta)$$
$$\psi_2 = Char((d)F_2', \eta).$$

The goal $\Gamma, \Delta \vdash^{()} F_1 \mid F_2 : {<}\tau{>}\phi'$ is reduced, using two cuts and the outermost induction hypothesis to a judgment of the form (4.89). As the previous subcase this is further reduced to a subgoal of the form (4.90). Observe now that

$$\psi_1 = \nu b \leftarrow Char(F_1', \eta) \tag{4.95}$$

while ψ_2 is unchanged from (4.92). Now, instead of using ($\rightarrow$-$\leftarrow$-COM) we use ($\rightarrow$-ν $\leftarrow$-COM) (observe that we can use alpha-conversion to identify the bound b in ψ_1 with the bound d in ψ_2) to reduce the current goal to the following:

$$\begin{aligned}&\Gamma,\Delta,y_1{:}Char(F_1'\{d/b\},\nu d.\eta),y_2{:}\bigvee\{d{=}e\wedge\cdots\}\vee(\bigwedge\{d{\neq}e|\cdots\}\wedge\cdots)\\ &\quad\vdash^{d} \quad y_1 \mid y_2 : \phi'\end{aligned} \tag{4.96}$$

and then further, using boolean reasoning, to

$$\Gamma, \Delta, y_1 : Char(F_1'\{d/b\}, \nu d.\eta), y_2 : Char(F_2', \nu d.\eta) \vdash^{d} y_1 \mid y_2 : \phi' \tag{4.97}$$

which is the required result.

7 (RES), (OPEN). Let $E = \nu a.F$ and assume that the inference of $E\eta \xrightarrow{l_\tau} E'$ has the shape

$$\frac{F\eta \xrightarrow{l_\tau} F'}{E\eta \xrightarrow{l_\tau} E'}$$

where $l_\tau \neq a$ and $l_\tau \neq \bar{a}$, and where $E' = \nu a.F'$. The reduction is simple, of $\Gamma, \Delta \vdash^{()} \nu a.F : \phi$ to $\Gamma, \Delta \vdash^{a} F : \phi$ using (NEW2).

8 (ID). Let $E = D(\vec{b})$ and assume that the inference of $E\eta \xrightarrow{l_\tau} E'$ has the shape

$$\frac{P\eta\{\vec{b}/\vec{a}\} \xrightarrow{l_\tau} E'}{E\eta \xrightarrow{l_\tau} E'}$$

because $D(\vec{a}) \triangleq P$. In this case the reduction is straightforward from (FIX).

This completes the case for ϕ a diamond formula. The case for ϕ a box formula is quite similar, as are the cases for input/output. These cases are therefore omitted. We thus need to pay attention to the proper discharge of hypotheses. These, however, are dealt with quite simply by observing that, as the structure of the model checker proof, and in particular the pattern of unfoldings of greatest fixed point formulas, is reflected in the structure of the compositional proof system proof, and in particular leaves discharged in the former proof will correspond to leaves that can be discharged in the latter. ∎

II

FROM A DESCRIPTIVE PERSPECTIVE

Chapter 5

A TUTORIAL INTRODUCTION TO SYMBOLIC MODEL CHECKING

David Déharbe
Departamento de Informática e Matemática Aplicada
Universidade Federal do Rio Grande do Norte
david@dimap.ufrn.br

Abstract Symbolic model checking is a powerful formal verification technique that, contrarily to theorem proving, requires no user assistance. It is able to verify that an implementation, modelled as a labelled finite-state transition graph, satisfies its specification, given as a set of terms in some temporal logic. This chapter introduces the basics of symbolic model checking. We first give the definition of Kripke structures, our model for finite-state transition graph. Temporal logic model checking, including the specification language CTL (Computation Tree Logic), a less powerful verification technique, is then defined. Symbolic model checking itself is then defined. Throughout this tutorial, we use as a running example the alternate bit protocol to illustrate the different concepts.

Keywords: formal methods, verification, model checking, temporal logic

1. Introduction

Model checking is a decision procedure to check that a given structure is a model of a given formula. [11] and [29] presented independently a fully automatic model checking algorithm for the branching time temporal logic CTL in finite-state transition systems. The verification was performed as a graph traversal, linear in both the size of the formula and in the size of the model. This algorithm has been used to verify systems of up to several million states and transitions, which is enough in practice only for small systems.

A major breakthrough has been achieved when [25] proposed to represent the state-transition graph and implement efficiently the search algorithms using binary decision diagrams [6] (BDDs), which are a concise representation for the propositional logic. In this algorithm, called symbolic model checking,

R.J.G.B. de Queiroz (ed.), Logic for Concurrency and Synchronisation, 215–237.

the BDDs are used to represent and operate on the characteristic functions of both sets of transitions and sets of states of the graph. Since sets are not explicitly enumerated, but represented by their characteristic functions, the size of the verified model is not bound by the memory of the computer carrying the verification. This turns possible the verification of systems that are several orders of magnitude larger than was previously achieved.

Outline: We present these techniques by first discussing Kripke structures, the state-transition model of preference in the literature on the subject (Section 2). We then introduce the temporal logic CTL and a corresponding decision procedure, temporal logic model checking (Section 3). We present the connection between propositional logic and Kripke structures, the binary decision diagram representation for propositional logic formulas, and symbolic model checking algorithms (Section 4). Finally, we conclude with a discussion on some of the ongoing research on the subject (Section 5).

2. Kripke structures

2.1 Definitions

Among the numerous concurrency models proposed in the last 20 years [33], Kripke structures are one of the most commonly used in the scope of temporal logic model checking. Kripke structures are interleaving, non-deterministic, state-based models and suitable to represent a wide-range of practical problems: hardware, protocols, software, to name some of the most important.

Informally, Kripke structures are finite-state transition systems, where the labelling is associated with states, instead of the more traditional labelling of transitions.

Definition 1 (Kripke structure) Let P be a finite set of boolean propositions. A Kripke structure over P is a quadruple $M = (S, T, I, L)$ where:

- S is a set of states (when S is finite, we say that M is a *finite Kripke structure*);
- $T \subseteq S \times S$ is a transition relation, such that $\forall s \in S \bullet \exists s' \in S, (s, s') \in T$;
- $I \subseteq S$ is the set of initial states;
- $L : S \rightarrow 2^P$ is a labeling function.

The labelling function L associates each state with a set of boolean propositions true in that state.

Example 2 (Kripke structure) The Kripke structure $ABPsender$ describes the sender component in the alternating bit protocol. The set of atomic propositions is $P = \{g, s, w, b\}$, and $ABPsender = (S_A, T_A, I_A, L_A)$ where:

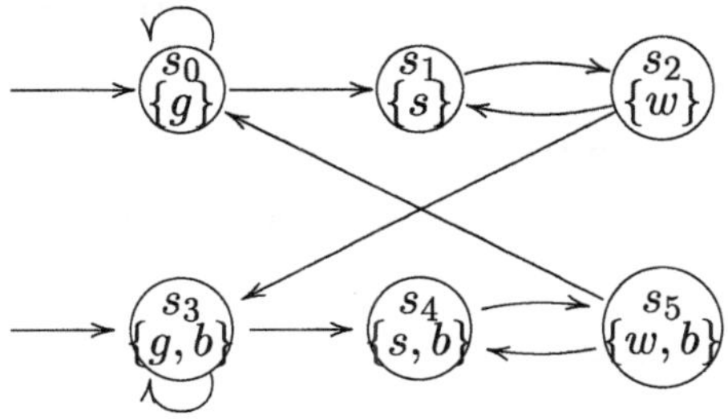

Figure 5.1. State transition diagram for Kripke structure $ABPsender$

- the set of states is such that $S_A = \{s_0, s_1, s_2, s_3, s_4, s_5\}$;
- the transition relation is such that

$$T_A = \{(s_0, s_0), (s_0, s_1), (s_1, s_2), (s_2, s_1), (s_2, s_3), ; \\ (s_3, s_3), (s_3, s_4), (s_4, s_5), (s_5, s_4), (s_5, s_0)\}$$

- the set of initial states is such that $I_A = \{s_0, s_3\}$;
- the labeling function is such that $L_A = \{s_0 \mapsto \{g\}, s_1 \mapsto \{s\}, s_2 \mapsto \{w\}, s_3 \mapsto \{g, b\}, s_4 \mapsto \{s, b\}, s_5 \mapsto \{w, b\}\}$.

The state transition diagram of $ABPsender$ is displayed in Figure 5.1.

In this protocol, the sender tags messages with a control bit alternatively set high and low. The receiver shall acknowledge each message, attaching the corresponding control value. The sender waits for the acknowledgement and checks that the control value is correct before sending another data message. If necessary, the same data is sent again. In this Kripke structure, b indicates the state of the control bit. g is set when the sender is getting the data from the user. s states that the sender is sending the message to the transmission medium. Finally, w indicates that the sender is waiting for the acknowledgement. The data messages are not modeled in this version of $ABPsender$.

A path π in the Kripke structure M is a possibly infinite sequence of states $(s_1, s_2, \ldots)$ such that $\forall i \bullet i \geq 1, (s_i, s_{i+1}) \in T$. $\pi(i)$ is the ith state of π. The set of states reachable from I, denoted RS, is the set of states s such that there exists a path to this state:

$$RS = \{s \in S \mid \exists \pi, ((\pi(1) \in I) \wedge \exists i \geq 1, (\pi(i) = s))\} \tag{5.1}$$

2.2 Computation tree

The computation tree of a Kripke structure is obtained by an operation similar to that of unfolding in CCS [27]. Consequently, computation trees form a special class of Kripke structures, such that the transition relation is acyclic, and only branches away from the initial states.

Definition 3 (Computation tree) Let P be a finite set of boolean propositions. A computation tree is a Kripke structure $M = (S, T, I, L)$ over P, such that:

- every state is reachable;
- Let T^+ the transitive closure of T. $\forall s \in S \bullet (s, s) \notin T^+$;
- $\forall s, s', s'' \in S \bullet (s', s) \in T \wedge (s'', s) \in T \Rightarrow s' = s''$.

For any Kripke structure M, it is possible to associate a computation tree M' such that the sets of states of M' is isomorphic to the set of the finite paths of M.

Definition 4 Let $M = (S, T, I, L)$ be a Kripke structure. The computation tree of M, denoted $ct(M)$, is the Kripke structure (S', T', I', L') such that:

- S' consists of all finite paths of M that start at initial states;
- $(\pi, \pi') \in T'$ iff $\pi = (s_1, \ldots s_n)$, $\pi' = (s_1, \ldots, s_n, s_{n+1})$ and $(s_n, s_{n+1}) \in T$.
- I' consists of all paths of M with only one (initial) state;
- For any path $\pi = (s_1, \ldots s_n)$ of M, $L'(\pi) = L(s_n)$.

Example 5 (Computation tree) Figure 5.2 depicts the initial part of the state transition diagram of the infinite Kripke structure $ct(ABPsender)$. For instance, state π_9 in $ct(ABPsender)$ corresponds to path $s_3 s_4 s_5$ in $ABPsender$. Note how the labeling function is preserved between the origin and destination of the transitions in the original Kripke structure and the corresponding computation tree.

3. Temporal logic model checking

3.1 Computation tree logic

Computation Tree Logic (CTL for short) is a logic to reason about Kripke structures and is interpreted over their computation tree. Within the framework of modal and temporal logics, CTL can be classified as a branching-time, discrete temporal logic.

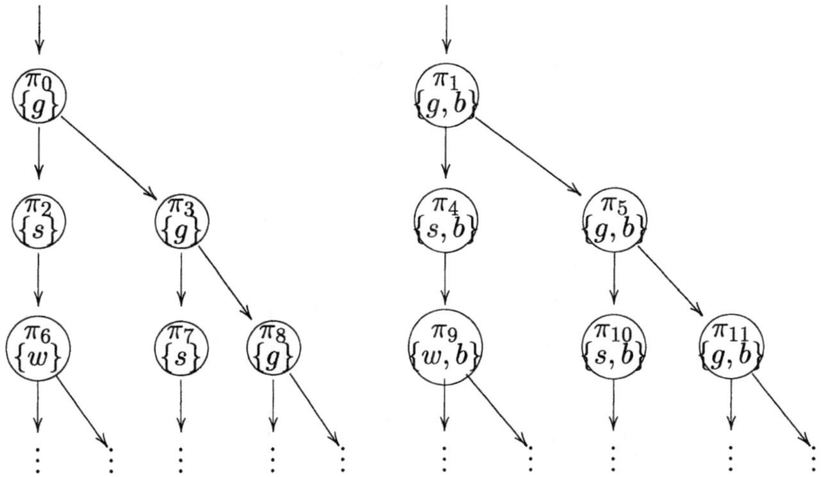

Figure 5.2. Partial state transition diagram of $ct(ABPsender)$

3.1.1 Syntax. The set $T_{\mathsf{CTL}}(P)$ of Computation Tree Logic (CTL for short) formulas over a set of propositions P is the smallest set such that $P \subseteq T_{\mathsf{CTL}}(P)$ and, if f and g are in $T_{\mathsf{CTL}}(P)$, then $\neg f$, $f \wedge g$, **EX**f, **AX**f, **EG**f, **AG**f, **EF**f, **AF**f, **E**[f**U**g], **A**[f**U**g] are in $T_{\mathsf{CTL}}(P)$.

Each temporal logic operator is composed of:

- a path quantifier: **E**, for some path, or **A**, for all paths;
- followed by a state quantifier: **E**, next state on the path, **U**, until, **G**, globally, or **F**, eventually.

3.1.2 Semantics. The semantics of CTL are defined with respect to a Kripke structure $M = (S, T, I, L)$ over a set of atomic propositions P. If f is in $T_{\mathsf{CTL}}(P)$, $M, s \models f$ means that f holds at state s of M.

Let f and g be in $T_{\mathsf{CTL}}(P)$, then

1 $M, s \models p$ iff $p \in L(s)$.

2 $M, s \models \neg f$ iff $M, s \not\models f$.

3 $M, s \models f \wedge g$ iff $M, s \models f$ and $M, s \models g$.

4 $M, s \models$ **EX**f iff there exists a state s' of M such that $(s, s') \in T$ and $s' \models f$. I.e., s has a successor where f is valid.

5 $M, s \models$ **EG**f iff there exists a path π of M such that $\pi(1) = s$ and $\forall i \geq 1 \bullet M, \pi(i) \models f$. I.e., s is at the start of a path where f holds globally.

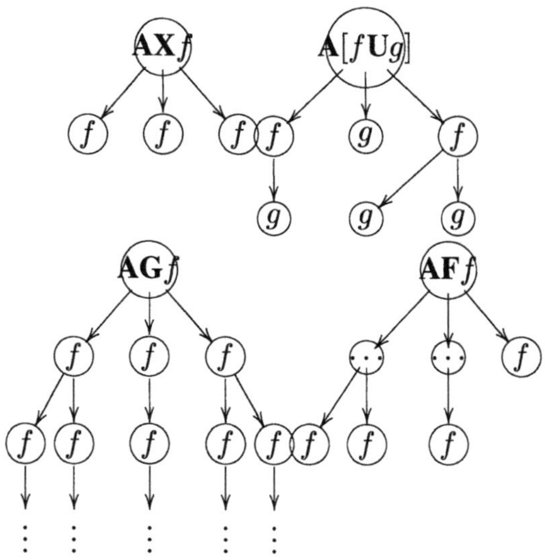

Figure 5.3. Illustration of universal CTL operators

6 $M, s \models \mathbf{E}[f\mathbf{U}g]$ iff there exists a path π of M such that $\pi(1) = s$ and $\exists i \geq 1 \bullet (M, \pi(i) \models g \wedge \forall j, i > j \geq 1 \bullet M, \pi(j) \models f)$. I.e., s is at the start of a path where g holds eventually and f holds until g becomes valid (nothing is said about f when g turns valid).

The other temporal logic operators can be defined in terms of **EX**, **EG** and **E**[**U**]:

$$\mathbf{AX}f = \neg\mathbf{EX}\neg f$$
$$\mathbf{AG}f = \neg\mathbf{EF}\neg f$$
$$\mathbf{AF}f = \neg\mathbf{EG}\neg f$$
$$\mathbf{EF}f = \mathbf{E}[\mathit{true}\mathbf{U}f]$$
$$\mathbf{A}[f\mathbf{U}g] = \neg\mathbf{E}[\neg g\mathbf{U}\neg f \wedge \neg g] \wedge \neg\mathbf{EG}\neg g$$

Figure 5.3 pictures the semantics of the four universal temporal operators (vertical dots indicate paths where f holds infinitely).

Definition 6 A formula f is valid in structure M if it is valid for all initial states:

$$M \models f \text{ iff } \forall s \in I \bullet M, s \models f.$$

Example 7 (CTL formulas) The following CTL formulas state properties of the $ABPsender$ Kripke structure:

- $\mathbf{AG}(s \vee w \vee g)$: The sender is always in one of the three states: getting from the sender, sending a message, or waiting for an acknowledgement.

- $\mathbf{EG}(\neg s \wedge \neg w)$: There is an execution path of the sender such that it is never sending or waiting.

3.2 Algorithm

Given a Kripke structure $M = (S, T, I, L)$ over a set of propositions P and a CTL formula f, its specification, the algorithm given in [12] repeatedly labels the states of M with the sub-formulas of f, starting with the sub-formulas of length 1 (i.e. atomic propositions in P) and finishing with f itself.

Figure 5.4 sketches this algorithm. Function Emc first calls the graph labeling function with the given formula and then checks that all initial states are labeled. Labelling function $Label$ recurses down the structure of the formula (function $args$, given a CTL formula f, returns the set of sub-formulas of f). The terminal case is when the formula is an atomic function. If the formula is a boolean function, each state is labeled according to the previous labeling of the function arguments. Finally, to deal with temporal operators, special-purpose labeling functions are invoked.

Function $LabelEX$ deals with $\mathbf{EX}f$ formulas and is given in Figure 5.5. For each transition $t = (s_1, s_2)$, if s_2 is labeled with f, s_1 has a successor where f is valid, and formula $\mathbf{EX}f$ is added to the label of the s_1.

Function $LabelEU$ handles $\mathbf{E}[f\mathbf{U}g]$ formulas (Figure 5.6). $\mathbf{E}[f\mathbf{U}g]$ is valid in a state if and only if there is a finite path starting in this state, where f is always valid but in the last state, where g is valid. First, each state already with g is also labeled with $\mathbf{E}[f\mathbf{U}g]$. Then, function $LabelEUaux$ is invoked and backtracks along the transitions while f appears in the labels of the states. Each state found along such paths is labeled with formula $\mathbf{E}[f\mathbf{U}g]$. To avoid infinite loops, this backtracking also stops as soon as it meets a state already labeled with $\mathbf{E}[f\mathbf{U}g]$.

Finally, function $LabelEG$ handles $\mathbf{EG}f$ formulas (Figure 5.7). $\mathbf{EG}f$ is valid in a state s if, and only if, there is an infinite path, starting at s and where f holds on each state. To detect such situations, it is necessary to find cycles in the transition graph along which f is always valid. This is the role of auxiliary routine $LabelEGaux$. $LabelEGaux$ has an additional parameter s, which is the state currently visited, and returns a boolean to indicate if $\mathbf{EG}f$ is valid in s. Additionally, two flags are associated with each state: $checked(s)$ and $mark(s)$. $checked(s)$ indicates if the algorithm has already computed if $\mathbf{EG}f$ is valid or not in state s. $mark(s)$ is true if the algorithm has not yet checked if $\mathbf{EG}f$ holds in s, and if s starts a finite path along which f always hold. $LabelEGaux$ does a depth-first search along the transition graph as long as:

```
function Emc(M: Kripke structure, f: CTL) : boolean
begin
    Label(M, f)
    for each s ∈ I do
        if f ∉ L(s) then return false
    return true
end function Emc

function Label(M: Kripke structure, f: CTL) : void
begin
    if f ∈ P then return;
    for each f_i ∈ args(f) do
        Label(M, f_i)
    case f of
        when ¬f_1: for each s ∈ S do
                       if f_1 ∉ L(s) then L(s) ← L(s) ∪ {f}
        when f_1 ∧ f_2: for each s ∈ S do
                       if f_1 ∈ L(s) ∧ f_2 ∈ L(s) then L(s) ← L(s) ∪ {f}
        when EX f_1: LabelEX(M, f_1)
        when EG f_1: LabelEG(M, f_1)
        when E[f_1 U f_2]: LabelEU(M, f_1, f_2)
    end case
end function Label
```

Figure 5.4. Labeling algorithm for CTL formulas

```
function LabelEX(M: Kripke structure, f: CTL) : void
begin
    for each t = (s_1, s_2) ∈ T do
        if f ∈ L(s_2) then L(s_1) ← L(s_1) ∪ {EX f}
end function LabelEX
```

Figure 5.5. Labeling algorithm for $\mathbf{EX}f$ formulas

1 It does not reach an already checked state s. If it does, it stops backtracking and returns a boolean to tell if $\mathbf{EG}f$ is valid in s or not.

2 It does not reach a state that is marked. If it does, then it has found a cycle where f is always valid. In this case it returns $true$.

3 It does not reach a state s where f does not hold. If it does, then the value returned is $false$ (this is implicit in this algorithm).

function $LabelEU(M$: Kripke structure, f: CTL, g: CTL) : void
begin
    for each $s \in S$ do
        if $g \in L(s)$ then
            $L(s) \leftarrow \{\mathbf{E}[f\mathbf{U}g]\} \cup L(s)$
            for each $t = (s', s) \in T$ do
                $LabelEUaux(M, f, g, s')$
end function $LabelEU$

function $LabelEUaux(M$: Kripke structure, f: CTL, g: CTL, s: state) : void
begin
    if $f \in L(s) \wedge \mathbf{E}[f\mathbf{U}g] \notin L(s)$ then
        $L(s) \leftarrow L(s) \cup \{\mathbf{E}[f\mathbf{U}g]\}$
        for each $(s', s) \in T$ do
            $LabelEUaux(M, f, s')$
end function $LabelEUaux$

Figure 5.6. Labeling algorithm for $\mathbf{E}[f\mathbf{U}g]$ formulas

4 Otherwise f is valid in s and $\mathbf{EG}f$ is potentially valid in s. The algorithm then calls itself recursively and checks if $\mathbf{EG}f$ is valid in one of the successors of f. As soon as one such successor is found, then formula $\mathbf{EG}f$ is added to the label set of s and the algorithm returns immediately $true$. If no such successor is found, then formula $\mathbf{EG}f$ is not added to the label set and the algorithm returns $false$.

Example 8 (Verification of $ABPsender$**)** To illustrate the model checking algorithm, we apply it to the verification of the $ABPsender$ (Figure 5.1). More specifically, we check that formula $\mathbf{EG}(\neg s \wedge \neg w)$ is verified by $ABPsender$, which is computed with the function call $Emc(ABPsender, \mathbf{EG}(\neg s \wedge \neg w))$.

The first step of the algorithm Emc (Figure 5.4) consists in labeling recursively the structure with the formulas and its sub-formulas. This is done by invoking algorithm *Label* with parameters $ABPsender$ and $\mathbf{EG}(\neg s \wedge \neg w)$). Algorithm *Label* first labels the states of the graph with each one of the sub-formulas of the specification that are valid in those states. Since these sub-formulas are all boolean, the application of *Label* is trivial and yields the following state labeling:

$$L(s_0)=\{g, \neg s, \neg w, \neg s \wedge \neg w\}$$
$$L(s_1)=\{w, \neg s\}$$
$$L(s_2)=\{s, \neg w\}$$

function $LabelEG(M$: Kripke structure, f: CTL) : void
begin
    for each $s \in S$ do
        $checked(s) \leftarrow false, mark(s) \leftarrow false$
    for each $s \in S$ do
        $LabelEGaux(M, f, s)$
end function $LabelEG$

function $LabelEGaux(M$: Kripke structure, f: CTL, s: state) : boolean is
begin
    if $\neg checked(s)$ then
        if $mark(s)$ then
            return $true$
        if $f \in L(s)$ then
            $mark(s) \leftarrow true$
            for each $(s, s') \in T$ do
                if $LabelEGaux(M, s', f)$ then
                    $L(s) \leftarrow L(s) \cup \{\mathbf{EG}f\}$
                    $checked(s) \leftarrow true$
                    return $true$
            $mark(s) \leftarrow false$
        $checked(s) \leftarrow true$
    return ($\mathbf{EG}f \in L(s)$)
end function $LabelEGaux$

Figure 5.7. Labeling algorithm for **EG**f formulas

$$L(s_3)=\{g, b, \neg s, \neg w, \neg s \wedge \neg w\}$$
$$L(s_4)=\{w, b, \neg s\}$$
$$L(s_5)=\{s, b, \neg w\}$$

Next, function *Label* invokes $LabelEG(ABPsender, \neg s \wedge \neg w)$ (Figure 5.7). The flags *checked* and $mark$ of each state are initialized to $false$. Then function $LabelEGaux$ is invoked on each state. We suppose that the first state to be inspected is s_0 and we trace the corresponding call $LabelEGaux(ABPsender, \neg g \wedge \neg w, s_0)$. Since s_0 is not yet checked and formula $\neg g \wedge \neg w$ belongs to $L(s_0)$, the $mark$ flag of s_0 is assigned $true$ and for each transition leaving s_0, function $LabelEGaux$ is called on the destination. Suppose transition (s_0, s_0) is chosen first. Then $LabelEGaux$ is invoked again on state s_0, tests the flag $mark$, which is now set, and returns $true$. The execution flows continues from the first invocation of $LabelEGaux$ and adds formula $\mathbf{EG}\neg s \wedge \neg w$ to set

$L(s_0)$, sets true the *checked* flag and returns true. This operation is repeated for each state of the structure. When *LabelEG* returns, the state labeling of *ABPsender* is:

$$
\begin{aligned}
L(s_0) &= \{g, \neg s, \neg w, \neg s \wedge \neg w, \mathbf{EG}(\neg s \wedge \neg w)\} \\
L(s_1) &= \{w, \neg s\} \\
L(s_2) &= \{s, \neg w\} \\
L(s_3) &= \{g, b, \neg s, \neg w, \neg s \wedge \neg w, \mathbf{EG}(\neg s \wedge \neg w)\} \\
L(s_4) &= \{w, b, \neg s\} \\
L(s_5) &= \{s, b, \neg w\}
\end{aligned}
$$

The second part of *Emc* executes then and checks that for each initial state of *ABPsender* the formula belongs to the label set, which is the case. Therefore $ABPsender \models \mathbf{EG}(\neg s \wedge \neg w)$.

3.3 Experimental results and conclusion

To illustrate the original model checking algorithm, a fully-functional implementation was demonstrated with a version of the complete alternating bit protocol that had a total of 251 states, with running times taking about 10 seconds for each formula to be verified [12]. After further optimizations, a parallel version of the model checking algorithm, implemented on a vector architecture, was able to verify a Kripke structure with 131,072 states and 67,108,864 transitions, its specification being a CTL formula with 113 sub-formulas. The time reported for this experiment was 225 seconds.

Despite these somehow impressive sounding results, in practice, the model checking presented above is not efficient enough to deal with industrial designs. In concurrent systems, the size of the state space grows exponentially with the number of components. For instance, the model of a sequential circuit with n flip-flops is a Kripke structure with potentially 2^n states: for $n = 32$, the order of magnitude of the number of potential states is 10^9. This phenomenon is known as the *state space explosion*, makes it practically impossible to represent exhaustively the set of states and the set of transitions of most systems.

4. Symbolic model checking

One possible way to avoid (or, at least, to delay) the state space explosion is to represent sets of states and transitions by their characteristic function rather than by enumeration. It is the purpose of Section 4.1 to explain how propositional logic may be used as a language to define and manipulate the characteristic functions. In Section 4.2, we present binary decision diagrams (BDDs), an efficient graph-based implementation of propositional logic. Finally, Section 4.3 details the symbolic version of the model checking algorithms presented in the previous section.

4.1 Kripke structures and propositional logic

4.1.1 Representing states and transitions. Let $M = (S, T, I, L)$ be a Kripke structure over $P = \{v_1, \ldots, v_n\}$. Let $\mathbf{v}$ denote $(v_1, \ldots v_n)$. The characteristic function of a state $s \in S$, denoted $[s]$, is defined as:

$$[s](\mathbf{v}) = \left(\left(\bigwedge_{v_i \in L(s)} v_i \right) \wedge \left(\bigwedge_{v_i \notin L(s)} \neg v_i \right) \right)$$

The definition of the characteristic function is extended to sets of states with the following definitions:

$$[\{\}](\mathbf{v}) = false$$
$$[\{x\} \cup X](\mathbf{v}) = [x](\mathbf{v}) \vee [X](\mathbf{v})$$

Let $P' = \{v'_1, \ldots v'_n\}$ be a set of fresh boolean propositions. The characteristic function of a transition $t = (s_1, s_2) \in T$, denoted $[t]$, is defined as:

$$[t](\mathbf{v}, \mathbf{v}') = [s_1](\mathbf{v}) \wedge [s_2](\mathbf{v}')$$

This definition can be extended to represent sets of transitions as for sets of states.

Example 9 (Characteristic function) In the Kripke structure $ABPsender$, the characteristic functions of the initial state s_0, of the transition (s_0, s_1) and of the initial states I are, respectively:

$$\begin{aligned}
[s_0] &= g \wedge \neg s \wedge \neg w \wedge \neg b \\
[(s_0, s_1)] &= (g \wedge \neg s \wedge \neg w \wedge \neg b) \wedge (\neg g' \wedge s' \wedge \neg w' \wedge \neg b') \\
[I] &= (g \wedge \neg s \wedge \neg w \wedge \neg b) \vee (g \wedge \neg s \wedge \neg w \wedge b) \\
&= (g \wedge \neg s \wedge \neg w) \\
[T] &= \ (b \underline{\vee} b') \wedge \neg g \wedge \neg s \wedge w \wedge g' \wedge \neg s' \wedge w' \\
&\quad \vee (b \leftrightarrow b') \wedge \ (g \wedge \neg s \wedge \neg w \wedge \neg w' \wedge (g' \underline{\vee} s')) \\
&\quad\quad \vee (\neg g \wedge s \wedge \neg w \wedge \neg g' \wedge \neg s' \wedge w') \\
&\quad\quad \vee (\neg s \wedge \neg g' \wedge s' \wedge \neg w' \wedge (g \underline{\vee} w))
\end{aligned}$$

To simplify notations, in the rest of the paper we will identify $[X]$ with X.

4.1.2 State space traversal. Let $M = (S, T, I, L)$ be a Kripke structure over P. The image of a set of states $X \subseteq S$ is the set of states that can be reached in one transition from X:

$$\{s \in S \mid \exists s' \in X, (s', s) \in T\}$$

The characteristic function of the image of X, denoted $Forward(M, X)$, is:

$$Forward(M, X)(\mathbf{v}')=\exists \mathbf{v}, X(\mathbf{v}) \wedge T(\mathbf{v}, \mathbf{v}') \qquad (5.2)$$

Conversely, the inverse image of a set of states $X \subseteq S$, is the set of states from which X can be reached in one transition:

$$\{s \in S \mid \exists s' \in X, (s, s') \in T\}$$

The characteristic function of the inverse image of a set of states X, denoted $Backward(M, X)$, is:

$$Backward(M, X)(\mathbf{v})=\exists \mathbf{v}', X(\mathbf{v}') \wedge T(\mathbf{v}, \mathbf{v}') \qquad (5.3)$$

4.2 Binary decision diagrams

Binary Decision Diagrams (BDDs for short) form a heuristically efficient data structure to represent formulas of the propositional logic. Let P be a totally ordered finite set of boolean propositions. Let f be a boolean formula over P, $bdd(f)$ is the BDD representing f, and $|bdd(f)|$ is the size of this BDD. [6] showed that BDDs are a canonical representation: two equivalent formulas are represented with the same BDD:

$$f \Leftrightarrow g \text{ iff } bdd(f) = bdd(g)$$

Moreover, most boolean operations can be performed efficiently with BDDs. Let $|b|$ denote the size of BDD b:

- $bdd(\neg f)$ is computed in constant time $O(1)$.
- $bdd(f \vee g)$ is realized in $O(|bdd(f)|.|bdd(g)|)$.
- $bdd(\exists x, f)$ is performed in $O(|bdd(f)|^2)$.

In this paper, we will use usual boolean operators to denote the corresponding operation on BDDs, e.g. $bdd(f) \vee bdd(g) = bdd(f \vee g)$.

We explain the basic principles of the BDD representation on an example. Fig. 5.8 presents the binary decision tree for the reachable states of Kripke structure $ABPsender$: $g \wedge \neg s \wedge \neg w$. The binary tree representation of a formula is exponential in the number of free boolean propositions in the formula.

The corresponding BDD is obtained by repeatedly applying the following rules:

- remove duplicate terminal vertices, i.e. after this operation there should be only two leafs, labeled respectively with 0 and 1;
- remove duplicate vertices bottom-up (two vertices are duplicate when they have the same label and their children are identical),

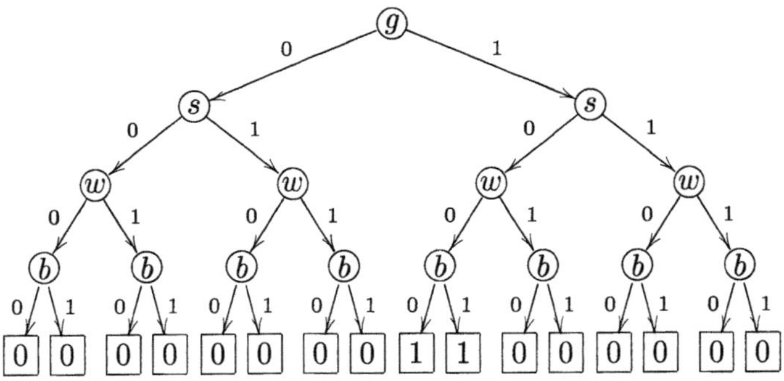

Figure 5.8. Binary tree for $g \wedge \neg s \wedge \neg w$.

- remove opposite vertices (two vertices are opposite when they are labeled with the same variable, and they have opposite children),
- remove redundant tests (a node is a redundant test if its children are identical).

Fig. 5.9 presents the BDD of the characteristic function for the reachable states of Kripke structure $ABPsender$, with variable ordering $g < s < w < b$. Dotted edges indicate that the function on the target node shall be negated. Therefore, the same BDD node is used to represent both a function and its negation (in Figure 5.9, the BDD represents $\neg g \vee s \vee w$ as well), it is interpreted differently according the type of the edge that is pointing to it. With variable ordering $g < g' < s < s' < w < w' < b < b'$, the BDD for the transition relation has 22 nodes.

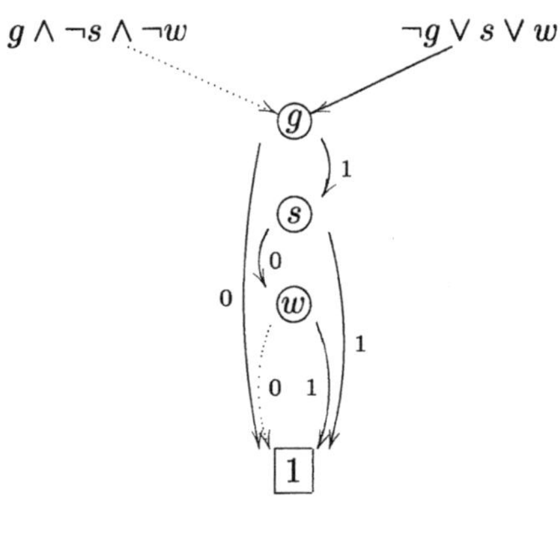

Figure 5.9. BDD for $g \wedge \neg s \wedge \neg w$.

[6] showed that some functions have an exponential BDD representation for any variable ordering, and that finding the optimum variable ordering is NP-hard. However, in practice, heuristic methods generally achieve a good variable ordering, when such ordering exists.

In a Kripke structure, states, transitions and sets thereof can be characterized with propositional logic formulas. These formulas can be represented and manipulated via their BDD representation. BDDs proved to be an efficient data structure to perform computations on large Kripke structures.

In the remainder, we will use the following operations on BDs:

- $BddFalse$, $BddTrue$ return the BDD for the boolean constants;
- $BddAtom$ takes a boolean proposition p as parameter and returns the BDD that represents p;
- $BddNot$ takes as parameter the BDD of a boolean formula f and returns the BDD of formula $\neg f$;
- $BddAnd$ (resp. $BddOr$) take as parameters the BDDs of two boolean formulas f and g and returns the BDD of $f \wedge g$ (resp. $f \vee g$).
- $BddImplies$ is a predicate that takes as parameters the BDDs of two boolean formulas f and g and checks wether f is a logical implication of g.

4.3 Algorithms

4.3.1 Fundamentals of lattices and fixpoints. A lattice is a set with a partial order on the elements of this set, a least element $\bot$, and a greatest element $\top$.

Let P be a non-empty finite set of atomic propositions. Let $M = (S, T, I, L)$ be a finite Kripke structure over P. We consider $(2^S, \subseteq)$ the lattice of subsets of S with set inclusion as the ordering. The empty set $\{\}$ and S are respectively the least and greatest elements of this lattice. Since a subset of S can be identified with its characteristic function, this lattice can also be interpreted as the lattice of characteristic functions, with boolean implication as ordering, *false* is the least element, and the characteristic function of S is the greatest element.

A function $\tau : 2^S \rightarrow 2^S$ is called a *predicate transformer*. τ is *monotonic* iff $P \subseteq Q$ implies $\tau(P) \subseteq \tau(Q)$. Also τ is $\cup$-*continuous* (resp. $\cap$-*continuous*) when $P_1 \subseteq P_2 \subseteq \ldots$ (resp. $P_1 \supseteq P2 \supseteq \ldots$) implies that $\tau(\cup_i P_i) = \cup_i \tau(P_i)$ (resp. $\tau(\cap_i P_i) = \cap_i \tau(P_i)$). [32] showed that if τ is monotonic, then τ has a least fixpoint, denoted $\mathbf{lfp} Z[\tau(Z)]$, and a greatest fixpoint, denoted $\mathbf{gfp} Z[\tau(Z)]$. Moreover, since the lattice is finite and τ is monotonic, it is also ∪-continuous and *cap*-continuous, and:

$$\mathbf{lfp}Z[\tau(Z)]{=}\cap\{Z \mid \tau(Z) = Z\} = \cup_i \tau^i(\mathit{false}) \tag{5.4}$$
$$\mathbf{gfp}Z[\tau(Z)]{=}\cup\{Z \mid \tau(Z) = Z\} = \cap_i \tau^i(\mathit{true}) \tag{5.5}$$

4.3.2 Fixpoint characterization of CTL operators.

CTL symbolic model checking uses the BDD representations of the characteristic functions of sets of states and transitions. The algorithm is based on the fixpoint characterization of the different temporal operators of CTL defined in [11]:

$$\mathbf{EG}f{=}\mathbf{gfp}Z[f \wedge \mathbf{EX}Z] \tag{5.6}$$
$$\mathbf{E}[f\mathbf{U}g]{=}\mathbf{lfp}Z[g \vee (f \wedge \mathbf{EX}Z)] \tag{5.7}$$

The symbolic model checking algorithm, named *Smc*, is shown in Figure 5.10, is a predicate that takes as arguments a Kripke structure M and a CTL formula f. It uses the auxiliary routine $SmcAux$, that returns the characteristic function of the states of M that satisfies f, and checks if f is valid in each initial state of M. $SmcAux$ itself relies on auxiliary routines $SmcEX$ (Figure 5.11), *SmcEU* (Figure 5.12) and *SmcEG* (Figure 5.13). All these routines are used to recurse over the syntactical structure of CTL formulas, and have as arguments M a Kripke structure, f a CTL formula, and as result the BDD of the characteristic function of the set of M states where f is valid.

function *Smc*(M: Kripke structure, f: CTL) : boolean
begin
    return BddImplies($M.I$, $SmcAux(M, f)$)
end function *Smc*
function $SmcAux$(M: Kripke structure, f: CTL) : BDD
begin
    case f of
        f is a boolean proposition: return BddAtom(f)
        $f = \neg f_1$: return BddNot($SmcAux(M, f_1)$)
        $f = f_1 \wedge f_2$: return BddAnd($SmcAux(M, f_1)$, $SmcAux(M, f_2)$)
        $f = \mathbf{EX}f_1$: return $SmcEX(M, f_1)$
        $f = \mathbf{EG}f_1$: return $SmcEG(M, f_1)$
        $f = \mathbf{E}[f_1\mathbf{U}f_2]$: return $SmcEU(M, f_1, f_2)$
end function $SmcAux$

Figure 5.10. Algorithm for symbolic model checking CTL formulas

$SmcEX(M, f)$ computes the states of M where $\mathbf{EX}f$ is valid (Figure 5.11). It first computes F, the BDD for the characteristic function of the set states of

function $SmcEX(M$: Kripke structure, f: CTL) : BDD
variables
    F: BDD
begin
    $F \leftarrow$SmcAux(M, f)
    return $Backward(M, F)$
end function $SmcEX$

Figure 5.11. Algorithm for $\mathbf{EX}f$ formulas

function $SmcEU(M$: Kripke structure, f: CTL, g: CTL) : BDD
variables
    Q, Q', F, G: BDD
begin
    $F \leftarrow SmcAux(M, f)$
    $G \leftarrow SmcAux(M, g)$
    $Q \leftarrow$ BddFalse
    $Q' \leftarrow$ BddOr(G, BddAnd(F, Backward(M, Q)))
    while $Q \neq Q'$ do
        $Q \leftarrow Q'$
        $Q' \leftarrow$ BddOr(G, BddAnd(F, Backward(M, Q)))
    end while
    return Q
end function $SmcEU$

Figure 5.12. Algorithm for $\mathbf{E}[f\mathbf{U}g]$ formulas

M where f is valid, and returns the inverse image of F. The inverse image is computed with $Backward$, a routine that implements Equation 5.3.

Similarly, $SmcEU(M, f, g)$ computes the states of M where $\mathbf{E}[f\mathbf{U}g]$ is valid (Figure 5.12). It first computes F and G, characterizing the states of M where f and g are valid, and then computes the least fixpoint defined in Equation 5.7.

$SmcEG(M, f)$ computes the states of M where $\mathbf{EG}f$ is valid (Figure 5.13). It first computes F, the BDD for the set of states of M where f is valid, and then computes the greatest fixpoint defined in Equation 5.6.

Example 10 (Symbolic verification of $ABPsender$**)** To illustrate the symbolic model checking algorithm, we apply it to the same verification as Example 9: we check that the formula $\mathbf{EG}(\neg s \wedge \neg w)$ is verified by $ABPsender$,

```
function SmcEG(M: Kripke structure, f: CTL) : BDD
variables
    Q, Q', F: BDD
begin
    F ← SmcAux(M, f)
    Q ← BddTrue
    Q' ← BddAnd(F, Backward(M, Q))
    while Q ≠ Q' do
        Q ← Q'
        Q' ← BddAnd(F, Backward(M, Q))
    end while
    return Q
end function SmcEG
```

Figure 5.13. Algorithm for **EG**f formulas

	F	Q	$Backward(Q)$	Q'	$Q = Q'$
Init.	$\neg s \wedge \neg w$	$true$	$(\neg g \wedge w \wedge \neg s) \vee$ $(g \wedge \neg w \wedge \neg s) \vee$ $(\neg g \wedge \neg w \wedge s)$	$g \wedge \neg w \wedge \neg s$	No
Iter. 1	$\neg s \wedge \neg w$	$g \wedge \neg w \wedge \neg s$	$(w \wedge \neg g \wedge \neg s) \vee$ $(g \wedge \neg w \wedge \neg s)$	$g \wedge \neg w \wedge \neg s$	Yes

Figure 5.14. Trace of function call $SmcEG(ABPsender, \mathbf{EG}\neg s \wedge \neg w)$

which is computed with the function call $Smc(ABPsender, \mathbf{EG}(\neg s \wedge \neg w))$ (Figure 5.10).

Most of the computation is carried out by the call $SmcEG(ABPsender, \neg s \wedge \neg w)$ (Figure 5.13). Figure 5.14 contains a trace for the values of the expressions Q, $Backward(ABPsender, Q)$, Q' and $Q = Q'$ on the while statement test at the different iterations of the fixpoint computation[1]. The result returned by the function call is the BDD for $g \wedge \neg w \wedge \neg s$ which is also that of the characteristic function for the set of initial states. Therefore the symbolic model checking returns a $true$ answer, stating that the formula $\mathbf{EG}\neg s \wedge \neg w$ is valid in the Kripke structure $ABPsender$.

4.3.3 Results and extensions to symbolic model checking.

Symbolic model checking has been used to verify a large variety of systems: hardware descriptions [19], software [2], protocols [26, 13]. The size of the Kripke structures used in these verification has been routinely much larger than 10^{20} [8].

An extremely useful feature of model checking is the possibility to compute counterexamples (or witnesses) when a universal formula is false (when an existential formula is true) [15]. For instance, the counterexample of an $\mathbf{AG}f$ formula is a path from an initial state to a state where f is not valid.

In practice, symbolic model checking is well-suited for the verification of the control components of a system. However it performs poorly with data parts. The reason is that BDDs are ill-suited to represent arithmetic expressions or other data-intensive operations. Practically this means that symbolic model checking cannot be used to uncover bugs such as the one found in the Pentium chip floating point division unit. An approach to verify this type of systems has been to combine model checking using other data structures than BDDs to represent the data parts of the system under verification. Word-level model checking [17] is an example of such approach. Word-level model checking uses functions mapping boolean vectors into the integers to model the system under verification. The internal representation of these functions is a combination of two different classes of data structures: multi-terminal binary decision diagrams (MTBDD) represent the control parts and binary moment diagrams [7] (BMD) represent the data parts.

Another limitation of symbolic model checking lies in the expressiveness of the specification logic CTL. Properties asserted in CTL are of a qualitative nature, for instance *if A happens then necessarily B happens in the future*. To express quantitative properties, such as *if A then necessarily B will happen between 4 and 8 time units in the future* it is necessary to nest several **X** operators into syntactically complex and error prone formulas. One possible solution is to write a preprocessor that converts formulas in a quantitative variant of CTL into an equivalent CTL formula and use the standard symbolic model checking algorithm [20]. Another solution is to develop special-purpose algorithms or model representations for this type of formulas. Some tools [9, 30] have an even more powerful capability of computing the lower and upper bound of all possible intervals between two given events. To consider also continuous-time systems it is necessary to develop completely different techniques based on timed automata [1].

5. Conclusion

We have introduced the main concepts necessary to get the reader familiar with the fundamentals of temporal logic model checking and symbolic model checking. A more advanced treatment of the subject is also presented in [16]. Symbolic model checking is the subject of intensive and diversified researches that all tend to cope with the state-space explosion. These approaches may be classified under the following categories:

Composition: The behavior of a system can be seen as the combination of the behavior of its parts. Therefore one may verify a system based on the verification of its components. The decomposition into elementary proofs and the combination of the results need to be controlled by a compositional proof system. In our case, we would have to combine a compositional proof system for CTL (or a subset thereof) with the symbolic model checking algorithms [21].

Abstraction: The verification of a property needs only consider the aspects of a system that are relevant to this property. Using techniques similar to abstract interpretation, it is possible to directly generate an abstract model of the system being verified on which the verification can take place [14, 23].

Alternative algorithms: More efficient algorithms may be found for some classes of CTL formulas [22, 18]. Also alternatives to the representation of propositional logic with BDDs are now being investigated (e.g. [4]).

BDD improvements: The verification engine basically consists in BDD-based data and operations. Improving BDD management and BDD uses has a direct impact on the performances of symbolic model checkers [31, 35, 34]. Important issues are variable ordering [3], more efficient image and inverse image algorithms [28], and intrinsic implementation issues [10, 5, 24].

Notes

1. Actually the values displayed in this table are that of the boolean formulas represented by Q and Q', instead of the less human-friendly BDDs.

References

[1] R. Alur, C. Courcoubetis, and D.L. Dill. Model-checking for real time systems. In 5th *Symposium on Logic in Computer Science*, pages 414–425, June 1990.

[2] R. J. Anderson, P. Beame, S. Burns, W. Chan, F. Modugno, D. Notkin, and R. Reese. Model checking large software specifications. In 4th *Symposium on the Foundations of Software Engineering*, pages 156–166. ACM/SIGSOFT, Oct. 1996.

[3] A. Aziz, S. Taşiran, and R.K. Brayton. Bdd variable ordering for interacting finite state machines. In *31st annual conference on Design Automation conference: DAC'94*, pages 283–288, 1994.

[4] A. Biere, A. Cimatti, E.M. Clarke, M. Fujita, and Y. Zhu. Symbolic model checking using sat procedures instead of bdds. In *36th Design Automation Conference: DAC'99*, number 1579 in Lecture Notes in Computer Science. Springer Verlag, 1999.

[5] K. Brace, R. Rudell, and R. Bryant. Efficient implementation of a bdd package. In *27th ACM/IEEEE Design Automation Conference*, pages 40–45, June 1990.

[6] R.E. Bryant. Graph-based algorithm for boolean function manipulation. *IEEE Transactions Computers*, C(35):1035–1044, 1986.

[7] R.E. Bryant and Y.-A. Chen. Verification of arithmetic circuits with binary moments diagrams. In *32nd ACM/IEEE Design Automation Conference: DAC'95*, pages 535–541, 1995.

[8] J.R. Burch, E.M. Clarke, K.L. Mc Millan, D.L. Dill, and J. Hwang. 10^{20} states and beyond. In *LICS'90:* 5th *annual IEEE symposium on logic in computer science*, pages 428–439, Philadelphia,PA,Etats-Unis, June 1990. IEEE.

[9] S. Campos and E. Clarke. The verus language: representing time efficiently with bdds. In *AMAST Workshop on Real-Time-Systems, Concurrent and Distributed Software*, 1997.

[10] Y.-A. Chen, B. Yang, and R. E. Bryant. Breadth-first with depth-first bdd construction: A hybrid approach. Technical Report CMU-CS-97-120, Carnegie Mellon University, March 1997.

[11] E.M. Clarke and E. A. Emerson. Design and synthesis of synchronization skeletons for branching time temporal logic. In *Logics of Programs: Workshop*, volume 131 of *Lecture Notes in Computer Science*, pages 52–71. Springer Verlag, 1981.

[12] E.M. Clarke, E.A. Emerson, and A.P. Sistla. Automatic verification of finite-state concurrent systems using temporal logic specifications. *ACM Transactions On Programming Languages and Systems*, 8(2):244–263, Apr. 1986.

[13] E.M. Clarke, O. Grumberg, H. Hirashi, S. Jha, D. Long, and K.L. McMillan. Verification of the future-bus+ cache coherence protocol. *Formal Methods in Systems Design*, 6(2):217–232, 1995.

[14] E.M. Clarke, O. Grumberg, and D.E. Long. *19th Annual Symposium on Principles of Programming Languages*, chapter Model checking and abstraction. 1990.

[15] E.M. Clarke, O. Grumberg, K.L. McMillan, and X. Zhao. Efficient generation of counterexamples and witnesses in symbolic model checking. In *32nd ACM/IEEE Design Automation Conference: DAC'95*, 1995.

[16] E.M. Clarke, O. Grumberg, and D. Peled. *Model Checking*. MIT Press, 2000.

[17] E.M. Clarke, M. Khaira, and X. Zhao. Word-level model checking - avoiding the pentium fdiv error. In *33rd ACM/IEEE Design Automation Conference: DAC'96*, pages 645–648, 1996.

[18] D. Déharbe and A. Martins Moreira. Symbolic model checking with fewer fixpoint computations. In *World Congress on Formal Methods: FM'99*, 1999.

[19] D. Déharbe, S. Shankar, and E. Clarke. Model checking vhdl with cv. In *FMCAD'98: Formal Methods in Circuit Automation Design*, number 1522 in Lecture Notes in Computer Science. Springer Verlag, 1998.

[20] J. Frl, J. Gerlach, and T. Kropf. An efficient algorithm for real-time model checking. In *European Design and Test Conference*, pages 15–21, 1996.

[21] O. Grumberg and D.E. Long. Model checking and modular verification. *ACM Transactions On Programming Languages and Systems*, 16(3):843–871, May 1994.

[22] H. Iwashita, T. Nakata, and F. Hirose. CTL model checking based on forward state traversal. In *ICCAD'96*, page 82, 1996.

[23] W. Lee, A. Pardo, J.-Y. Jang, G. Hachter, and F. Somenzi. Tearing based automatic abstraction for ctl model checking. In *International Conference on Computer-Aided Design: ICCAD'96*, page 76, 1996.

[24] D.E. Long. Design of a cache-friendly bdd library. In *ICCAD'98*, pages 639–645, 1998.

[25] K.L. McMillan. *Symbolic Model Checking*. Kluwer Academic Publishers, 1993.

[26] K.L. McMillan and J. Schwalbe. *Shared Memory Multi-Processing*, chapter Formal Verification of the Gigamax Cache Coherency Protocol. MIT Press, 1992.

[27] A.R.G. Milner. *Calculus of communicating systems*, volume 92 of *Lecture Notes in Computer Science*. Springer-Verlag, 1980.

[28] A. Narayan, A.J. Isles, J. Jain, R.K. Brayton, and A.L. Sangiovanni-Vincentelli. Reachability analysis using partitioned-robdds. In *1997 IEEE/ACM international conference on Computer-aided design: ICCAD'97*, pages 388–393. 1997.

[29] J.-P. Queille and J. Sifakis. Specification and verification of concurrent systems in CESAR. In *Procs.* 5th *international symposium on programming*, volume 137 of *Lecture Notes in Computer Science*, pages 244–263. Springer Verlag, 1981.

[30] J. Ruf and T. Kropf. Symbolic model checking for a discrete clocked temporal logic with intervals. In *Advances in Hardware Design and Verification – Proceedings of the International Conference on Correct Hardware and Verification Methods: CHARME'97*, pages 146–163, 1997.

[31] F. Somenzi. *CUDD: CU decision diagrams package – release 2.3.0*. Department of Electrical and Computer Engineering, University of Colorado at Boulder, September 1998.

[32] A. Tarski. A lattice-theoretical fixpoint theorem and its applications. *Pacific J. Math*, pages 285–309, 1955.

[33] Glynn Winskel and Mogens Nielsen. *Handbook of Logic in Computer Science. Vol. 4: Semantic Modelling*, chapter Models for Concurrency, pages 1–148. Oxford Science Publications, 1995.

[34] B. Yang. *Optimizing Model Checking based on BDD Characterization.* PhD thesis, School of Computer Science – Carnegie Mellon University, May 1999. Available as research report CMU-CS-99-129.

[35] B. Yang, R.E. Bryant, D.R. O'Hallaron, A. Biere, O. Coudert, G. Janssen, R.K. Ranjan, and F. Somenzi. A performance study of bdd-based model checking. In *Formal Methods in Computer-Aided Design: FMCAD'98*, number 1522 in Lecture Notes in Computer Science. Springer Verlag, 1998.

Chapter 6

MODAL LOGICS FOR FINITE GRAPHS

Mario R. F. Benevides
Instituto de Matemática and COPPE
Universidade Federal do Rio de Janeiro, Brazil
mario@cos.ufrj.br

Abstract We present modal logics for four classes of finite graphs: finite directed graphs, finite acyclic directed graphs, finite undirected graphs and finite loopless undirected graphs. For all these modal proof theories we discuss soundness and completeness results with respect to each of these classes of graphs. Moreover, we investigate whether some well-known properties of undirected graphs are modally definable or not: k-colouring, planarity, connectivity and properties that a graph is Eulerian or Hamiltonian. Finally, we present an axiomatization for colouring and prove that it is sound and complete with respect to the class of finite k-colourable graphs. One of most interesting feature of this approach is the use of the axioms of Dynamic Logic together with the Löb axiom to ensure acyclicity.

Keywords: Dynamic Logic, Löb axiom, Graphs

1. Introduction

Graphs are among the most frequently used structures in Computer Science. In this discipline, usually many important concepts admit a graph representation, and sometimes a graph lies at the very kernel of the model of computation used. This happens, for instance, in the field of distributed systems, where the underlying model of computation is built on top of a graph. In addition to this central role, in distributed systems graphs are also important as tools for the description of resource sharing problems, scheduling problems, deadlock issues, and so on. The case of distributed systems is also particularly appealing from the standpoint of the use of graphs as modeling tools because it illustrates well two different levels at which graph properties have to be described. One is the "local" level, encompassing properties that hold for vertices or constant-size vertex-neighborhood. The other level is "global" and comprises properties that

R.J.G.B. de Queiroz (ed.), Logic for Concurrency and Synchronisation, 239–267.

hold for the graph as a whole, as for example k-colouring, planarity, acyclicity, connectivity, and so on.

In this work, we present a description language and a proof theory to express and reason about properties that hold at vertices of finite graphs. It is inspired by [2, 1], where a proof theory is presented to reason and express properties of finite trees. We prove soundness and completeness using techniques used in Dynamic logic [4] and in [2, 1]. One of most interesting features of these approaches is the use of the axioms of Dynamic Logic together with the Löb axiom to ensure acyclicity. We propose five proof theories and discuss their completeness w.r.t. five classes of finite graphs.

First, we present modal logics for four classes of finite graphs: finite directed graphs, finite acyclic directed graphs, finite undirected graphs and finite loopless undirected graphs. For all these modal proof theories we discuss soundness and completeness with respect to each of these classes of graphs. Moreover, we investigate, for a mono-modal language, whether some well-known properties of undirected graphs are modally definable or not: k-colouring, planarity, connectivity and properties that a graph is Eulerian or Hamiltonian. Finally, we present an axiomatization for colouring and prove that it is sound and complete with respect to the class of finite k-colourable graphs.

A finite directed graph G is a pair (V, E), where V is a finite set of vertices and $E \subseteq V \times V$ is a set of edges. G is said to be an *undirected* graph if E is a symmetric relation. We call a graph G *loopless* if E is irreflexive. A *path* in a graph G is a sequence of vertices $< v_1, v_2, \cdots, v_n >$, where $< v_i, v_{i+1} >\in E$, for $0 < i < n$. A *cycle* is a path where $v_1 = v_n$. A graph G is said to be *acyclic* if there is no cycle in it, otherwise it is *cyclic*. If $G = (V, E)$ is an undirected graph and $< v_i, v_j >\in E$, we say that v_i and v_j are *adjacent* to each other. If G is a directed graph we say that v_i is *adjacent to* v_j and v_j is *adjacent from* v_i. The *out-degree* of a vertex is the number of vertices adjacent to it.

The binary relations R_t and R_f are used to express the fact that for every edge $< v_i, v_j >\in E$, v_i is adjacent to v_j ($v_i R_t v_j$) and v_j is adjacent from v_i ($v_j R_f v_i$). Figure 6.1 illustrates this fact.

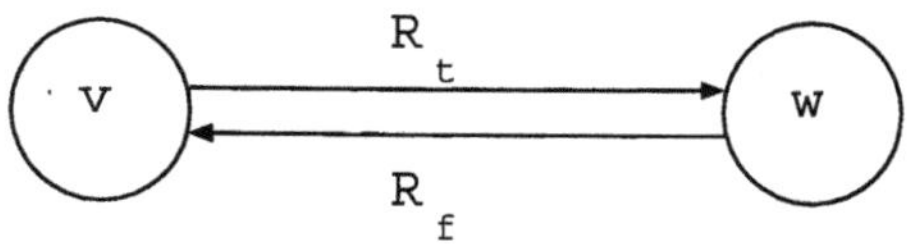

Figure 6.1. vR_tw and wR_fv.

In Section 2 we present a modal proof theory and discuss its completeness with respect to the class of finite directed graphs. In Section 3, we present a modal logic that is an extension of the one presented in Section 2 with Löb axioms and prove its completeness with respect to the class of finite acyclic directed graphs. In Sections 4 and 5, we extend the two modal systems presented in the previous sections and discuss their completeness with respect to the class of finite undirected graphs and finite loopless undirected graphs. In Section 6, we investigate the issue of whether some well-known properties of undirected graphs are modally definable or not: k-colouring, planarity, connectivity, and the Eulerian and Hamiltonian properties. In Section 7, we present a axiomatization for colouring and prove that it is sound and complete with respect to the class of finite k-colourable graphs.

2. Finite directed graphs

2.1 Language and models

The language is a multi-modal language with four modal operators: $< t >$, $< f >$, $< t+ >$ and $< f+ >$, and their duals ($[x]$ for $x \in \{t, f, t+, f+\}$). A formula $< x > A$ for $x \in \{t, f\}$ is true at a vertex w if for some vertex v wR_xv and A is true at v. A formula $< x+ > A$ for $x \in \{t, f\}$ is true at a vertex w if there exist a seguence of verteces $v_1, \cdots, v_n$, $n \geq 2$, such that $w = v_1$, $v = v_n$ and $v_iR_xv_{i+1}$ for $1 \leq i < n$, and A is true at v.

Let G be the finite graph shown in Figure 6.2 (we are only representing the relation R_t, R_f is its converse). In order to illustrate the use of the language the following formulas are true at vertex w supposing A is true at vertex v: $G, w \models < t > A$, $G, w \models < t >< t >< t > A$, $G, w \models < t >< f >< t > A$, $G, w \models < t+ > A$ and $G, w \models < t+ > [t]\bot$.

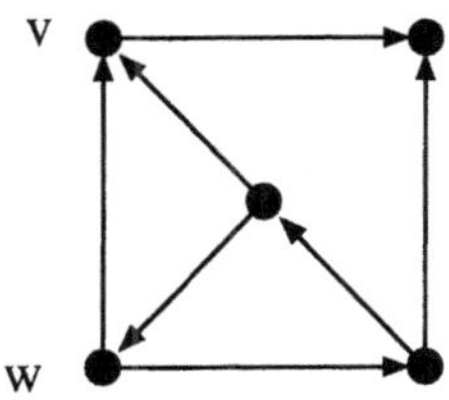

Figure 6.2. Graph G, where a formula A is true at vertex v.

Definition 1 The *directed graph language* is a multi-modal language consisting of a set Φ of countably many propositional symbols (the elements of Φ are denoted by p, q, ..), the booleans connectives $\neg$ and $\wedge$ and four modal operators: $< t >$, $< f >$, $< t+ >$ and $< f+ >$, and the formulas are defined as follows:

$$A ::= p \mid \top \mid \neg A \mid A_1 \wedge A_2 \mid < t > A \mid < f > A \mid < t+ > A \mid < f+ > A$$

We freely use the standard boolean abbreviations $\vee$, $\rightarrow$ and $\leftrightarrow$ and also the following abbreviations for the duals: $[x]A := \neg < x > \neg A$ for $x \in \{t, f, t+, f+\}$.

Definition 2 A *finite directed graph* is a tuple DG $=< V, R_t, R_f >$ where
V - is a finite set of vertices;
R_t and R_f - are binary relations over S, i.e., $R_t, R_f \subseteq S \times S$

And the following condition must be satisfied:

- R_t and R_f are converse relations;

Definition 3 A *model* for a directed graph language is a pair $\mathcal{M} =< DG, \mathbf{V} >$, where DG is a directed graph and $\mathbf{V}$ is a valuation function mapping proposition symbols into subsets of V, i.e., $\mathbf{V} : \Phi \longmapsto Pow(V)$.

The notion of satisfaction is defined as follows:

Definition 4 Let $\mathcal{M} =< DG, \mathbf{V} >$ be a model. The notion of *satisfaction* of a formula A in a model $\mathcal{M}$ at a vertex v, notation $\mathcal{M}, v \models A$ can be inductively defined as follows:

1 $\mathcal{M}, v \models p$ iff $v \in \mathbf{V}(p)$

2 $\mathcal{M}, v \models \neg A$ iff $\mathcal{M}, v \not\models A$

3 $\mathcal{M}, v \models A \wedge B$ iff $\mathcal{M}, v \models A$ and $\mathcal{M}, v \models B$

4 $\mathcal{M}, v \models < t > A$ iff there exists a $w \in V$, vR_tw and $\mathcal{M}, w \models A$

5 $\mathcal{M}, v \models < f > A$ iff there exists a $w \in V$, vR_fw and $\mathcal{M}, w \models A$

6 $\mathcal{M}, v \models < t+ > A$ iff there exists a $w \in V$, vR_t^+w and $\mathcal{M}, w \models A$

7 $\mathcal{M}, v \models < f+ > A$ iff there exists a $w \in V$, vR_f^+w and $\mathcal{M}, w \models A$.

Here R_t^+ and R_f^+ denote the transitive closure of R_t and R_f respectively.

If $\mathcal{M}, v \models A$ for every vertex v, we say that A is *valid in the model* $\mathcal{M}$, notation $\mathcal{M} \models A$. And if A is valid in all models $\mathcal{M}$ we say that A is *valid*, notation $\models A$.

2.2 Proof theory

The proof theory presented below is based in the one presented in [2, 1]. The axioms are all classical tautologies (1), the distribution axioms for the modalities t, f, $t+$ and $f+$ (2), the converse axioms for t and f (3 and 4) and the Segerberg axioms (5 and 6). The inference rules are Modus Ponens, Universal Generalization and Substitution.

In order to make the proof theory presented below more elegant, we introduce some abbreviations for the reflexive and transitive closures and for some special constants:

$< x* > A := A \vee < x+ > A$ and its dual $[x*]A := \neg < x* > \neg A$; for $x \in \{t, f\}$

Special constants:

$sink := [t]\bot$ and $source := [f]\bot$

Axioms

1 *All tautologies*

2 $[x](A \to B) \to ([x]A \to [x]B)$ for $x \in \{t, f, t+, f+\}$

3 $A \to [t] < f > A$

4 $A \to [f] < t > A$

5 $[x+]A \leftrightarrow [x][x*]A$ for $x \in \{t, f\}$

6 $[x+](A \to [x]A) \to ([x]A \to [x+]A)$ for $x \in \{t, f\}$

Inference Rules

M.P. $A, A \to B/B$

U.G. $A/[t]A$, $A/[t+]A$, $A/[f]A$ and $A/[f+]A$

SUB. $A/\sigma A$
where σ is a map uniformly substituting formulas for propositional variables.

A formula A is said to be a *theorem* of a set of formulas Γ, notation $\Gamma \vdash A$ iff there exists a sequence $A_0, A_1, ..., A_n$ of formulas such that A_i is either an axiom or was obtained by applying an inference rule to formulas of $\{A_0, A_1, ..., A_{i-1}\}$ and A is the last item A_n. We say that a set formula Γ is *inconsistent* iff $\Gamma \vdash \bot$ otherwise Γ is said to be *consistent*. A formula A is consistent iff $\{A\}$ is consistent.

2.3 Soundness and completeness results

The modal theory presented above is clearly a fragment of PDL with converse and only four actions $t, f, t+$ and $f+$. It is shown in [6] that PDL with converse has finite model property and completeness of our logic follows from the completeness of PDL. From the canonical model construction [3, 4, 2] it is easy to prove that for every consistent formula we can build a finite model which satisfies it and, moreover, this canonical model is a finite directed graph.

3. Finite acyclic directed graphs

In this section, we extend the proof theory presented in Section 2 with Löb axioms for modalities $[t+]$ and $[f+]$. In order to prove completeness we make a filtration and then using a breadth-first search we assign levels to atoms. We present a construction that takes the result of the filtration and the levels and yields a finite acyclic model.

The language is the same as for finite directed graphs.

Definition 5 A *finite acyclic directed graph* is a tuple
DG $=< V, R_t, R_f >$ where
V - is a finite set of vertices;
R_t and R_f - are binary relations over S, i.e., $R_t, R_f \subseteq S \times S$
And the following conditions must be satisfied:
- R_t and R_f are converse relations;
- R_t and R_f are conversely well-founded (this means that there is no cycle).

The notions of model and satisfaction are the same as for finite directed graphs.

3.1 Proof theory

In order to obtain a proof theory for the modal logic for finite acyclic directed graphs, denoted by **DAG**, we add Löb axiom for t (7) and f (8) and axioms for the constants $sink$ and $source$ (9 and 10) to the proof theory **DG** presented in the previous section.

Axioms

1 *All tautologies*

2 $[x](A \rightarrow B) \rightarrow ([x]A \rightarrow [x]B)$ for $x \in \{t, f, t+, f+\}$

3 $A \rightarrow [t] < f > A$

4 $A \rightarrow [f] < t > A$

5 $[x+]A \leftrightarrow [x][x*]A$ for $x \in \{t, f\}$

6 $[x+](A \rightarrow [x]A) \rightarrow ([x]A \rightarrow [x+]A)$ for $x \in \{t, f\}$

7 $[t+]([t+]A \rightarrow A) \rightarrow [t+]A)$

8 $[f+]([f+]A \rightarrow A) \rightarrow [f+]A)$

9 $< f* > source$

10 $< t* > sink$

Inference Rules

M.P. $A, A \to B/B$

U.G. $A/[t]A$, $A/[f]A$, $A/[t+]A$ and $A/[f+]A$

SUB. $A/\sigma A$
where σ is a map uniformly substituting formulas for propositional variables.

Intuitively, axiom 7 and 8 (Löb) assure that relations R_t and R_f have no cycles and axioms 9 and 10 say that from every vertex we can always reach a sink via R_t-transitions and a source via R_f-transitions.

3.2 Canonical models

The canonical model construction is analogous to the one used for PDL. First, we define the Fisher–Ladner closure $C_{FL}(\Gamma)$ for a set Γ of formulas and then we define the set of atoms of Γ $At(\Gamma)$.

Definition 6 (Fisher–Ladner closure) Let Γ be a set of formulas. The *closure* of Γ, notation $C_{FL}(\Gamma)$, is the smallest set of formulas satisfying the following conditions:

1 $C_{FL}(\Gamma)$ is closed under subformulas;

2 if $< t+ > A \in C_{FL}(\Gamma)$, then $< t > A \in C_{FL}(\Gamma)$

3 if $< t+ > A \in C_{FL}(\Gamma)$, then $< t >< t+ > A \in C_{FL}(\Gamma)$

4 if $< f+ > A \in C_{FL}(\Gamma)$, then $< f > A \in C_{FL}(\Gamma)$

5 if $< f+ > A \in C_{FL}(\Gamma)$, then $< f >< f+ > A \in C_{FL}(\Gamma)$

6 $< t > \top$ and $< f > \top \in C_{FL}(\Gamma)$

7 $< t* > sink$ and $< f* > source \in C_{FL}(\Gamma)$

8 if $A \in C_{FL}(\Gamma)$ and A is not of the form $\neg B$, then $\neg A \in C_{FL}(\Gamma)$.

It is easy to verify that if Γ is a finite set of formulas, then the closure $C_{FL}(\Gamma)$ of Γ is also finite.

Definition 7 Let Γ be a set of formulas. A set of formulas $\mathcal{A}$ is said to be an *atom of* Γ if it is a maximal consistent subset of $C_{FL}(\Gamma)$. The set of all atoms of Γ is denoted by $At(\Gamma)$.

Lemma 8 *Let Γ be a set of formulas. If $A \in C_{FL}(\Gamma)$ and A is consistent then there exists an $\mathcal{A} \in At(\Gamma)$ such that $A \in \mathcal{A}$.*

PROOF. We can construct the atom $\mathcal{A}$ as follows. First, we enumerate the elements of $C_{FL}(\Gamma)$ as $\alpha_1, \cdots, \alpha_n$. We start the construction making $\mathcal{A}_1 = \{A\}$, then for $1 < i < n$, we know that $\vdash \bigwedge \mathcal{A}_i \leftrightarrow (\bigwedge \mathcal{A}_i \wedge \alpha_{i+1}) \vee (\bigwedge \mathcal{A}_i \wedge \neg\alpha_{i+1})$ is a tautology and therefore either $\mathcal{A}_i \wedge \alpha_{i+1}$ or $\mathcal{A}_i \wedge \neg\alpha_{i+1}$ is consistent. We take $\mathcal{A}_{i+1}$ as the union of $\mathcal{A}_i$ with the consistent member of the previous disjunction. At the end, we make $\mathcal{A} = \mathcal{A}_n$. ∎

The canonical relations are defined based on the notion of consistency.

Definition 9 Let Γ be a set of formulas. The *canonical relations over* Γ $S_t^\Gamma, S_f^\Gamma, S_{t+}^\Gamma$ and S_{f+}^Γ on $At(\Gamma)$ are defined as follows:

$\mathcal{A} S_x^\Gamma \mathcal{B}$ iff $\bigwedge \mathcal{A} \wedge < x > \bigwedge \mathcal{B}$ is consistent, for $x \in \{t, f, t+, f+\}$;

Definition 10 Let Γ be a set of formulas. The *canonical model over* Γ is a tuple $\mathbf{G}^\Gamma =< At(\Gamma), S_t^\Gamma, S_f^\Gamma, S_t^{\Gamma+}, S_f^{\Gamma+}, \mathbf{V}^\Gamma >$, where for all propositional symbols p and for all atoms $\mathcal{A} \in At(\Gamma)$ we have

- $\mathbf{V}^\Gamma(p) = \{\mathcal{A} \in At(\Gamma) \mid p \in \mathcal{A}\}$;
- $S_t^\Gamma, S_f^\Gamma, S_t^{\Gamma+}$ and $S_f^{\Gamma+}$ are the canonical relations and their transitive closure.

We say that $\mathbf{V}^\Gamma$ is the *canonical valuation*. For the sake of clarity we avoid using the Γ subscripts.

The following lemma ensures that the canonical model behaves well.

Lemma 11 *Let Γ be a set of formulas and $\mathcal{A} \in At(\Gamma)$. Then for all formulas $< x > A \in C_{FL}(\Gamma)$, where $x \in \{t, f, t+, f+\}$, we have*

$< x > A \in \mathcal{A}$ iff there exists $\mathcal{B} \in At(\Gamma)$ such that $\mathcal{A} S_x \mathcal{B}$ and $A \in \mathcal{B}$.

PROOF.
- Proof for $x = t$.

  $\Rightarrow$: Suppose $< t > A \in \mathcal{A}$. By Definition 7, we have that $\bigwedge \mathcal{A} \wedge < t > A$ is consistent. By a modal reasoning, using the tautology $\vdash A \leftrightarrow ((A \wedge \alpha) \vee (A \wedge \neg\alpha))$, we have that either $\bigwedge \mathcal{A} \wedge < t > (A \wedge \alpha)$ is consistent or $\bigwedge \mathcal{A} \wedge < t > (A \wedge \neg\alpha)$ is consistent. So, by the appropriate choice of α, for all formulas $\alpha \in C_{FL}$, we can construct an atom $\mathcal{B}$ such that $A \in \mathcal{B}$ and $\bigwedge \mathcal{A} \wedge < t > (A \wedge \bigwedge \mathcal{B})$ is consistent and by Definition 9 $\mathcal{A} S_t \mathcal{B}$.

  $\Leftarrow$: Suppose there is a $\mathcal{B}$ such that $A \in \mathcal{B}$ and $\mathcal{A} S_t \mathcal{B}$. Then $\bigwedge \mathcal{A} \wedge < t > \bigwedge \mathcal{B}$ is consistent and also $\bigwedge \mathcal{A} \wedge < t > A$ is consistent. But $< t > A \in C_{FL}$ and by maximality $< t > A \in \mathcal{A}$.

- Proofs for $x \in \{f, t+, f+\}$ are analogous to the previous case. ■

It is important to notice that Lemma 11 is proved only for relations S_x, where $x \in \{t, f, t+, f+\}$. But, we would like to have it for relations S_t^+ and S_f^+. In order to prove this, we must show that the relations S_{x+} are included in relations S_x^+ (for $x \in \{t, f\}$). This is what Lemma 12 does.

Lemma 12 *Let $\mathcal{A}, \mathcal{B} \in At(\Gamma)$. Then*

if $\mathcal{A}S_{x+}\mathcal{B}$ then $\mathcal{A}S_x^+\mathcal{B}$ for $x \in \{t, f\}$;

PROOF. Suppose $\mathcal{A}S_{x+}\mathcal{B}$ for $x \in \{t, f\}$. Let $\mathbf{C} = \{\mathcal{C} \in At(\Gamma) \mid \mathcal{A}S_x^+\mathcal{C}\}$ and $\mathbf{C}_\vee^\wedge = (\bigwedge \mathcal{C}_1 \vee \cdots \vee \bigwedge \mathcal{C}_n)$. It is not difficult to see that $\mathbf{C}_\vee^\wedge \wedge <x>\neg\mathbf{C}_\vee^\wedge$ is inconsistent. By a simple modal reasoning using Segerberg axiom (axiom 6), we have $\vdash \bigwedge \mathcal{A} \rightarrow [x+]\mathbf{C}_\vee^\wedge$. By supposition, $\bigwedge \mathcal{A} \wedge <x+> \bigwedge \mathcal{B}$ is consistent and so is $\bigwedge \mathcal{B} \wedge \mathbf{C}_\vee^\wedge$. Therefore, for at least one $\mathcal{C} \in \mathbf{C}$, we know that $\bigwedge \mathcal{B} \wedge \bigwedge \mathcal{C}$ is consistent. By maximality, we have that $\mathcal{B} = \mathcal{C}$. And by the definition of $\mathbf{C}_\vee^\wedge$, we have $\mathcal{A}S_x^+\mathcal{B}$. ■

Lemma 13 *Let $\mathcal{A}, \mathcal{B} \in At(\Gamma)$. Then, for $x \in \{t, f\}$;*

- *$< x+ > A \in \mathcal{A}$ iff for some $\mathcal{B}$, $\mathcal{A}S_x^+\mathcal{B}$ and $A \in \mathcal{B}$;*

PROOF.

- Proof for $x = t$

  $\Rightarrow$: Suppose $< t+ > A \in \mathcal{A}$. By Lemma 11, there is an atom $\mathcal{B}$ such that $\mathcal{A}S_{t+}\mathcal{B}$ and by Lemma 12 we have $\mathcal{A}S_t^+\mathcal{B}$.

  $\Leftarrow$: Suppose that for some $\mathcal{B}$, $\mathcal{A}S_t^+\mathcal{B}$ and $A \in \mathcal{B}$. Then, for some n $\mathcal{A} = \mathcal{A}_1 S_t \cdots S_t \mathcal{A}_n = \mathcal{B}$. We can prove by induction on $1 \leq k \leq n$.

  $k = 1$: $\mathcal{A}S_t\mathcal{B}$ and $A \in \mathcal{B}$. By Lemma 11, $< t > A \in \mathcal{A}$. From axiom 5, we know that $\vdash < t > A \rightarrow < t+ > A$ and by the definition of C_{FL} and maximality we have $< t+ > A \in \mathcal{A}$.

  $k > 1$: By the induction hypothesis $< t+ > A \in \mathcal{A}_2$ and $< t >< t+ > A \in \mathcal{A}_1$. From axiom 5, we know that $\vdash < t > A \rightarrow < t+ > A$ and by the definition of C_{FL} and maximality we have $< t+ > A \in \mathcal{A}$.

- Proof for $x = f$ is analogous to the previous case. ■

Another nice property of canonical models is that the relations S_t and S_f are converse relations.

Lemma 14 *Let $\mathbf{G}$ be a canonical model over Γ. Then S_t and S_f are converse relations. That is, for every $\mathcal{A}, \mathcal{B} \in At(\Gamma)$, $\mathcal{A}S_t\mathcal{B}$ iff $\mathcal{B}S_f\mathcal{A}$.*

PROOF. $\Rightarrow$: Suppose $\mathcal{A}S_t\mathcal{B}$ and it is not the case that $\mathcal{B}S_f\mathcal{A}$. So, $\bigwedge \mathcal{B} \wedge < f > \bigwedge \mathcal{A}$ is inconsistent, and $\vdash \bigwedge \mathcal{B} \to \neg < f > \bigwedge \mathcal{A}$. By generalization, $\vdash [t] \bigwedge \mathcal{B} \to [t] \neg < f > \bigwedge \mathcal{A}$ (*).
By hypothesis, $\bigwedge \mathcal{A} \wedge < t > \bigwedge \mathcal{B}$ is consistent (**).
From (*) and (**) $\bigwedge \mathcal{A} \wedge < t > \neg < f > \bigwedge \mathcal{A}$ is also consistent. By axiom 6, $\vdash \bigwedge \mathcal{A} \to [t] < f > \bigwedge \mathcal{A}$, and hence $< t > (< f > \bigwedge \mathcal{A} \wedge \neg < f > \bigwedge \mathcal{A})$ is consistent, which is a contradiction.

$\Leftarrow$: is analogous to the previous case. ∎

The only problem is that finite canonical models are not acyclic directed graphs. Our strategy now is to construct a finite acyclic graph out of our canonical models and then prove soundness and completeness w.r.t. the class of finite acyclic graphs.

3.3 Completeness for finite acyclic directed graphs

In order to prove completeness, we first assign levels to atoms using the Löb axiom. Second, we propose a construction that takes finite canonical models and levels and yields a finite acyclic model. Third, we prove that this construction halts and the finite acyclic model is indeed a finite acyclic graph. Finally, we prove completeness by showing that for every consistent formula we can build a finite model which satisfies it and, moreover, this model is a finite acyclic graph.

The techniques used in this section to prove completeness with respect to the class of finite acyclic directed graphs is based on [2, 1], which the presented paper extends. First, they only use Löb axiom in one direction (daughter-of relation, d), but not for its converse (mother-of relation, u). In our approach, we have to take Löb in both direction. Second, while, in their construction, unsatisfied points (atoms which contains a formula of the form $< x > A$, for $x \in \{t, f\}$, and do not x-sees another atom which contains A), in the d-relation, are cared for, no such precautions are taken for unsatisfied points in the u-relation. On the other hand, our construction takes care of unsatisfied points in both direction (t- and f-). Finally, they have only one definition of levels for the d-relation, while we use a breadth-first search to assign levels to atoms, which is a unique assignment, that can be used in both directions.

In the left-hand side picture, of Figure 6.3, a cyclic graph (model) obtained by the filtration defined in Section 3.2 is shown. If our construction only treats unsatisfied points w.r.t. the t-relation, we can obtain an acyclic graph (model), like the one on the right-hand side, which contains no unsatisfied points w.r.t. the t-relation but contains unsatisfied points w.r.t. f-relation.

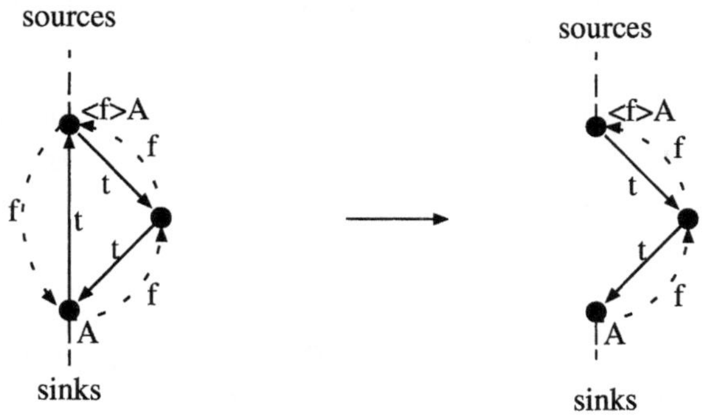

Figure 6.3. *f*-unsatisfied point

Next lemma plays an important role in the proof of the following lemmas. Intuitively, it says that from a sink (source) vertex we cannot make any S_t-transition (S_f-transition).

Lemma 15 *Let $\mathcal{A} \in At(\Gamma)$. Then*

i. $sink \in \mathcal{A}$ iff there is no formula $< t > A \in \mathcal{A}$ or $< t+ > A \in \mathcal{A}$;

ii. $source \in \mathcal{A}$ iff there is no formula $< f > A \in \mathcal{A}$ or $< f+ > A \in \mathcal{A}$.

PROOF. i. $\Rightarrow$: Suppose $sink \in \mathcal{A}$.

- Suppose $< t > A \in \mathcal{A}$, by Lemma 11, there exists $\mathcal{B} \in At(\Gamma)$ such that $\mathcal{A}S_t\mathcal{B}$ and $A \in \mathcal{B}$. As $[t]\perp \in \mathcal{A}$, then $\perp \in \mathcal{B}$, which is a contradiction, because atoms are maximal consistent sets.
- Suppose $< t+ > A \in \mathcal{A}$. By axiom 5 and closure conditions 2 and 3, we have either $< t > A \in \mathcal{A}$ or $< t >< t+ > A \in \mathcal{A}$. In both cases, there exists $\mathcal{B} \in At(\Gamma)$ such that $\mathcal{A}S_t\mathcal{B}$. As $[t]\perp \in \mathcal{A}$, then $\perp \in \mathcal{B}$, which is a contradiction.

$\Leftarrow$: Suppose there is no formula $< t > A \in \mathcal{A}$ or $< t+ > A \in \mathcal{A}$. In particular, $< t > \top \notin \mathcal{A}$, by closure conditions 6 and 8 and maximality $\neg < t > \top \in \mathcal{A}$ and $[t]\perp \in \mathcal{A}$. Therefore, $sink \in \mathcal{A}$.

ii. is analogous to the previous case. ∎

Lemma 16 *Let **G** be a canonical model over Γ and $SI_\Gamma = \{\mathcal{A} \in At(\Gamma) \mid sink \in \mathcal{A}\}$ and $SO_\Gamma = \{\mathcal{A} \in At(\Gamma) \mid source \in \mathcal{A}\}$. Then $SI_\Gamma \neq \emptyset$ and $SO_\Gamma \neq \emptyset$.*

PROOF. $< t* > sink, < f* > source \in C_{FL}$ by Definition 6 (7), as C_{FL} is closed under subformula $sink, source \in C_{FL}$. Using Lemma 8, there exists atoms $\mathcal{A}, \mathcal{B} \in At(\Gamma)$ such that $sink \in \mathcal{A}$ and $source \in \mathcal{B}$. ∎

Next lemma ensures that the following definition of levels is correct. Here, the Löb axiom plays an important role.

Lemma 17 *Let* Γ *be a set of formulas,* $At(\Gamma)\backslash SO_\Gamma$ *be a non-empty set and* $\mathbf{A} = \{\mathcal{A}_1, ..., \mathcal{A}_n\}$ *and* $\mathbf{B} = \{\mathcal{B}_1, ..., \mathcal{B}_m\}$ *be disjoint non-empty sets such that* $\mathbf{A} \cup \mathbf{B} = At(\Gamma)\backslash SO_\Gamma$*. Then for some* $\mathcal{A} \in \mathbf{A}$ *we have,*

$$\bigwedge \mathcal{A} \wedge [t][t*](\bigwedge \mathcal{B}_1 \vee \cdots \vee \bigwedge \mathcal{B}_m) \text{ is consistent.}$$

PROOF. Let $SO_\Gamma = \{S_1, \cdots, S_k\}$ and $\mathbf{A}_\vee^\wedge = (\bigwedge \mathcal{A}_1 \vee \cdots \vee \bigwedge \mathcal{A}_n)$, $\mathbf{B}_\vee^\wedge = (\bigwedge \mathcal{B}_1 \vee \cdots \vee \bigwedge \mathcal{B}_m)$ and $\mathbf{SO_\Gamma}_\vee^\wedge = (\bigwedge S_1 \vee \cdots \vee \bigwedge S_k)$

We know that $\mathbf{A}_\vee^\wedge$ is consistent and as $source \notin \mathcal{A}_i$, for all $1 \leq i \leq n$, then $< t+ > \mathbf{A}_\vee^\wedge$ is also consistent. But , $< t+ > (\mathbf{A}_\vee^\wedge \vee \mathbf{SO_\Gamma}_\vee^\wedge)$ is consistent. Using the counter positive of Löb axiom and Segerberg axiom we obtain
$(\mathbf{A}_\vee^\wedge \vee \mathbf{SO_\Gamma}_\vee^\wedge) \wedge [t][t*]\neg(\mathbf{A}_\vee^\wedge \vee \mathbf{SO_\Gamma}_\vee^\wedge)$ is consistent.

Using the tautology $\phi \leftrightarrow ((\phi\wedge\psi)\vee(\phi\wedge\neg\psi))$ we can prove $\vdash (\mathbf{A}_\vee^\wedge \vee \mathbf{SO_\Gamma}_\vee^\wedge \vee \mathbf{B}_\vee^\wedge)$. Therefore, $(\mathbf{A}_\vee^\wedge \vee \mathbf{SO_\Gamma}_\vee^\wedge) \wedge [t][t*]\mathbf{B}_\vee^\wedge$ is consistent, and $\mathbf{A}_\vee^\wedge \wedge [t][t*]\mathbf{B}_\vee^\wedge$ is consistent too. And, for some $\mathcal{A} \in \mathbf{A}$ we also have $\bigwedge \mathcal{A} \wedge [t][t*]\mathbf{B}_\vee^\wedge$ is consistent. ∎

Lemma 18 *Let* Γ *be a set of formulas,* $At(\Gamma)\backslash SI_\Gamma$ *be a non-empty set and* $\mathbf{A} = \{\mathcal{A}_1, ..., \mathcal{A}_n\}$ *and* $\mathbf{B} = \{\mathcal{B}_1, ..., \mathcal{B}_m\}$ *be disjoint non-empty sets such that* $\mathbf{A} \cup \mathbf{B} = At(\Gamma)\backslash SI_\Gamma$ *and for* $\mathcal{A}_i \in \mathbf{A}$ $(1 \leq i \leq n)$ $source \notin \mathcal{A}_i$*. Then for some* $\mathcal{A} \in \mathbf{A}$ *we have,*

$$\bigwedge \mathcal{A} \wedge [f][f*](\bigwedge \mathcal{B}_1 \vee \cdots \vee \bigwedge \mathcal{B}_m) \text{ is consistent.}$$

PROOF. Analogous to the proof of Lemma 17. ∎

Definition 19 Let Γ be a set of formulas and $At(\Gamma)$ a non-empty set. We define the levels of $At(\Gamma)$ as follows

$L_0 := \{\mathcal{A} \in At(\Gamma) | sink \in \mathcal{A}\}$

$V_0 := L_0$

$i := 0$

while $(At(\Gamma)\backslash SO_\Gamma)\backslash V_i$ is not empty **do**

$V_i := \bigcup_{0 \leq j \leq i} L_j$ for $i \geq 0$
$L_{i+1} := \{\mathcal{A} \in At(\Gamma)\backslash SO_\Gamma | \mathcal{A} \notin V_i$ and $\mathcal{A} S_t \mathcal{B}$ for some $\mathcal{B} \in V_i\}$
$i := i + 1$

halt and assign the level L_i to all atoms in SO_Γ

Lemma 20 *The above construction halts at level L_{max} after a finite number of steps.*

PROOF. Lemma 17 ensures that each level L_i is non-empty, and as $At(\Gamma)$ is finite and each atom $\mathcal{B}$ belongs to a unique level and at each iteration the set V_i increases and $At(\Gamma)\backslash SO_\Gamma$ decreases, eventually $At(\Gamma)\backslash V_i$ will be empty and level L_{max} will be assign to the atoms in SO_Γ. ∎

Next two lemmas ensure that the above definition of levels behaves well.

Lemma 21 *Let $\mathcal{A}, \mathcal{B} \in At(\Gamma)$. If $\mathcal{A} \in L_{i+1}$ ($i \geq 0$) and $< t > A \in \mathcal{A}$ then for some $\mathcal{B} \in L_m$ where $m \leq i$, $\mathcal{A}S_t\mathcal{B}$ and $A \in \mathcal{B}$.*

PROOF. Suppose $\mathcal{A} \in L_{i+1}$ ($i \geq 0$) and $< t > A \in \mathcal{A}$. We have two possibilities:

i) $\mathcal{A}$ is a source and by Definition 19 $\mathcal{A} \in L_{max}$. Suppose for the sake of contradiction that there is no $\mathcal{B} \in L_m$ where $m \leq i+1$, such that $A \in \mathcal{B}$ and $\mathcal{A}S_t\mathcal{B}$. By Lemma 11, there is a C such that $\mathcal{A}S_t\mathcal{C}$ and $A \in \mathcal{C}$. By supposition, $\mathcal{C} \in L_{max}$ and therefore $\mathcal{C}$ is a source. By Lemma 14, $\mathcal{C}S_f\mathcal{A}$ and $< f > \top \in \mathcal{C}$. But, this is a contradiction with Lemma 15.

ii) $\mathcal{A}$ is not a source. Let $V_i = \{\mathcal{B}_1, \cdots, \mathcal{B}_n\}$ and $\mathbf{B}^{\wedge}_{\vee} = (\bigwedge \mathcal{B}_1 \vee \cdots \vee \bigwedge \mathcal{B}_n)$. Suppose that it is not the case that for some $\mathcal{B} \in V_i$, $\mathcal{A}S_t\mathcal{B}$ and $A \in \mathcal{B}$. Then, for all $\mathcal{B}_i \in V_i$, $1 \leq i \leq n$, $\bigwedge \mathcal{A} \wedge < t > \bigwedge \mathcal{B}_i$ is inconsistent, and hence $\vdash \bigwedge \mathcal{A} \rightarrow [t]\neg \bigwedge \mathcal{B}_i$. By modal reasoning $\vdash \bigwedge \mathcal{A} \rightarrow [t]\neg \mathbf{B}^{\wedge}_{\vee}$. By Lemma 17 and Definition 19, we have $\bigwedge \mathcal{A} \rightarrow [t+]\mathbf{B}^{\wedge}_{\vee}$ is consistent and therefore $\bigwedge \mathcal{A} \rightarrow [t]\mathbf{B}^{\wedge}_{\vee}$ is also consistent. As $< t > A \in \mathcal{A}$, then $< t > (\mathbf{B}^{\wedge}_{\vee} \wedge \neg \mathbf{B}^{\wedge}_{\vee})$ is consistent, which is a contradiction. ∎

Lemma 22 *Let $\mathcal{A}, \mathcal{B} \in At(\Gamma)$. If $\mathcal{A} \in L_i$ ($i \geq 0$) and $< f > A \in \mathcal{A}$ then for some $\mathcal{B} \in L_m$ where $m > i$, $\mathcal{A}S_f\mathcal{B}$ and $A \in \mathcal{B}$.*

PROOF. Suppose $\mathcal{A} \in L_i$ ($i \geq 0$) and $< f > A \in \mathcal{A}$. We have two possibilities:

i) $\mathcal{A}$ is a sink and by Definition 19 $\mathcal{A} \in L_0$. Suppose for the sake of contradiction that there is no $\mathcal{B} \in L_m$ where $m \geq i$, such that $A \in \mathcal{B}$ and $\mathcal{A}S_f\mathcal{B}$. As $< f > A \in \mathcal{A}$, by Lemma 11, there is a C such that $\mathcal{A}S_f\mathcal{C}$ and $A \in \mathcal{C}$. By supposition, $\mathcal{C} \in L_0$ and therefore $\mathcal{C}$ is a sink. By Lemma 14, $\mathcal{C}S_t\mathcal{A}$ and $< t > \top \in \mathcal{C}$. But, this is a contradiction with Lemma 15.

ii) $\mathcal{A}$ is not a sink. Let $W_i = At(\Gamma)\backslash V_i$, $W_i = \{\mathcal{B}_1, \cdots, \mathcal{B}_n\}$ and $\mathbf{B}^{\wedge}_{\vee} = (\bigwedge \mathcal{B}_1 \vee \cdots \vee \bigwedge \mathcal{B}_n)$. Suppose that it is not that case that for some $\mathcal{B} \in W_i$, $\mathcal{A}S_f\mathcal{B}$ and $A \in \mathcal{B}$. Then, for all $\mathcal{B}_i \in W_i$, $1 \leq i \leq n$,

$\bigwedge \mathcal{A} \wedge < f > \bigwedge \mathcal{B}_i$ is inconsistent, and hence $\vdash \bigwedge \mathcal{A} \rightarrow [f]\neg \bigwedge \mathcal{B}_i$. By modal reasoning $\vdash \bigwedge \mathcal{A} \rightarrow [f]\neg \mathbf{B}^{\wedge}_{\vee}$. From Definition 19 and Lemma 14, we can see that the same definition of levels could have been done using the relation S_f, so by Lemma 18 and Definition 19, we have $\bigwedge \mathcal{A} \rightarrow [f+]\mathbf{B}^{\wedge}_{\vee}$ is consistent and therefore $\bigwedge \mathcal{A} \rightarrow [f]\mathbf{B}^{\wedge}_{\vee}$ is also consistent. As $< f > A \in \mathcal{A}$, then $< f > (\mathbf{B}^{\wedge}_{\vee} \wedge \neg \mathbf{B}^{\wedge}_{\vee})$ is consistent, which is a contradiction. ∎

Definition 23 Let $\mathcal{A} \in L_i$ and $V_i := \bigcup_{0 \leq j \leq i} L_j$ for $i \geq 0$. For $(0 < i \leq max)$, we define the set *t-target* of $\mathcal{A}$ as $\{\mathcal{B} \in V_{i-1} | \mathcal{A} S_t \mathcal{B}\}$. For $(0 \leq i < max)$, we define the set *f-target* of $\mathcal{A}$ as $\{\mathcal{B} \in At(\Gamma) \backslash V_i | \mathcal{A} S_f \mathcal{B}\}$.

Lemma 24 *Let* $\mathcal{A} \in L_i$.

i. for $0 < i \leq max$*, then the* t-target *of* $\mathcal{A}$ *is non-empty;*

ii. for $0 \leq i < max$*, then the* f-target *of* $\mathcal{A}$ *is non-empty;*

PROOF. i. Suppose $\mathcal{A} \in L_{i+1}$ $(i \geq 0)$, then $sink \notin \mathcal{A}$, and by Lemma 15 for some $A < t > A \in \mathcal{A}$. By Lemma 21, there exists a $\mathcal{B} \in L_m$ where $m \leq i+1$, $\mathcal{A} S_t \mathcal{B}$ and $A \in \mathcal{B}$. By the Definition 23 $\mathcal{B} \in target$ of $\mathcal{A}$ and therefore, the target of $\mathcal{A}$ is not empty.

ii. analogous to the previous case. ∎

Definition 25 Let $\mathbf{G} =< V, R_t, R_f >$ a finite acyclic directed graph where $V = At(A)$ for some formula A. We say that a vertex $\mathcal{A}$ is

- *t-unsatisfied* if for some $< t > B \in \mathcal{A}$, there is no vertex $\mathcal{B}$ such that $\mathcal{A} R_t \mathcal{B}$ and $B \in \mathcal{B}$.
- *f-unsatisfied* if for some $< f > B \in \mathcal{A}$, there is no vertex $\mathcal{B}$ such that $\mathcal{A} R_f \mathcal{B}$ and $B \in \mathcal{B}$.
- *unsatisfied* if it is *t-unsatisfied* or *f-unsatisfied.*

Now we are ready to construct a finite acyclic graph out of canonical models. Intuitively, we start with the source vertices at step 1, and then for every vertex $\mathcal{A}$, which has a formula $< t > A$ ($< f > A$) and there is no vertex $\mathcal{B}$ such that $A \in \mathcal{B}$ and $\mathcal{A} R_t \mathcal{B}$ ($\mathcal{A} R_f \mathcal{B}$), we add the pair $\mathcal{A} R_t \mathcal{B}_i$ ($\mathcal{A} R_f \mathcal{B}_i$) for every $\mathcal{B}_i$ in the set t-target (f-target) of $\mathcal{A}$. By Lemma 24, we know that t-target (f-target) of $\mathcal{A}$ is non-empty and by Lemma 21 (Lemma 22) for at least one $\mathcal{B}_i$ $A \in \mathcal{B}_i$.

Constructing a finite acyclic graph

Let A be a consistent formula, $At(A)$ be the set of atoms over A and $\mathbf{G} =< V, R_t, R_f >$ a finite acyclic directed graph where $V = At(A)$.

step 1:

$V^1 := SO$
$R_t^1 := \emptyset$
$R_f^1 := \emptyset$

step n+1:

if there are no unsatisfied vertices

then

$V := V^n$
$R_t := R_t^n$
$R_f := R_f^n$
halt

else case

t-unsatisfied: choose an t-unsatisfied vertex $\mathcal{A}$, so there is a formula $< t > A \in \mathcal{A}$, by Lemma 24 $\mathcal{A}$ has a non-empty t-target set $\{\mathcal{B}_1, \cdots, \mathcal{B}_k\}$

$V^{n+1} := V^n \cup \{\mathcal{B}_1, \cdots, \mathcal{B}_k\}$
$R_t^{n+1} := R_t^n \cup \{(\mathcal{A}, \mathcal{B}_1), \cdots, (\mathcal{A}, \mathcal{B}_k)\}$
$R_f^{n+1} := R_f^n \cup \{(\mathcal{B}_1, \mathcal{A}), \cdots, (\mathcal{B}_k, \mathcal{A})\}$

f-unsatisfied: choose an f-unsatisfied vertex $\mathcal{A}$, so there is a formula $< f > A \in \mathcal{A}$, by Lemma 24 $\mathcal{A}$ has a non-empty f-target set $\{\mathcal{B}_1, \cdots, \mathcal{B}_k\}$

$V^{n+1} := V^n \cup \{\mathcal{B}_1, \cdots, \mathcal{B}_k\}$
$R_f^{n+1} := R_f^n \cup \{(\mathcal{A}, \mathcal{B}_1), \cdots, (\mathcal{A}, \mathcal{B}_k)\}$
$R_t^{n+1} := R_t^n \cup \{(\mathcal{B}_1, \mathcal{A}), \cdots, (\mathcal{B}_k, \mathcal{A})\}$

Lemma 26 *Let* **G** *be directed graph constructed by the construction presented above. For all $\mathcal{A}$ and for $x \in \{t, f\}$, if $< x > A \in \mathcal{A}$ then there exists a $\mathcal{B}$ in the x-target of $\mathcal{A}$ such that $A \in \mathcal{B}$.*

PROOF.

- Proof for $x = t$

  Suppose $< t > A \in \mathcal{A}$ and $\mathcal{A} \in L_{i+1}(i \geq 0)$. By Lemma 21, there exists a $\mathcal{B} \in L_m(m < i + 1)$ such that $A \in \mathcal{B}$ and $\mathcal{A}S_t\mathcal{B}$. By the Definition 23 of t-target, $\mathcal{B}$ is in the t-target of $\mathcal{A}$.

- Proof for $x = f$ is analogous to the previous case. ■

Lemma 27 *The above construction halts after a finite number of steps.*

PROOF. For each unsatisfied vertex $\mathcal{A}$, we add at most k vertices

$\{\mathcal{B}_1, \cdots, \mathcal{B}_k\}$ corresponding to its t-target/f-target set. By Lemma 26, all unsatisfied formulas are going to become satisfied by these additions, and as each $\mathcal{B}_1, \cdots, \mathcal{B}_k$ belongs to a strictly lower/upper level, and the number of levels is finite we can only add new vertices finitely many times. Hence, all branches reach a sink/source eventually, where, by Lemma 15, there is no t-unsatisfied/f-unsatisfied formulas. Therefore, the construction halts after a finite number of steps. ∎

Lemma 28 *Let* **G** *be a directed graph constructed by the construction presented above. For all vertices $\mathcal{A}, \mathcal{B} \in V$ and for $x \in \{t, f\}$,*

if $\mathcal{A}R_x\mathcal{B}$ then $\mathcal{A}S_x\mathcal{B}$.

PROOF. This proof is straightforward from the definition of target (Definition 23) and from the construction. ∎

Next lemma ensures that the graph generated by the construction behaves well.

Lemma 29 *Let* **G** *be a directed graph constructed by the construction presented above. For all $\mathcal{A} \in V$ and for $x \in \{t, f\}$,*

if $< x > A \in \mathcal{A}$ then there exists a $\mathcal{B} \in V$ such that $A \in \mathcal{B}$ and $\mathcal{A}R_x\mathcal{B}$.

PROOF.

- Proof for $x = t$:

  Suppose $< t > A \in \mathcal{A}$. Also, suppose for the sake of contradiction that there is no $\mathcal{B}$ such that $A \in \mathcal{B}$ and $\mathcal{A}R_t\mathcal{B}$. We have two possibilities:

  i) $\mathcal{A}$ is a sink, and by Lemma 15, for all formulas A, $< t > A \notin \mathcal{A}$, which is a contradiction;

  ii) $\mathcal{A}$ is an unsatisfied vertex, which is a contradiction, because the construction only halts when there is no unsatisfied vertex, and by Lemma 27 it always halts.

- Proof for $x = f$ is analogous to the previous case. ∎

Lemma 30 *Let* **G** *be a directed graph constructed by the construction presented above. For all $\mathcal{A} \in V$ and for $x \in \{t, f\}$,*

if $< x+ > A \in \mathcal{A}$ then there exists a $\mathcal{B} \in V$ such that $A \in \mathcal{B}$ and $\mathcal{A}R_x^+\mathcal{B}$.

PROOF.

- Proof for $x = t$:

  Suppose $< t+ > A \in \mathcal{A}$, by Lemma 15 $\mathcal{A}$ is not a sink. Suppose, for the sake of contradiction, that for all $\mathcal{A}'$, $\mathcal{A}R_t^+\mathcal{A}'$ and $A \notin \mathcal{A}'$.

  Also, suppose, for the sake of contradiction, that for some $\mathcal{B}$, $\mathcal{A}S_t\mathcal{B}$ and $A \in \mathcal{B}$. By Lemma 11, $< t > A \in \mathcal{A}$ and by Lemma 29, there exists a $\mathcal{B}' \in V$ such that $\mathcal{A}R_t\mathcal{B}'$ and $A \in \mathcal{B}'$, which is a contradiction. Therefore, for all $\mathcal{B}$, such that $\mathcal{A}S_t\mathcal{B}$, $A \notin \mathcal{B}$. By Lemma 11, $< t > A \notin \mathcal{A}$, so by maximality and Definition 6 $< t >< t+ > A \in \mathcal{A}$ and by Lemma 29, for some $\mathcal{A}_1 \in V$, $< t+ > A \in \mathcal{A}_1$ and $\mathcal{A}R_t\mathcal{A}_1$.

We can apply the same argument to $\mathcal{A}_1$ yielding $\mathcal{A}_2$ such that $\mathcal{A}R_t\mathcal{A}_1R_t\mathcal{A}_2$ and $< t+ > A \in \mathcal{A}_2$. If we apply the same argument many times we can obtain $\mathcal{A}R_t\mathcal{A}_1R_t\mathcal{A}_2\cdots$. But, it can only be applied finitely many times since G is a finite acyclic graph (Lemma 31). Hence, it must end up in a sink $\mathcal{A}_j$ and we have $< t+ > A \in \mathcal{A}_j$, but by Lemma 15 $< t+ > A \notin \mathcal{A}_j$ which is a contradiction. Therefore, $< t+ > A \notin \mathcal{A}$;

- Proof for $x = f$ is analogous to the previous case. ∎

Lemma 31 **G** *is a finite acyclic directed graph.*

PROOF.

- **G** is finite because $At(\Gamma)$ is finite;
- R_t and R_f are converse relation. This follows straightforward by the construction of these relations;
- **G** is acyclic. By the construction, for all paths $< \mathcal{A}_1, \cdots, \mathcal{A}_n >$, where $\mathcal{A}_1 \in SO$ and $\mathcal{A}_n \in SI$, and for all vertices $\mathcal{A}_i$ and $\mathcal{A}_j$ in this path, if $i < j$ then $level(\mathcal{A}_i) > level(\mathcal{A}_j)$. Therefore we cannot have cycles. ∎

Lemma 32 (Truth lemma for finite acyclic directed graph) *Let* **G** *be a finite acyclic directed graph constructed over* Γ *by the construction presented above and* $\mathcal{M} = (G, \mathbf{V})$ *a model according to Definition 3. For all atoms* $\mathcal{A}$ *and all* $A \in C_{FL}(\Gamma)$, $\mathcal{M}, \mathcal{A} \models A$ *iff* $A \in \mathcal{A}$.

PROOF. By induction on the structure of A

- Atomic formulas: straightforward from the definition of **V** (Definition 3);
- Boolean operators: straightforward from the definition of **V** (Definition 3);
- Modality $< t >$

  $\Rightarrow$: Suppose $\mathcal{M}, \mathcal{A} \models < t > A$, then there exists $\mathcal{A}'$ such that $\mathcal{A}R_t\mathcal{A}'$ and $\mathcal{M}, \mathcal{A}' \models A$. By Lemma 28 $\mathcal{A}S_t\mathcal{A}'$ and by the induction hypothesis $A \in \mathcal{A}'$, by Lemma 11 we have $< t > A \in \mathcal{A}$;

  $\Leftarrow$: Suppose $\mathcal{M}, \mathcal{A} \not\models < t > A$. We have two possibilities:

    i) $\mathcal{A}$ is a sink, and by Lemma 15, for all formulas A, $< t > A \notin \mathcal{A}$;

ii) For all $\mathcal{A}'$, $\mathcal{A}R_t\mathcal{A}'$ and $\mathcal{M}, \mathcal{A}' \not\models A$. By the induction hypothesis $A \notin \mathcal{A}'$, by Lemma 29 we have $< t > A \notin \mathcal{A}$;

- Modality $< f >$: Analogous to the case of modality $< t >$;

- Modality $< t+ >$

$\Rightarrow$: Suppose $\mathcal{M}, \mathcal{A} \models < t+ > A$, then there exists $\mathcal{A}'$ such that $\mathcal{A}R_t^+\mathcal{A}'$, i.e., there is a finite sequence of $\mathcal{A}R_t, \cdots, R_t\mathcal{A}'$ and $\mathcal{M}, \mathcal{A}' \models A$. By the induction hypothesis $A \in \mathcal{A}'$. Applying Lemma 28 finitely many times we have $\mathcal{A}S_t, \cdots, S_t\mathcal{A}'$, by Lemma 13 we have $< t+ > A \in \mathcal{A}$;

$\Leftarrow$: Suppose $\mathcal{M}, \mathcal{A} \not\models < t+ > A$. We have two possibilities:

i) $\mathcal{A}$ is a sink, and by Lemma 15, for all formulas A, $< t+ > A \notin \mathcal{A}$;

ii) For all $\mathcal{A}'$, $\mathcal{A}R_t^+\mathcal{A}'$ and $\mathcal{M}, \mathcal{A}' \not\models A$. By the induction hypothesis $A \notin \mathcal{A}'$, by Lemma 30 we have $< t+ > A \notin \mathcal{A}$;

- Modality $< f+ >$: Analogous to the case of modality $< t+ >$. ■

Theorem 33 (Completeness for finite acyclic directed graph) *The modal logic for directed graphs is complete with respect to the class of finite acyclic directed graph.*

PROOF. For every formula A we can build a canonical model and then we use the construction to build a finite acyclic directed graph $\mathbf{G}_A$. Using Definition 3 we can obtain a model $\mathcal{M} = (\mathbf{G}_A, \mathbf{V}_A)$, by Lemma 8 there exist an atom $\mathcal{A} \in At(A)$ such that $A \in \mathcal{A}$, and by the truth Lemma 32 $\mathcal{M}, \mathcal{A} \models A$. Therefore, our modal system is complete with respect to the class of finite acyclic directed graphs. ■

Theorem 34 (Soundness for finite acyclic directed graph) *The modal logic for directed graphs is sound with respect to the class of finite acyclic directed graph.*

PROOF. The proof of soundness is analogous to the proof of soundness for dynamic logic, it can be found in [4]. It is not difficult to see that every finite acyclic directed graph is a model for **DAG**. ■

4. Undirected graphs

Now, we extend the language with a new modality $\Box$ and the proof theory with the following definition:

Definition 35 $\Box A \leftrightarrow [t]A \wedge [f]A$

And its dual $\Diamond A := \neg\Box\neg A$

We denote by **UG** the proof theory presented in Section 2.2 plus Definition 35. It is easy to verify that the following facts can be demonstrated in **UG**:

Dual $\vdash \Diamond A \leftrightarrow < t > A \vee < f > A$

Distribution $\vdash \Box(A \to B) \to \Box A \to \Box B$.

Symmetry $\vdash A \to \Box\Diamond A$.

Universal Generalization $\vdash A / \vdash \Box A$.

We extend our frame defined in Definition 2 with a new relation R expressing the that we can either make R_t or R_f transitions.

Definition 36 A *finite undirected graph* is a tuple DG $=< V, R_t, R_f, R >$ where

V - is a finite set of vertices;

R_t and R_f - are binary relations over S, i.e., $R_t, R_f \subseteq S \times S$

$R = R_t \cup R_f$

And the following condition must be satisfied:

- R_t and R_f are converse relations;

UG is a definitional extension of the modal logic presented in Section 2 and thus soundness and completeness follow. It is important to notice that a frame DG $=< V, R_t, R_f, R >$ is a finite undirected graph only with respect to the R relation.

5. Loopless undirected graphs

In order to obtain a proof theory for finite loopless undirected graphs we extend the language presented in Section 3.1 **DAG** with a new modality $\Box$ and the Definition 37 presented below. We denote by **IUG** the proof theory presented in Section 2.2 plus the following definitions:

Definition 37 $\Box A \leftrightarrow [t]A \wedge [f]A$

And its dual $\Diamond A := \neg\Box\neg A$

It is easy to verify that the following facts can be demonstrated in **IUG**:

Dual $\vdash \Diamond A \leftrightarrow < t > A \vee < f > A$

Distribution $\vdash \Box(A \rightarrow B) \rightarrow \Box A \rightarrow \Box B.$

Symmetry $\vdash A \rightarrow \Box\Diamond A.$

Universal Generalization $\vdash A / \vdash \Box A.$

Definition 5, of frame, can be extented with a new relation R expressing the fact that we can either make R_t or R_f transitions.

Definition 38 A *finite loopless undirected graph* is a tuple IDG $=< V, R_t, R_f, R >$ where

V - is a finite set of vertices;
R_t and R_f - are binary relations over S, i.e., $R_t, R_f \subseteq S \times S$
$R = R_t \cup R_f$

And the following condition must be satisfied:
- R_t and R_f are converse relations;
- R_t and R_f are conversely well-founded (this means that there is no cycle).

IUG is just a definitional extension of the modal logic presented in Section 3 it is straightforward to prove soundness and completeness. It is important to notice that a frame IDG $=< V, R_t, R_f, R >$ is a finite loopless undirected graph only with respect to the R relation.

6. Modal definability

In this section, we investigate if some well-known graph properties are modally definable or not. Among this properties are: coloring, connectivity, Eulerian graphs, Hamiltonians graphs and planarity. We use a mono-modal language.

We start presenting som⌣ ʷ⌣ll-known facts about modal definability, and then we prove some theorems for graphs properties using these facts.

Definition 39 Let $\mathcal{F} = (W, R)$ and $\mathcal{F}' = (W', R')$ be two frames. A function $f : W \rightarrow W'$ is a *bounded morphism* if

i) f is a homomorphism (if wRv, then $R'f(w), f(v)$);

ii) if $R'f(w), v'$, then there is a v s.t. wRv and $f(v) = v'$;

We use $\mathcal{F} \Rightarrow \mathcal{F}'$ if $\mathcal{F}'$ is a bounded morphic image of $\mathcal{F}$.

Definition 40 Let $\mathcal{F}_i = (W_i, R_t)$ $(i \in I)$ be a collection of frames. The *disjoint union* $\uplus\mathcal{F}_i = (W, R)$ is defined

W - the disjoint union of W_i;

R - the disjoint union of R_t.

Theorem 41 *Let $\mathcal{F} = (W, R)$ and $\mathcal{F}' = (W', R')$ be two frames such that $\mathcal{F} \Rightarrow \mathcal{F}'$. Then*

If $\mathcal{F} \models \phi$, then $\mathcal{F}' \models \phi$

Theorem 42 *Let $\mathcal{F}_i = (W_i, R_t)$ $(i \in I)$ be a collection of frames. And $\uplus \mathcal{F}_i = (W, R)$ their* disjoint union.

If $\mathcal{F}_i \models \phi$ for every $(i \in I)$, then $\uplus \mathcal{F}_i \models \phi$

It is important to notice that according to our definition an undirected graph $G = (V, E)$ is just a directed graph which relation E is symmetric. As a convention, whenever we draw an undirected graph we omit the arrows as shown in the Figure 6.4.

Figure 6.4. Abbreviation for undirected graphs

6.1 Colouring

Let G be an undirected graph and k be the number of colours available. We say that G is k-colourable iff every vertex of G is painted with a colour and every edge of G has endpoints of different colours.

Theorem 43 *Colouring is not modally definable.*

PROOF.

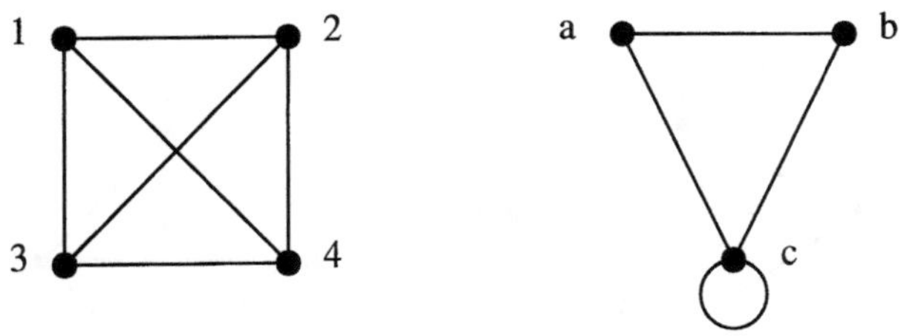

Figure 6.5. Square 1,2,3,4 is 4-colourable and triangle a,b,c is not.

From Figure 6.5, let $f = \{(1, a), (2, b), (3, c), (4, c)\}$. It is straightforward to prove that f is a bounded morphism. By Theorem 41, colouring is not preserved under taking bounded morphic image, therefore it is not modally definable. ∎

6.2 Hamiltonian graphs

A connected undirected graph G is said to be Hamiltonian if there exists a cycle which passes exactly once through each vertex of G.

Theorem 44 *The class of Hamiltonian graphs is not modally definable.*

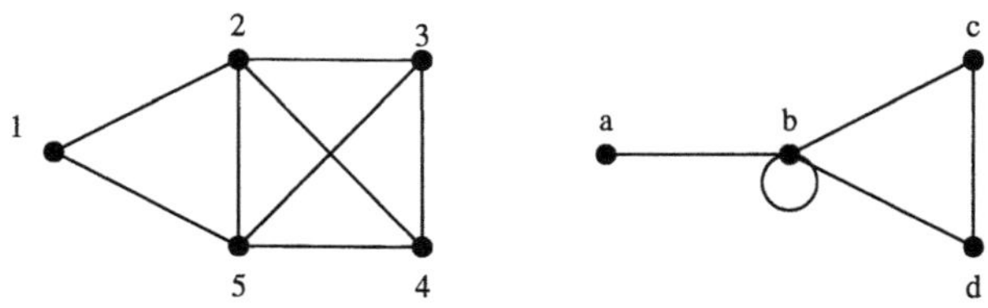

Figure 6.6. Graph 1,2,3,4,5 is Hamiltonian and graph a,b,c,d is not.

PROOF. From Figure 6.6, let $f = \{(1,a),(2,b),(3,c),(4,d),(5,b)\}$. It is straightforward to prove that f is a bounded morphism. By Theorem 41, Hamiltonian property is not preserved under taking bounded morphic image, therefore it is not modally definable. ∎

6.3 Eulerian graphs

A connected undirected graph G is said to be Eulerian if there exists a closed path (cycle) which includes every edge of G exactly once and if edge $< v, w >$ belongs to the path, then edge $< w, v >$ cannot belong to the path.

Theorem 45 *A connected graph G is Eulerian iff the out-degree of every vertex of G is even.*

Theorem 46 *The class of Eulerian graphs is not modally definable.*

PROOF.

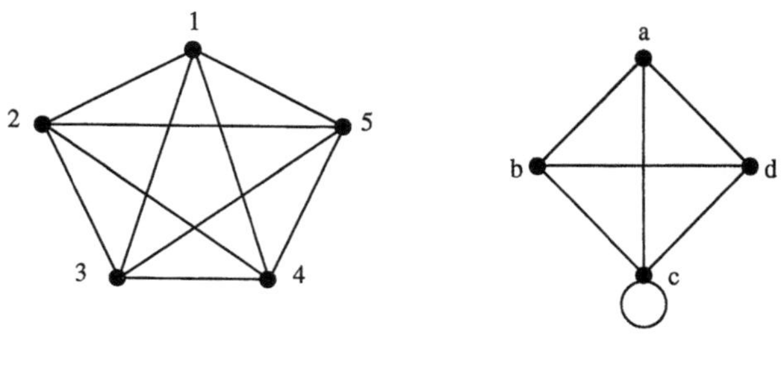

Figure 6.7. Graph 1,2,3,4,5 is Eulerian and graph a,b,c,d is not.

From Figure 6.7, let $f = \{(1,a),(2,b),(3,c),(4,c),(5,d)\}$. It is straightforward to prove that f is a bounded morphism. By Theorem 41, Eulerian property is not preserved under taking bounded morphic image, therefore it is not modally definable. ∎

6.4 Connectivity

A graph G is said to be connected if for every two vertices there is a path connecting them.

Theorem 47 *Connectivity is not modally definable.*

PROOF. The disjoint union of connected graphs is not a connected graph. By Theorem 42, connectivity is not preserved under taking disjoint union, therefore it is not modally definable. ∎

6.5 Planarity

A planar graph is an undirected graph drawn in the plane such that no two edges intersect geometrically except at a vertex to which they are both incident.

A planar graph is a graph which is isomorphic to a planar graph.

Let G be a graph and $< v, w >$ be one edge of G. A *edge subdivision of* $< v, w >$ is a path $< v, u_1. \cdots, u_n, w >$ $(n \geq 0)$ where each u_i has degree two. A graph G_1 is a *subdivision of a graph* G_2 if G_1 can be obtained from G_2 by making a sequence of edge subdivision in G_2. [1]

Theorem 48 (Kuratowski) *A graph is planar iff it contains no subgraph which is a subdivision of* K_5 *or* $K_{3,3}$.

Theorem 49 *Planarity is not modally definable.*

PROOF.
Let $f : i^n \to i$ be a morphism from the graph presented in Figure 6.8 into K_5 (Figure 6.10). It is straightforward to prove that f is a bounded morphism. By Theorem 41, planarity is not preserved under taking bounded morphic image, therefore it is not modally definable. ∎

7. k-Colourable graphs

The problem with colouring is due to the fact that the frame can have reflexive points. But in our modal logic for finite loopless undirected graphs **IUG** frames are irreflexive. In this section, we present an extension of **IUG** which is sound and complete w.r.t the class of k-colourable graphs.

In order to obtain a proof theory for k-Colourable graphs we first extend our modal language with k 0-ary operators $\{\xi_1, \cdots, \xi_k\}$. And then, we add the following axioms to the proof theory presented in Section 5 **IUG**. In Section 7.1, we prove that this new axiomatic is sound and complete with respect to the

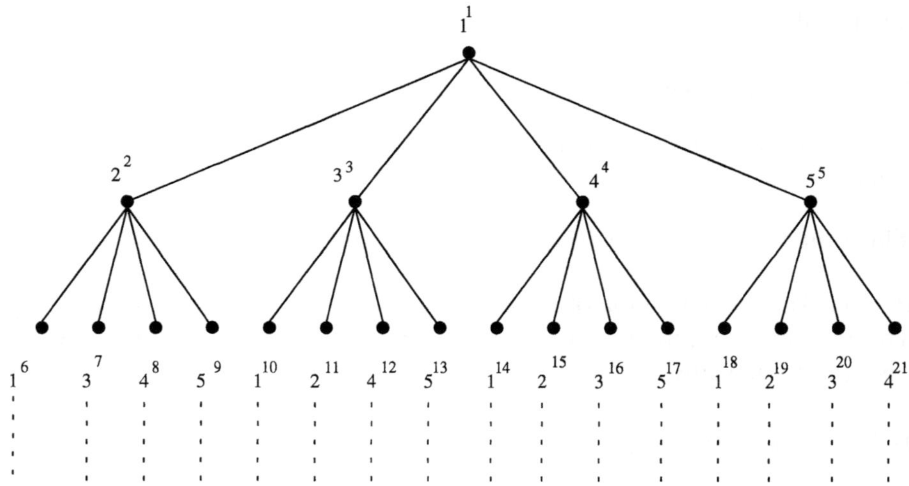

Figure 6.8. Planar Graph.

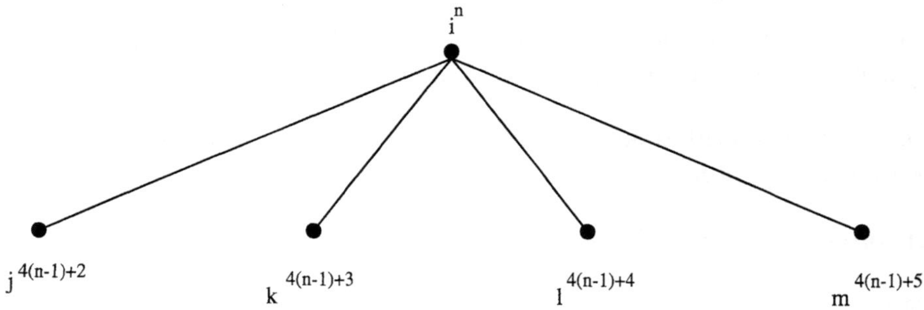

Figure 6.9. Formation rule for the graph presented in Figure 6.8 .

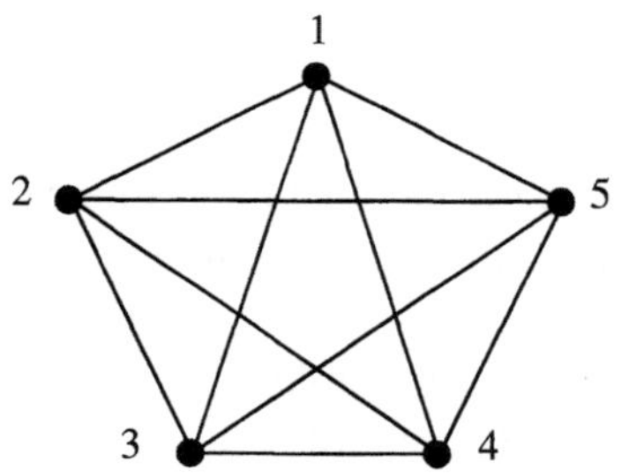

a

Figure 6.10. Graph **K5**.

class of finite k-Colourable graphs. The proof methodology is analogous to the one used in the previous sections. We denote by k-**CG** the proof theory presented in Section 2.2 plus the following axioms:

axioms

M_0: $\xi_1 \vee \cdots \vee \xi_k$

M_1: $\neg(\xi_i \wedge \xi_j)$ $i, j = 1, ..., k$ and $i \neq j$

M_2: $\xi_i \rightarrow \Box\neg\xi_i$ $i = 1, ..., k$ (or $\Diamond\xi_i \rightarrow \neg\xi_i$)

Definition 50 k-**CG** $\equiv$ **UG** + M_0 + M_1 + M_2

7.1 Completeness for k-colourable graphs

We extend our frame defined in Definition 5 with a family of k new unary relations $C_i x$, expressing the fact that vertex x is coloured with colour i.

Definition 51 A *k-colourable graph* is a tuple
kCG $=< V, R_t, R_f, R, C_1, \cdots, C_k >$ where

V - is a finite set of vertices;
R_t, R_f, and R - are binary relations over S as defined in Definition 38
C_i - are unary relations over S (for $i = 1, \cdots, k$)

And the following frame conditions must be satisfied:

A_0: $\forall x C_1 x \vee \cdots \vee C_k x$

A_1: $\forall x \neg(C_i x \wedge C_j x)$ $i, j = 1, ..., k$ and $i \neq j$

A_2: $\forall x \forall y (xRy \wedge C_i x) \rightarrow \neg C_i y$ $i = 1, ..., k$

Frame conditions A_0 and A_1 express the fact that every vertex has a unique colour and A_2 that adjacent vertices have different colours.

The definition of a model is exactly the same as in Definition 3 and in the definition of satisfaction 4 we add the following condition:

Definition 52 (satisfaction) 1,2,3,4,5,6 and 7 as in Definition 4;

8. $\mathcal{M}, v \models \xi_i$ iff $C_i v$

Definition 53 (Fisher–Ladner closure) Let Γ be a set of formulas. The *closure* of Γ, notation $C_{FL}(\Gamma)$, is the smallest set of formulas satisfying the following conditions:

1,2,3,4,5,6,7,8 and 9 as in Definition 6;

10. $\xi_i \in C_{FL}(\Gamma)$.

The definitions of canonical relations, and canonical model are exactly the same as presented in Section 4.

It is important to notice that, although the modal formulas M_0, M_1 and M_2 are all Sahlqvist [5] formulas, it does not imply that their first order correspondent (A_0, A_1 and A_2) are valid in the result of our construction.[2]

Now we are ready to construct a finite loopless undirected k-colourable graph out of canonical models.

Constructing a k-colourable graph

Let A be a consistent formula, $At(A)$ be the set of atoms over A and **IUG** $=< V, R_t, R_f, R >$ a finite loopless undirected graph constructed according to Section 5. We can construct a k-colourable graph

k-CG $=< V, R_t, R_f, R, C_1, \cdots, C_k >$ by defining the relations $C_1, \cdots, C_k$ as follows:

Definition 54 For all $\mathcal{A} \in At(A)$, $\mathcal{A} \in C_i$ iff $\xi_i \in \mathcal{A}$.

Lemma 55 *The above construction halts after a finite number of steps.*

PROOF. This proof is straightforward from the fact that the construction for finite loopless undirected graph halts by Lemma 27 and that $\mathcal{A}$ and $At(A)$ are finite. ■

Lemma 56 *Let $\mathcal{A} \in At(A)$. Then $\xi_i \in \mathcal{A}$ iff $\neg\xi_j \in \mathcal{A}$, for every $j \neq i$.*

PROOF. $\Rightarrow$: suppose $\xi_i \in \mathcal{A}$ and for the sake of contradiction suppose for some $j \neq i$ $\xi_j \in \mathcal{A}$. As $\mathcal{A}$ is consistent, we cannot have $\xi_i \in \mathcal{A}$ and $\xi_j \in \mathcal{A}$, because that would contradict axiom M_1.

$\Leftarrow$: suppose $\neg\xi_j \in \mathcal{A}$, for every $j \neq i$. By maximality of $\mathcal{A}$ and by axiom M_0, $\xi_i \in \mathcal{A}$. ■

Lemma 57 *Let $\mathcal{A}, \mathcal{B} \in At(A)$. Then*

i. If $\mathcal{A}S_t\mathcal{B}$, then $\xi_j \in \mathcal{A}$ iff $\xi_n \in \mathcal{B}$, for $j \neq n$;

ii. If $\mathcal{A}S_f\mathcal{B}$, then $\xi_j \in \mathcal{A}$ iff $\xi_n \in \mathcal{B}$, for $j \neq n$.

iii. If $\mathcal{A}S\mathcal{B}$, then $\xi_j \in \mathcal{A}$ iff $\xi_n \in \mathcal{B}$, for $j \neq n$.

PROOF. i. $\Rightarrow$: suppose $\xi_j \in \mathcal{A}$ and suppose $\xi_j \in \mathcal{B}$, by Lemma 11(1) $< t > \xi_j \in \mathcal{A}$ which is inconsistent with axiom M_2.

$\Leftarrow$: suppose $\xi_n \in \mathcal{B}$, by Lemma 11(1) $< t > \xi_n \in \mathcal{A}$. As $\mathcal{A}$ is consistent, by axiom M_2 $\neg\xi_n \in \mathcal{A}$. By Lemma 56 for some $j \neq n$ $\xi_j \in \mathcal{A}$.

ii. Analogous to the proof of i.

iii. suppose $\mathcal{A}S\mathcal{B}$, by the definition of S relation $\mathcal{A}S\mathcal{B}$ iff $\mathcal{A}S_t\mathcal{B}$ or $\mathcal{A}S_f\mathcal{B}$. In both case, by i. and ii. we have $\xi_j \in \mathcal{A}$ iff $\xi_n \in \mathcal{B}$, for $j \neq n$. ∎

Lemma 58 *Let $\mathcal{A}, \mathcal{B} \in At(A)$. Then*

If $\mathcal{A}R\mathcal{B}$, then $\xi_j \in \mathcal{A}$ iff $\xi_n \in \mathcal{B}$, for $j \neq n$.

PROOF. Suppose $\mathcal{A}R\mathcal{B}$. By Lemma 28 $\mathcal{A}S\mathcal{B}$.But by Lemma 57 (iii.) we have $\xi_j \in \mathcal{A}$ iff $\xi_n \in \mathcal{B}$, for $j \neq n$. ∎

Lemma 59 **G** *is a finite loopless undirected k-colourable graph.*

PROOF.

- From Section 5, **G** is finite loopless undirected graph;
- By Lemma 56 we know that each vertex has precisely one colour;
- By Lemma 58 we know that adjacent vertices have different colours.

Therefore, **G** is a k-colourable graph. ∎

Lemma 60 (Truth lemma for k-colourable graphs) *Let* **G** *be a k-colourable graph over Γ and $\mathcal{M} = (G, \mathbf{V})$ a model according to Definitions 51, 3. For all atoms $\mathcal{A}$ and all $A \in C_{FL}(\Gamma)$, $\mathcal{M}, \mathcal{A} \models A$ iff $A \in \mathcal{A}$.*

PROOF. By induction on the structure of A

- Atomic formulas, Boolean operators and modalities $< t >, < f >, < t+ >, < f+ >$ and $\Diamond$ are analogous to Lemma 32;
- Modalities ξ_i, for $i = 1, \cdots, k$

  $\Leftrightarrow$: Suppose $\mathcal{M}, \mathcal{A} \models \xi_i$, by the Definition 52 of satisfaction $\mathcal{M}, \mathcal{A} \models \xi_i$ iff $C_i\mathcal{A}$, by the construction $C_i\mathcal{A}$ iff $\xi_i \in \mathcal{A}$. ∎

Theorem 61 (Completeness for k-colourable graphs) *The modal logic for k-colourable graphs k-***CG** *is complete with respect to the class of finite loopless undirected k-colourable graphs.*

PROOF. For every formula A we can build a canonical model and then using Definition 51 and 3 we can obtain a model $\mathcal{M} = (\mathbf{G}_A, \mathbf{V}_A)$, by Lemma 8 there exist an atom $\mathcal{A} \in At(A)$ such that $A \in \mathcal{A}$, and by the truth Lemma 60 $\mathcal{M}$,A$\models A$. Therefore, our modal system is complete with respect to the class of finite loopless undirected k-colourable graphs. ∎

8. Conclusions

We have presented five modal systems and shown that they are sound and complete to the classes of finite directed graphs, finite directed acyclic graphs, finite undirected graphs, finite undirected loopless graphs and k-colourable finite loopless graphs.

One of most interesting features of these approaches is the use of the axioms of Dynamic Logic together with the Löb axiom to ensure acyclicity.

We have also investigated, for a mono-modal language, whether some well-known properties of undirected graphs are modally definable or not: planarity, connectivity and properties that a graph is Eulerian or Hamiltonian. Although all results are negative, it is well-known that modal definability is sensitive to the language used. So, if we use a richer language many properties which are not modally definable for the mono-modal language can become definable.

This work opens up possibilities to investigate some interesting related problems. First, we would like to extend our completeness results for a full Dynamic logic with Löb axiom. Also, we would like to investigate axiomatization for other classes of finite graphs, for instance: planar graphs, connected graphs, Eulerian graphs and Hamiltonian graphs. And finally, it would be interesting to see if the use of universal modality and the D operator could help to overcome some of the negative results of Section 6.

Acknowledgments

The author has received partial support from the Brazilian Research Agencies CNPq and FAPERJ.

Notes

1. It is important to notice that if $G = (V, E)$ is an undirected graph, a subdivision $< a, u, b >$ of an edge $< a, b > \in E$ is obtained by adding edges $< a, u >, < u, a >, < u, b >$ and $< b, u >$ to E and removing edges $< a, b >$ and $< b, a >$ from E.

2. It is well-known that Sahlqvist theorem does not hold for models obtained by filtration. A classical example is transitivity ($\Box A \rightarrow \Box\Box A$), it is a Sahlqvist formula but if we make a filtration the model yielded is not necessary transitive.

References

[1] P. Blackburn and W. Meyer-Viol. Linguistic, logic and finite trees. *Bulletin of the IGPL*, 2:2–29, 1994.

[2] P. Blackburn, W. Meyer-Viol, and M. de Rijke. *A Proof System for Finite Trees*. Lecture Notes in Computer Science 1092. Springer-Verlag, Berlin, 1996. Computer Science Logic 95.

[3] M. J. Fisher and R. F. Ladner. Propositional dynamic logic of regular programs. *Journal of Computer and System Sciences*, 18, 1979.

[4] R. Goldblatt. *Logics of Time and Computation.* CSLI Lecture Notes 7. CSLI, Stanford, 1992.

[5] H. Sahlqvist. Completeness and correspondence in first and second-order semantics for modal logic. In S. Kanger, editor, *Proceedings of the third Scandinavian Logic Symposium.* North-Holland, Amsterdam, 1975.

[6] D. Vakarelov. *Filtration Theorem for Dynamic Algebras with Test and Inverse Operators.* Lecture Notes in Computer Science 148. Springer-Verlag, Berlin, 1983. Logics of Programs and their Applications.

[7] J. van Benthem. Correspondence theory. In D. M. Gabbay and F. Guenthner, editors, *Handbook of Philosophical Logic, vol. 2.* D. Reidel Publishing Company, Dordrecht, 1984.

[8] J. van Benthem. *Modal Logic and Classical Logic.* Bibliopolis, Italy, 1985.

Chapter 7

BISIMULATION AND LANGUAGE EQUIVALENCE

Colin Stirling
Division of Informatics
Edinburgh University, Scotland
cps@dcs.ed.ac.uk

Abstract We contrast bisimulation equivalence and language equivalence. There are two threads. First is that because bisimulation is more intensional, results in language and automata theory can be recast for bisimulation. The second thread is the contrast between definability of language equivalence and bisimulation equivalence. Bisimulation equivalence is definable as a "simple" formula in first-order logic with fixed points. We show that language equivalence is not definable as an unconditional projection of simple least fixed point.

Keywords: bisimulation, formal languages

1. Introduction

One way to understand an interactive system is firmly rooted in language theory, that a system is its set of runs (or words). Properties of systems are described in a linear time temporal logic. Relationships between automata, language theory and logic are then utilised, such as the theory of ω-regular languages and Büchi automata.

An alternative viewpoint is that an interactive system should be understood as its capability for interacting with other systems. Language and automata theory then have less relevance because a more intensional account of system behaviour is needed than that given by sets of words. Bisimulation equivalence has a pivotal role within this approach.

Bisimulation is a rich concept which appears in various areas of theoretical computer science. Besides its origin for understanding concurrency, it was independently developed in the context of modal logic. In this paper we make

R.J.G.B. de Queiroz (ed.), Logic for Concurrency and Synchronisation, 269–284.

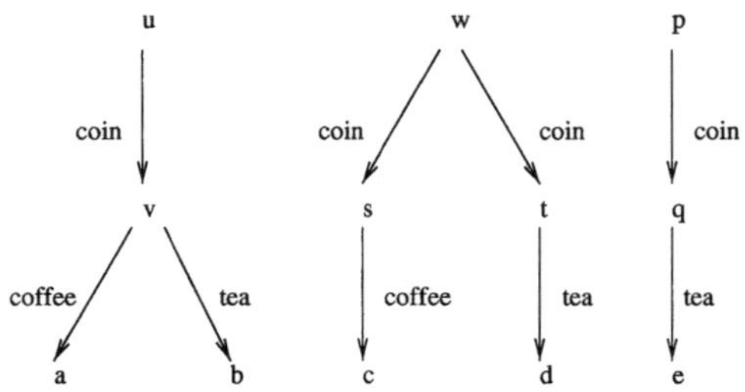

Figure 7.1. Simple transition graphs

some contrasts between bisimulation equivalence and language equivalence. There are two threads. First is that because bisimulation is more intensional, results in language and automata theory can be recast for bisimulation. The second thread is the contrast between definability of language equivalence and bisimulation equivalence. Bisimulation equivalence is definable as a "simple" formula in first-order logic with fixed points. Language equivalence is not definable as an unconditional projection of a simple least fixed point. This should be contrasted with a known normal form result for least fixed point logic: any least fixed point definable relation is definable as a projection of a simple least fixed point under equality conditions on its components. It should be noted that undefinability of language equivalence in least fixed point logic per se would actually imply P $\neq$ NP. In Section 2 we consider the two origins of bisimulation. In Section 3 we describe some results which contrast bisimulation equivalence and language equivalence on automata. The final two sections discuss logics and the undefinability result.

2. Background

Labelled transition systems are commonly encountered in operational semantics of programs and systems. They are just labelled graphs. A transition system is a pair $G = (S, \{\xrightarrow{a}: a \in A\})$ where S is a non-empty set (of states), A is a non-empty set (of labels) and for each $a \in A$, $\xrightarrow{a}$ is a binary relation on S. We write $s \xrightarrow{a} s'$ instead of $(s, s') \in \xrightarrow{a}$. Sometimes there is extra structure in a transition system, a set of atomic colours Q, such that each colour $q \subseteq S$ (the subset of states with colour q).

Consider the transition systems pictured in Figure 7.1. Here u and w are simple vending machines, and p is a person who wishes to obtain tea. The language accepted by u, $\{\text{coin coffee}, \text{coin tea}\}$, is the same as that accepted by w. When p is placed in parallel with w, $p||w$, then deadlock is possible before a tea action because of the joint transition $p||w \xrightarrow{\text{coin}} q||s$: we assume here that parallel composition requires both components to do the same action. In contrast $p||u \xrightarrow{\text{coin}} q||v$, and tea is then the only next possible action. Therefore the language accepted by $p||w$, $\{\text{coin}, \text{coin tea}\}$, is different from the language accepted by $p||u$, $\{\text{coin tea}\}$. Consequently language equivalence is not a congruence for interacting "automata". Because of this Milner and others sought a more intensional notion of equivalence which would be preserved by communicating automata.

Bisimulations were introduced by Park [16] as a small refinement of the behavioural equivalence originally defined by Hennessy and Milner between basic CCS processes (whose behaviours are transition systems).

Definition 1 A binary relation R between states of a transition system is a *bisimulation* just in case whenever $(s, t) \in R$ and $a \in A$,

1 if $s \xrightarrow{a} s'$ then $t \xrightarrow{a} t'$ for some t' such that $(s', t') \in R$ and

2 if $t \xrightarrow{a} t'$ then $s \xrightarrow{a} s'$ for some s' such that $(s', t') \in R$.

In the case of an enriched transition system with colours there is an extra clause in the definition of a bisimulation that it preserves colours: if $(s, t) \in R$ then

0. for all colours q, $s \in q$ iff $t \in q$

Simple examples of bisimulations are the identity relation and the empty relation. Two states of a transition system s and t are *bisimulation equivalent* (or *bisimilar*), written $s \sim t$, if there is a bisimulation relation R with $(s, t) \in R$. The machines u and w in Figure 7.1 are not bisimulation equivalent. The transition $u \xrightarrow{\text{coin}} v$ cannot be matched either by $w \xrightarrow{\text{coin}} s$ because s does not have a tea transition or by $w \xrightarrow{\text{coin}} t$ because t does not have a coffee transition.

Transition systems are models for basic process calculi, such as ACP, CCS and CSP. Bisimulation equivalence is a congruence for all the operators of these calculi. By permitting more general operators, whose rules for transitions belong to a general format, bisimulation equivalence turns out to be the least congruence induced by language equivalence [9]. Models for richer process calculi capturing value passing, mobility, causality, time, probability and locations have been developed. The basic notion of bisimulation has been generalised, often in a variety of different ways, to cover these extra features. Bisimulation also has a nice categorical representation via co-algebras due to Aczel, see for example [18], which allows a very general definition. It is an interesting question whether all the different brands of bisimulation are instances of this categorical account.

It is common to identify a root of a transition system as a start state. Above we defined a bisimulation on states of the same transition graph. Equally we could have defined it between states of different transition systems. When

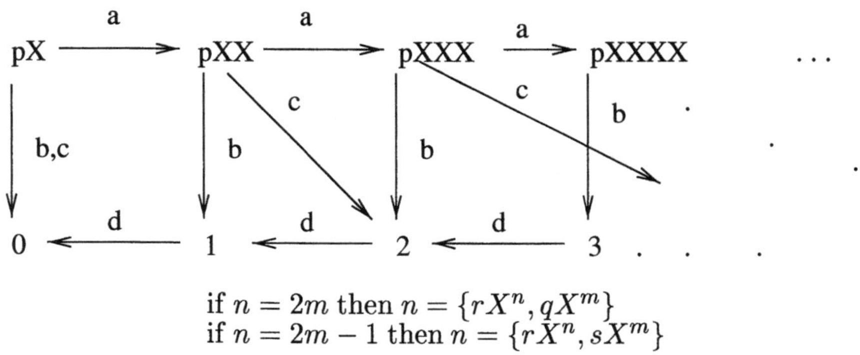

Figure 7.2. Quotiented transition graph

transition systems are rooted we can then say that two systems are bisimilar if their roots are. A family Δ of rooted transition graphs is said to be *closed under bisimulation equivalence* when the following holds.

$$\text{if } G \in \Delta \text{ and } G \sim G' \text{ then } G' \in \Delta$$

Given a rooted transition system there is a "smallest" transition system which is bisimilar to it: this is its *canonical* transition graph which is the result of first removing any states which are not reachable from the root, and then identifying bisimilar states (using quotienting). For instance Figure 7.2 is the canonical graph of Figure 7.3.

An alternative perspective on bisimulation closure is from the viewpoint of properties of transition systems. Properties whose transition systems are bisimulation closed are said to be *bisimulation invariant.* Over rooted transition graphs property Φ is bisimulation invariant when the following holds.

$$\text{if } G \models \Phi \text{ and } G \sim G' \text{ then } G' \models \Phi$$

(By $G \models \Phi$ we mean that Φ is true of the transition graph G.) On the whole, "counting" properties are not bisimulation invariant, for example "has 32 states" or "has an even number of states". In contrast temporal properties are bisimulation invariant, for instance "will eventually do an a-transition" or "is never able to do a b-transition". Other properties such as "has an Hamiltonian circuit" or "is 3-colourable" are also not bisimulation invariant. Later we shall be interested in parameterised properties, that is properties of arbitrary arity. We say that an n-ary property $\Phi(x_1, \ldots, x_n)$ on transition systems is bisimulation invariant when the following is true.

$$\begin{array}{l}\text{if } G \models \Phi[s_1, \ldots, s_n] \text{ and } t_1, \ldots, t_n \text{ are states of } G' \text{ and} \\ t_i \sim s_i \text{ for all } i : 1 \leq i \leq n \text{ then } G' \models \Phi[t_1, \ldots, t_n]\end{array}$$

(By $G \models \Phi[s_1, \ldots, s_n]$ we mean that $\Phi(x_1, \ldots, x_n)$ is true of G when x_i is interpreted as state s_i for each i.) An example of a property which is not bisimulation invariant is "$x_1, \ldots, x_n$ is a cycle", and an example of a bisimulation invariant property is "x_1 is language equivalent to x_2". The notions of bisimulation closure and invariance have appeared independently in a variety of contexts, see for instance [2, 3, 4, 5, 15].

Bisimulation was first introduced in the context of modal logic by Van Benthem [2] to give an account of which subfamily of first-order logic is definable in modal logic. Let M be the following modal logic where a ranges over A:

$$\Phi \ ::= \ \mathtt{tt} \ \mid \ \neg\,\Phi \ \mid \ \Phi_1 \vee \Phi_2 \ \mid \ \langle a \rangle \Phi$$

The inductive stipulation below defines when a state s of a transition graph G has a modal property Φ, written $s \models_G \Phi$, however we drop the index G.

$$\begin{array}{lll} s \models \mathtt{tt} & & \\ s \models \neg\,\Phi & \text{iff} & s \not\models \Phi \\ s \models \Phi \vee \Psi & \text{iff} & s \models \Phi \text{ or } s \models \Psi \\ s \models \langle a \rangle \Phi & \text{iff} & \exists t.\, s \stackrel{a}{\longrightarrow} t \text{ and } t \models \Phi \end{array}$$

In the context of an enriched transition system one adds propositions q for each colour $q \in Q$ to the logic, with semantic clause: $s \models q$ iff $s \in q$. Modal formulas are bisimulation invariant: if $s \models \Phi$ and $s \sim t$ then $t \models \Phi$ for any modal Φ.

First-order logic, FOL, over transition systems contains binary relations E_a for each $a \in A$ (and monadic predicates $q(x)$ for each colour q if extended transition systems are under consideration). Formulas of FOL have the following form.

$$\Phi \ ::= \ xE_a y \ \mid \ x = y \ \mid \ \neg\,\Phi \ \mid \ \Phi_1 \wedge \Phi_2 \ \mid \ \forall x.\Phi$$

A formula $\Phi(x_1, \ldots, x_n)$ with at most the free variables $x_1, \ldots, x_n$ will be true or false of a transition system G and states $\widetilde{s} = s_1, \ldots, s_n$ in the usual way. For example

$$\begin{array}{lll} G \models x_i E_a x_j[\widetilde{s}] & \text{iff} & s_i \stackrel{a}{\longrightarrow} s_j \\ G \models \forall x_{n+1}.\Phi[\widetilde{s}] & \text{iff} & \forall s'.\, G \models \Phi[\widetilde{s}, s'] \end{array}$$

Not all first-order formulas are bisimulation invariant. An example is the formula $\exists y.\exists z.(x \stackrel{a}{\longrightarrow} y \wedge x \stackrel{a}{\longrightarrow} z \wedge y \neq z)$ which says "x has at least two different a-transitions".

Van Benthem introduced bisimulation to identify which formulas $\Phi(x)$ of FOL are equivalent to modal formulas (to M formulas) [3]. A formula $\Phi(x)$ is equivalent to a modal formula Φ' provided that for any G and for any state s, $G \models \Phi[s]$ iff $s \models_G \Phi'$. Van Benthem proved the following characterisation.

$$
\begin{array}{lllll}
q\epsilon \overset{d}{\longleftarrow} sX \overset{d}{\longleftarrow} qX & \overset{d}{\longleftarrow} & sXX \overset{d}{\longleftarrow} qXX & \overset{d}{\longleftarrow} \ldots \\
\uparrow c & & \uparrow c & & \uparrow c \quad \vdots \\
pX \quad \overset{a}{\longrightarrow} & pXX & \overset{a}{\longrightarrow} & pXXX \overset{a}{\longrightarrow} \ldots \\
\downarrow b & & \downarrow b & & \downarrow b \quad \vdots \\
r\epsilon \quad \overset{d}{\longleftarrow} & rX & \overset{d}{\longleftarrow} & rXX \overset{d}{\longleftarrow} \ldots
\end{array}
$$

Figure 7.3. A pushdown automaton

Proposition 2 *A FOL formula $\Phi(x)$ is equivalent to an M formula iff $\Phi(x)$ is bisimulation invariant.*

3. Caucal's hierarchy

Bisimulation equivalence is a very fine equivalence between states. An interesting line of enquiry is to re-consider classical results in automata and language theory, replacing language equivalence with bisimulation equivalence. These results concern definability, closure properties and decidability/undecidability.

Grammars can be viewed as generators of transition systems. Let Γ be a finite family of nonterminals and assume that A is a finite set (of terminals). A basic transition has the form $\alpha \xrightarrow{a} \beta$ where α, $\beta \in \Gamma^*$ and $a \in A$. A state is then any member of Γ^*, and the transition relations on states are defined as the least relations containing the basic transitions and satisfying the following prefix rule.

$$\text{if } \alpha \xrightarrow{a} \beta \text{ then } \alpha\delta \xrightarrow{a} \beta\delta$$

Given a state α we can define its rooted transition system whose states are just the ones reachable from α. An example is a pushdown automaton over the alphabet $A = \{a, b, c, d\}$ whose basic transitions are as follows (where ϵ is the empty stack sequence).

$$\{pX \xrightarrow{a} pXX,\ pX \xrightarrow{c} q\epsilon,\ pX \xrightarrow{b} r\epsilon,\ qX \xrightarrow{d} sX,\ sX \xrightarrow{d} q\epsilon,\ rX \xrightarrow{d} r\epsilon\}$$

The transition graph generated by pX is pictured in Figure 7.3. For any $n \geq 0$ the transition $qXX^n \xrightarrow{d} sXX^n$ is derived from the basic transition $qX \xrightarrow{d} sX$ using the prefix rule when β is X^n. In this example the set of nonterminals is divided into two, states Q' and stack elements Γ. Each basic transition belongs to $(Q' \times \Gamma) \times A \times (Q' \times \Gamma^*)$.

In the table below is the "Caucal hierarchy" of transition graph descriptions, which depends on how the family of basic transitions is specified. In each case we assume a finite family of rules. Type 3 captures regular grammars (whose graphs are finite-state), Type 2 captures context-free grammars in Greibach normal form, and Type $1\frac{1}{2}$, in fact, captures pushdown automata. For Type 0 and below this means that in each case there are finitely many basic transitions. In the other two cases R_1 and R_2 are regular expressions over Γ. The idea is that each rule $R_1 \xrightarrow{a} \beta$ stands for the possibly infinite family of basic transitions $\{\alpha \xrightarrow{a} \beta : \alpha \in R_1\}$ and $R_1 \xrightarrow{a} R_2$ stands for the family $\{\alpha \xrightarrow{a} \beta : \alpha \in R_1 \text{ and } \beta \in R_2\}$. For instance a Type -1 rule of the form $X^*Y \xrightarrow{a} Y$ includes for each $n \geq 0$ the basic transition $X^nY \xrightarrow{a} Y$.

	Basic Transitions
Type -2	$R_1 \xrightarrow{a} R_2$
Type -1	$R_1 \xrightarrow{a} \beta$
Type 0	$\alpha \xrightarrow{a} \beta$
Type $1\frac{1}{2}$	$\alpha \xrightarrow{a} \beta$ where $\lvert\alpha\rvert = 2$ and $\lvert\beta\rvert > 0$
Type 2	$X \xrightarrow{a} \beta$
Type 3	$X \xrightarrow{a} Y$ or $X \xrightarrow{a} \epsilon$

This hierarchy is implicit in Caucal's work on understanding context-free graphs, and understanding when a graph has a decidable monadic second-order theory [5, 4, 6]. With respect to language equivalence, the hierarchy collapses to just two levels, the regular and the context-free. The families between Type 2 and Type -2 are equivalent: for every G of Type -2 and root α there is a G' of Type 2 and root α' such that the language of α is the same as the language of α'.

The standard textbook transformation from pushdown automata to context-free grammars (Type $1\frac{1}{2}$ to Type 2) does not preserve bisimulation equivalence. In fact, with respect to bisimilarity pushdown automata are richer than context-free grammars. Caucal [5] shows that there is not a Type 2 transition graph and root α which is bisimulation equivalent to pX of Figure 7.3. He also shows that Type 0 transition systems coincide (up to isomorphism) with Type $1\frac{1}{2}$. There is a strict hierarchy between Type 0 and Type -2. Therefore, with respect to bisimulation equivalence there are five levels in the hierarchy. An interesting consideration is to what extent this hierarchy is closed under canonical transition graphs. Figure 7.2 is clearly not a Type 0 graph but it is the canonical graph for Figure 7.3, and therefore this shows that Type 0 is not closed under canonical graphs, see [4] for further details and results.

Baeten, Bergstra and Klop proved that bisimulation equivalence is decidable for a subset of Type 2 transition systems[1] [1]. The decidability result was generalised in [7] to encompass all Type 2 graphs. Groote and Hüttel proved that other standard equivalences (traces, failures, simulation, 2/3-bisimulation etc..,) on Type 2 graphs are all undecidable using reductions from the undecidability of language equivalence for these graphs [10]. The most recent result is by Sénizergues [20], who shows that bisimulation equivalence is decidable for transition systems somewhere between Type 0 and -1. This result is a small generalisation of his formidable proof of decidability of language equivalence for DPDA [19, 20]. Using ideas developed in concurrency theory (tableaux methods) we have simplified his proof of the DPDA result [22]. This leaves as an open question whether bisimulation equivalence is also decidable for Type -1 and Type -2 systems.

4. Richer logics

Modal logic M of Section 2 is not very expressive. For instance it cannot express temporal properties, such as safety or liveness properties, of transition systems. Such properties have been found to be very useful when analysing the behaviour of concurrent systems. Modal mu-calculus, μM, introduced by Kozen [13], has the required extra expressive power. The new constructs over and above those of M are

$$\Phi ::= X \mid \ldots \mid \mu X.\Phi$$

where X ranges over a family of propositional variables, and in the case of $\mu X.\Phi$ there is a restriction that all free occurrences of X in Φ are within the scope of an even number of negations (to guarantee monotonicity).

The semantics of M is extended to encompass these extra constructs. The inductive definition of satisfaction stipulates when a state s of a transition system has the property $\Phi(X_1, \ldots, X_n)$ when each X_i is interpreted as the set of states S_i, written $s \models \Phi[\widetilde{S}]$, and the semantic clauses for the modal fragment are as before (except for the presence of the state sets).

$$\begin{array}{lll} s \models X_i[\widetilde{S}] & \text{iff} & s \in S_i \\ s \models \mu X_{n+1}.\Phi[\widetilde{S}] & \text{iff} & \forall S'.\,\text{if } (\forall t \in S'.\, t \models \Phi[\widetilde{S}, S']) \text{ then } s \in S' \end{array}$$

The stipulation for the fixed point follows directly from the Tarski-Knaster theorem, as a least fixed point is the intersection of all prefixed points. (Again we would add atomic formulas q if we are interested in extended transition systems.)

Second-order propositional modal logic, 2M, is defined as an extension of M as follows.

$$\Phi ::= X \mid \ldots \mid \Box\Phi \mid \forall X.\Phi$$

The modality $\Box$ is the reflexive and transitive closure of $\bigcup\{[a] \,:\, a \in A\}$, and is included so that 2M includes μM. As with modal mu-calculus we define when $s \models \Phi[\widetilde{S}]$. The new clauses are:

$$\begin{array}{lll} s \models \Box\Phi[\widetilde{S}] & \text{iff} & \forall t. \forall w \in A^*. \text{ if } s \xrightarrow{w} t \text{ then } t \models \Phi[\widetilde{S}] \\ s \models \forall X_{n+1}.\Phi[\widetilde{S}] & \text{iff} & \forall S'.\ s \models \Phi[\widetilde{S}, S'] \end{array}$$

There is a straightforward translation of μM into 2M. Let Tr be this translation. The important case is the fixed point: $\text{Tr}(\mu X.\Phi) = \forall X.(\Box(\text{Tr}(\Phi) \to X) \to X)$.

Formulas of M and closed formulas of μM are bisimulation invariant. This is not true in the case of 2M, for it is too rich for characterising bisimulation: for instance, a variety of "counting" properties are definable, such as "has at least two different a-transitions", expressible as $\exists X.(\langle a \rangle X \wedge \langle a \rangle \neg X)$. This means that two bisimilar states need not have the same 2M properties.

FOL over transition graphs is also not rich enough for capturing interesting properties. One extension of first-order logic is monadic second-order logic, 2OL, with the extra formulas

$$\Phi \ ::= \ X(x) \ | \ \ldots \ | \ \forall X.\Phi$$

The semantic clauses are generalised as follows:

$$\begin{array}{lll} G \models X_i(x_j)[\widetilde{s}, \widetilde{S}] & \text{iff} & s_j \in S_i \\ G \models \forall X_{n+1}.\Phi[\widetilde{s}, \widetilde{S}] & \text{iff} & \forall S'. G \models \Phi[\widetilde{s}, \widetilde{S}, S'] \end{array}$$

Formulas of 2OL need not be bisimulation invariant. An interesting question is the relationship between μM and 2OL. Van Benthem's result was generalised by Janin and Walukiewicz [12] as follows.

Proposition 3 *A 2OL formula $\Phi(x)$ is equivalent to a closed μM formula iff $\Phi(x)$ is bisimulation invariant.*

One corollary of this result is that the bisimulation invariant closed formulas of 2M has the same expressive power as the closed formulas of μM.

A different extension of FOL is first-order logic with fixed points, μFOL, where there is the following extra formulas

$$\Phi \ ::= \ X(x_1, \ldots, x_k) \ | \ \ldots \ | \ (\mu X(x_1, \ldots, x_k).\, \Phi)(y_1, \ldots, y_k)$$

In the case of $(\mu X(\ldots).\, \Phi)(\ldots)$, there is the same restriction as in μM that all free occurrences of X in Φ lie within the scope of an even number of negations. The interpretation of a predicate X_i with arity k is a set of k-tuples, a subset of S^k. The semantic clauses are therefore as follows (where we use the notation S_i for sets of tuples).

The new semantic clauses are (where $\widetilde{s'}$ is $s_{k+1}, \ldots, s_n$)

$$\begin{array}{lll} G \models X_j(x_{i1}, \ldots, x_{ik})[\widetilde{s}, \widetilde{S}] & \text{iff} & (s_{i1}, \ldots, s_{ik}) \in S_j \\ G \models (\mu X_{n+1}(\ldots).\Phi)(x_1, \ldots, x_k)[\widetilde{s}, \widetilde{S}] & \text{iff} & \forall S'. \text{if } (\forall (t_1, \ldots, t_k) \in S'. \\ & & \quad G \models \Phi[\widetilde{t}, \widetilde{s'}, \widetilde{S}, S']) \text{ then} \\ & & \quad (s_1, \ldots, s_k) \in S' \end{array}$$

When the alphabet A is finite, bisimulation equivalence is definable in μFOL as a dyadic greatest fixed point formula

$$(\nu Z(x,y).(Z(y,x) \wedge \bigwedge_{a \in A} (\forall x'.\exists y'.xE_a x' \rightarrow yE_a y' \wedge Z(x',y'))))(\ldots)$$

where $(\nu Z(\ldots)\Phi)(\ldots)$ is $\neg((\mu Z(\ldots)\Phi(\neg Z))(\ldots))$. An interesting open question is how to characterise the bisimulation invariant sublogic of μFOL.

5. Finite model theory

Finite model theory is concerned with relationships between complexity classes and logics over finite structures. It is interesting to consider bisimulation invariance in the context of finite model theory. Rosen showed that Proposition 2 (in Section 2) remains true with the restriction to finite transition systems [17]. It is an open question whether Proposition 3 also remains true under this restriction.

Part of the interest in relationships between μM and 2M or 2OL with respect to finite transition systems is that within 2M and 2OL one can define NP-complete problems: examples include 3-colourability on finite connected undirected graphs. Consider such a graph. If there is an edge between two states s and t let $s \xrightarrow{a} t$ and $t \xrightarrow{a} s$. So in this case $A = \{a\}$, and 3-colourability is given by:

$$\exists X.\, \exists Y.\, \exists Z.\, (\Phi \wedge \Box((X \rightarrow [a]\neg X) \wedge (Y \rightarrow [a]\neg Y) \wedge (Z \rightarrow [a]\neg Z)))$$

where Φ, which says that every vertex has a unique colour, is

$$\Box((X \wedge \neg Y \wedge \neg Z) \vee (Y \wedge \neg Z \wedge \neg X) \vee (Z \wedge \neg X \wedge \neg Y))$$

In contrast, μM formulas over finite transition systems can only express PTIME properties.

An interesting open question is whether there is a logic which captures exactly the PTIME properties of transition systems. Otto has shown that there is a logic for the PTIME properties that are bisimulation invariant [15]. The right setting is μFOL over canonical transition systems (where $=$ is $\sim$, and a linear ordering on states is thereby definable).

We now consider emaciated finite transition systems whose set A is a singleton. That is now $G = (S, \longrightarrow)$ where S is finite. We write $s \xrightarrow{n} t$, $n \geq 0$, if there is a sequence of transitions of length n from s to t (and by convention $s \xrightarrow{0} s$). A state is terminal if it has no transitions. The language of state s is therefore the set of words $L(s) = \{i \geq 0 \,:\, s \xrightarrow{i} t \text{ and } t \text{ is terminal}\}$. Consequently, s and s' are language equivalent if $L(s) = L(s')$. The property "x is language equivalent to y" as was noted earlier is bisimulation invariant. The definition in μFOL of bisimulation equivalence of the previous section remains correct when transition systems are finite. However it is unlikely that language equivalence is definable in μFOL because of the following result, proved in [23].

Proposition 4 *Language equivalence on (canonical) emaciated finite transition graphs is co-NP complete.*

Hence language equivalence over finite transition systems is definable in μFOL iff PTIME = NP. Dawar offers a different route to this observation [8].

A classical result (due to Immermann, Gurevich and Shelah) in a slightly normalised form is:

Proposition 5 *A μFOL formula $\Psi(y_1, \ldots, y_n)$ over finite transition systems is equivalent to a formula of the form $\exists u.\,((\mu Z(x_1, \ldots, x_m).\,\Phi)(y_1, \ldots, y_n, u, \ldots, u))$ where Φ is first-order and contains at most $x_1, \ldots, x_m$ free.*

The argument places in the application $(\ldots)$ from $n+1$ to m are all filled by the same element u. This allows for the arity of the defining fixed point m to be larger than the arity of the μFOL formula n. Consequently, if one can prove that "y is language equivalent to z" is not definable by a μFOL formula in normal form, $\exists u.\,(\mu Z(x_1, \ldots, x_m).\,\Phi\,(y, z, u, \ldots, u))$, then this would show that PTIME is different from NP.

The result below has the consequence that language equivalence is not definable by a normal formula of the form $\exists u.\,(\mu Z(x_1, x_2, x_3).\,\Phi\,(y, z, u))$. We present the theorem in the most general form possible, that language equivalence is not definable as an unconditional projection of a simple fixed point.

Theorem 6 *Language equivalence is not definable by a normal formula of the form $(\mu Z(x_1, \ldots, x_n).\,\Phi)(\ldots)$ (over finite transition systems).*

That is, language equivalence is not definable as an unconditional projection of a simple fixed point. The rest of the paper is devoted to its proof. One popular method for showing non-definability is to use games. Here we use a variant method which introduces "proofs" of formulas. A sufficiently concrete account of when a formula is true of a transition system is given by a tableau proof. The aim is to provide a proof system $G \vdash \Psi(s_1, \ldots, s_n)$ for showing $G \models \Psi[s_1, \ldots, s_n]$ when Ψ is a formula of the form $(\mu Z(x_1, \ldots, x_n).\,\Phi)(\ldots)$ containing the single fixed point $\mu Z(\ldots)$: it is straightforward to extend the proof system to formulas with multiple fixed points. The property checker is a tableau system, a goal directed proof system. Assume that the starting formula is $(\mu Z(x_1, \ldots, x_n).\,\Phi)(\widetilde{x})$ and let $\{s_1/x_1, \ldots, s_n/x_n\}$ be the simultaneous substitution of each s_i for x_i. We assume that in Φ all negations are moved inwards in the usual way. The tableau rules are therefore as follows.

$$\frac{G \vdash (\mu Z(x_1, \ldots, x_n).\, \Phi)(\widetilde{x})(s_1, \ldots, s_n)}{G \vdash Z(s_1, \ldots, s_n)}$$

$$\frac{G \vdash Z(s_1, \ldots, s_n)}{G \vdash \Phi\{s_1/x_1, \ldots, s_n/x_n\}}$$

$$\frac{G \vdash \exists x.\Psi}{G \vdash \Psi\{s/x\}}\ s \in S$$

$$\frac{G \vdash \forall x.\Psi}{G \vdash \Psi\{s_1/x\} \ \ \ldots \ \ G \vdash \Psi\{s_k/x\}}\ S = \{s_1, \ldots, s_k\}$$

$$\frac{G \vdash \Psi_1 \wedge \Psi_2}{G \vdash \Psi_1 \quad G \vdash \Psi_2}$$

$$\frac{G \vdash \Psi_1 \vee \Psi_2}{G \vdash \Psi_1} \qquad \frac{G \vdash \Psi_1 \vee \Psi_2}{G \vdash \Psi_2}$$

To test if $G \models (\mu Z(x_1, \ldots, x_n).\, \Phi)(\widetilde{x})[s_1, \ldots, s_n]$ one tries to develop a proof of $G \vdash (\mu Z(x_1, \ldots, x_n).\, \Phi)(\widetilde{x})(s_1, \ldots, s_n)$ by building a tableau, a finite proof tree whose root is labelled with this initial sequent. The sequents labelling the immediate successors of a node are determined by an application of one of the rules. One keeps building a proof until we reach a terminal sequent.

A terminal sequent has one of the following forms

1 $G \vdash s = t$ or $G \vdash s \neq t$

2 $G \vdash sEt$ or $G \vdash \neg(sEt)$

3 $G \vdash Z(s_1, \ldots, s_n)$ and in the proof tree above this sequent there is the same sequent $G \vdash Z(s_1, \ldots, s_n)$.

Terminal sequents of type 1 or 2 which are true are successful. A tableau proof is successful if all of its leaves are successful, as shown by the next result whose proof is straightforward.

Lemma 7 *$G \models (\mu Z(x_1, \ldots, x_n).\, \Phi)(\widetilde{x})[s_1, \ldots, s_n]$ iff there is a successful tableau whose root is $G \vdash (\mu Z(x_1, \ldots, x_n).\, \Phi)(\widetilde{x})(s_1, \ldots, s_n)$.*

PROOF. [of Theorem 6] Suppose $(\mu Z(x_1, \ldots, x_n).\, \Phi)(\widetilde{x})$ defines language equivalence. That is for any transition graph G and states $s_1, \ldots, s_n$ of S

$$G \models (\mu Z(x_1, \ldots, x_n).\, \Phi)(\widetilde{x})[s_1, \ldots, s_n] \text{ iff } L(s_1) = L(s_2)$$

Assume $0 < t < k < k+t < a$

$$\begin{array}{ll} e_0 \longrightarrow \ldots \longrightarrow e_k \longrightarrow \ldots \longrightarrow e_{k+t} \longrightarrow \ldots \longrightarrow e_a & \text{and } e_0 \longrightarrow e_t \\ l_0 \longrightarrow \ldots \longrightarrow l_k \longrightarrow \ldots \longrightarrow l_{k+t} \longrightarrow \ldots \longrightarrow l_a & \text{and } e_k \longrightarrow e_{k+t} \\ l'_1 \longrightarrow \ldots \longrightarrow l'_k \longrightarrow \ldots \longrightarrow l'_{k+t} \longrightarrow \ldots \longrightarrow l'_a & \text{and } l'_k \longrightarrow l'_{k+t} \end{array}$$

Figure 7.4. Ingredients of the graphs G and G'

We assume that Φ is in prenex normal form $Q_1 x_{i1} \ldots Q_m x_{im} \Psi$ where Ψ is in DNF: each clause in Ψ contains atomic formulas of the form $x = y$, $x \neq y$, xEy, $\neg(xEy)$ and $Z(y_1, \ldots, y_n)$ where the variables x, y range over the set $X = \{x_1, \ldots, x_n, x_{i1}, \ldots, x_{im}\}$ (where x_{ij} could be the same as x_k).

Consider the transition systems in Figure 7.4. The e vertices have an "early" branching point whereas the l and l' vertices have a "late" branching point. Let G be the graph whose vertices are e_i and l_i, and let G' be the similar graph whose vertices are e_i and l'_i. Notice that $L(e_0) = L(l_0)$ but $L(e_i) \neq L(l_i)$ when $i : 0 < i < k+1$. Moreover $L(e_0) \neq L(l'_1)$. We assume that $k > n2^{m+1}$ where m is the number of quantifiers and n is the arity of Z, and we assume that $t < k$ and $t > 1$.

There is a tableau proof of $G \vdash (\mu Z(x_1, \ldots, x_n). \Phi)(\widetilde{x})(e_0, l_0, \widetilde{v})$ by Lemma 7, where $\widetilde{v}$ is any sequence of vertices $v_3, \ldots, v_n$. Consider any $\widetilde{v}$ such that there is a shortest depth tableau proof of $G \vdash Z(e_0, l_0, \widetilde{v})$. The argument proceeds by showing that there is also a proof of $G' \vdash Z(e_0, l'_1, \widetilde{v}')$. First we define the elements $\widetilde{v}'$.

Assume $\widetilde{v} = v_3 \ldots v_n$. We define $\widetilde{v}' = v'_3 \ldots v'_n$ as follows. If v_i is e_j then $v'_i = e_j$ and if $v_i = l_{k+j}$ then $v'_i = l'_{k+j}$. Otherwise $v_i = l_j$ and $j < k$. Consider all such v_i in decreasing order, say $l_{j1} \ldots l_{jb}$ (where $b < (n-1)$). If $j1 \geq (k - 2^{m+1})$ then the corresponding element is l'_{j1} otherwise it is l'_{j1+1}: and in the second case all the other corresponding elements are l'_{js+1} for $1 < s \leq b$. Assume then that for the first element l_{j1} the index $j1 \geq (k - 2^{m+1})$. Consider now the second element l_{j2}. If the index $j2 \geq (j1 - 2^{m+1})$ then the corresponding element is l'_{j2} otherwise it is l'_{j2+1} and again in this second case the rest of the corresponding elements are l'_{js+1} for $2 < s \leq b$. Repeat this construction as long as $js \geq j(s-1) - 2^{m+1}$ where the corresponding element is l'_{js}, and otherwise it is l'_{js+1} and the rest of the corresponding elements are l'_{jz+1} for $s < z \leq b$.

Consider the sequence of elements $l_k\, l_{j1} \ldots l_{jb}\, l_0$ in G and the corresponding sequence $l'_k\, l'_{j1'} \ldots l'_{jb'}\, l'_1$ in G'. The index ji' in the second sequence is either ji or $ji+1$. Let p be the pivot point in the sequence where all indices to the left also occur in the first sequence and all indices to the right including that of p are 1 plus the index of the corresponding element in the first sequence.

Note that the difference in the index between element p and the next element to the left is greater than 2^{m+1}.

We now show how a proof of $G \vdash Z(e_0, l_0, \widetilde{v})$ can be used to develop a proof of $G' \vdash Z(e_0, l'_1, \widetilde{v}')$. The goal $G \vdash Z(e_0, l_0, \widetilde{v})$ reduces to the subgoal $G \vdash (Q_1 x_{i1} \ldots Q_m x_{im} \Psi)(e_0, l_0, \widetilde{v})$. Similarly the goal $G' \vdash Z(e_0, l'_1, \widetilde{v}')$ reduces to the subgoal $G' \vdash (Q_1 x_{i1} \ldots Q_m x_{im} \Psi)(e_0, l'_1, \widetilde{v}')$.

We consider the quantifiers in turn, and at each stage the pivot point may become updated.

Suppose $Q_1 x_{i1} = \forall x_{i1}$. The goal $G \vdash (Q_1 x_{i1} \ldots Q_m x_{im} \Psi)(e_0, l_0, \widetilde{v})$ reduces to subgoals one for each $u \in G$, $G \vdash ((Q_2 x_{i2} \ldots Q_m x_{im} \Psi)(e_0, l_0, \widetilde{v}))$ $\{u/x_{i1}\}$. For each such subgoal we associate a subgoal of the second proof which has the following form $G' \vdash ((Q_2 x_{i2} \ldots Q_m x_{im} \Psi)(e_0, l'_1, \widetilde{v}'))\{u'/x_{i1}\}$, so that all $u' \in G'$ are dealt with. If $u = e_i$ then $u' = e_i$. Let j be the index of the pivot element p. If $u = l_i$ and $i > j + 2^m$ then $u' = l'_i$ otherwise $u' = l'_{i+1}$. In the circumstance that $i \geq j$ but $j + 2^m \geq i$ then the pivot element p is updated to that of u'.

The argument is similar when $Q_1 x_{i1} = \exists x_{i1}$. Now there is only one subgoal $G \vdash ((Q_2 x_{i2} \ldots Q_m x_{im} \Psi)(e_0, l_0, \widetilde{v}))\{u/x_{i1}\}$. The corresponding subgoal $G' \vdash ((Q_2 x_{i2} \ldots Q_m x_{im} \Psi)(e_0, l'_1, \widetilde{v}'))\{u'/x_{i1}\}$ is chosen as above, and again the pivot element may be updated.

The argument continues for the remaining quantifiers. Suppose the goal is $G \vdash ((Q_c x_{ic} \ldots Q_m x_{im} \Psi)(e_0, l_0, \widetilde{v}))\{u_1/x_{i1}, \ldots, u_{c-1}/x_{i(c-1)}\}$, and the corresponding goal is $G' \vdash ((Q_c x_{ic} \ldots Q_m x_{im} \Psi)(e_0, l'_1, \widetilde{v}'))\{u'_1/x_{i1}, \ldots, u'_{c-1}/x_{i(c-1)}\}$ in the second proof. If $Q_c x_{ic}$ is $\forall x_{ic}$ then we proceed as above except if j is the index of the current pivot element p and $u_c = l_i$ and $i > j + 2^{(m+1)-c}$ then $u'_c = l'_i$ otherwise $u' = l'_{i+1}$. The pivot element is updated to u_c when $i \geq j$ but $j + 2^{(m+1)-c} \geq i$.

As quantifiers are eliminated the sequence of elements $l_k \, l_{j1} \ldots l_{jb} \, l_0$ in G and the corresponding sequence $l'_k \, l'_{j1'} \ldots l'_{jb'} \, l'_1$ in G' may be expanded, and the pivot element in the second sequence updated. However at each stage, after eliminating quantifier $Q_c x_{ic}$, the difference in the index between the pivot element p and the next element to the left is greater than $2^{(m+1)-c}$.

Finally all the quantifiers are removed, and a subgoal of the first proof has the form $G \vdash \Psi(e_0, l_0, \widetilde{v}, \widetilde{u})$ and in the second proof has the form $G' \vdash \Psi(e_0, l'_1, \widetilde{v}', \widetilde{u}')$. The formula Ψ is in DNF. Assume that $\Psi = \bigvee \Psi_i$ where each Ψ_i is a conjunction of atomic formulas. Suppose the subgoal of $G \vdash \Psi(e_0, l_0, \widetilde{v}, \widetilde{u})$ is $G \vdash \Psi_i(e_0, l_0, \widetilde{v}, \widetilde{u})$, then the corresponding subgoal is $G' \vdash \Psi_i(e_0, l'_1, \widetilde{v}', \widetilde{u}')$. Consider any atomic formula B in Ψ_i. If B has the form $x = y$ or xEy then because of the construction of $\widetilde{v}'$ and $\widetilde{u}'$ it follows that $G \models B[e_0, l_0, \widetilde{v}, \widetilde{u}]$ iff $G' \models B[e_0, l'_1, \widetilde{v}', \widetilde{u}']$. The only other possible atomic sentences have the form $Z(y_1, \ldots, y_n)$. Suppose a subgoal of $G \vdash \Psi_i(e_0, l_0, \widetilde{v}, \widetilde{u})$ is $G \vdash Z(w_1, \ldots w_n)$. It follows that w_1 and w_2 are not e_0 and l_0 (for otherwise the proof of $G \vdash Z(e_0, l_0, \widetilde{w})$ must be shorter than that of $G \vdash Z(e_0, l_0, \widetilde{v})$ contrary to assumption). Hence either $w_1 = w_2$ or $w_1 = e_j$ and $w_2 = l_j$ where $j > k$. But then there must also be a successful proof for $G' \vdash Z(w'_1, \ldots, w'_n)$ as $L(w'_1) = L(w'_2)$. ∎

Acknowledegment

I would like to thank the referee for comments and wording of the main result of this paper.

Notes

1. They proved it for the normed subfamily. G is normed if for every state s there is a word w such that $s \xrightarrow{w} \epsilon$.

References

[1] J. Baeten, J. Bergstra and J. Klop. Decidability of bisimulation equivalence for processes generating context-free languages. *Journal of Association of Computing Machinery* 40:653–682, 1993.

[2] J. van Benthem. Correspondence theory. In *Handbook of Philosophical Logic*, Vol. II, ed. Gabbay, D. and Guenthner, F., 167–248, Reidel, 1994.

[3] van Benthem, J. Exploring Logical Dynamics. *CSLI Publications*, Stanford, 1996.

[4] O. Burkart, D. Caucal and B. Steffen. Bisimulation collapse and the process taxonomy. *Lecture Notes in Computer Science* 1119:247–262, 1996.

[5] D. Caucal. On the regular structure of prefix rewriting. *Theoretical Computer Science* 106:61–86, 1992.

[6] D. Caucal. On infinite transition graphs having a decidable monadic theory. *Lecture Notes in Computer Science* 1099:194–205, 1996.

[7] S. Christensen, H. Hüttel and C. Stirling. Bisimulation equivalence is decidable for all context-free processes. *Information and Computation* 121:143–148, 1995.

[8] A. Dawar. A restricted second-order logic for finite structures, *Information and Computation* 143:154–174, 1998.

[9] J. Groote. Transition system specifications with negative premises. *Theoretical Computer Science* 118:263–299, 1993.

[10] Groote, J., and Hüttel, H. Undecidable equivalences for basic process algebra. *Information and Computation* 115:354–371, 1994.

[11] M. Hennessy and R. Milner. Algebraic laws for nondeterminism and concurrency. *Journal of Association of Computer Machinery* 32:137–162, 1985.

[12] D. Janin and I. Walukiewicz. On the expressive completeness of the propositional mu-calculus with respect to the monadic second order logic. *Lecture Notes in Computer Science* 1119:263–277, 1996.

[13] D. Kozen. Results on the propositional mu-calculus. *Theoretical Computer Science* 27:333–354, 1983.

[14] R. Milner. *Communication and Concurrency*. Prentice Hall, 1989.

[15] M. Otto. Bisimulation-invariant ptime and higher-dimensional μ-calculus. *Theoretical Computer Science*, 224:73–113, 1999.

[16] D. Park. Concurrency and automata on infinite sequences. *Lecture Notes in Computer Science* 154:561–572, 1981.

[17] E. Rosen. Modal logic over finite structures. *Journal of Logic, Language and Information* 6:427–439, 1997.

[18] J. Rutten. A calculus of transition systems (towards universal coalgebra). In *Modal Logic and Process Algebra*, ed. Ponse, A., De Rijke, M. and Venema, Y. *CSLI Publications*, 187–216, Stanford. 1995.

[19] G. Sénizergues. The equivalence problem for deterministic pushdown automata is decidable. *Lecture Notes in Computer Science* 1256:671–681, 1997.

[20] G. Sénizergues. L(A) = L(B)? decidability results from complete formal systems. *Theoretical Computer Science* 251:1–166, 2001

[21] G. Sénizergues. Decidability of bisimulation equivalence for equational graphs of finite out-degree. *Procs IEEE FOCS 98*, 120–129, 1998.

[22] C. Stirling. Decidability of DPDA equivalence. *Theoretical Computer Science*, 255:1–31, 2001.

[23] L. Stockmeyer and A. Meyer. Word problems requiring exponential time. *Procs. 5th ACM STOC*, 1–9, 1973.

TRENDS IN LOGIC

1. G. Schurz: *The Is-Ought Problem.* An Investigation in Philosophical Logic. 1997 ISBN 0-7923-4410-3
2. E. Ejerhed and S. Lindström (eds.): *Logic, Action and Cognition.* Essays in Philosophical Logic. 1997 ISBN 0-7923-4560-6
3. H. Wansing: *Displaying Modal Logic.* 1998 ISBN 0-7923-5205-X
4. P. Hájek: *Metamathematics of Fuzzy Logic.* 1998 ISBN 0-7923-5238-6
5. H.J. Ohlbach and U. Reyle (eds.): *Logic, Language and Reasoning.* Essays in Honour of Dov Gabbay. 1999 ISBN 0-7923-5687-X
6. K. Došen: *Cut Elimination in Categories.* 2000 ISBN 0-7923-5720-5
7. R.L.O. Cignoli, I.M.L. D'Ottaviano and D. Mundici: *Algebraic Foundations of many-valued Reasoning.* 2000 ISBN 0-7923-6009-5
8. E.P. Klement, R. Mesiar and E. Pap: *Triangular Norms.* 2000 ISBN 0-7923-6416-3
9. V.F. Hendricks: *The Convergence of Scientific Knowledge.* A View From the Limit. 2001 ISBN 0-7923-6929-7
10. J. Czelakowski: *Protoalgebraic Logics.* 2001 ISBN 0-7923-6940-8
11. G. Gerla: *Fuzzy Logic.* Mathematical Tools for Approximate Reasoning. 2001 ISBN 0-7923-6941-6
12. M. Fitting: *Types, Tableaus, and Gödel's God.* 2002 ISBN 1-4020-0604-7
13. F. Paoli: *Substructural Logics: A Primer.* 2002 ISBN 1-4020-0605-5
14. S. Ghilardi and M. Zawadowki: *Sheaves, Games, and Model Completions.* A Categorical Approach to Nonclassical Propositional Logics. 2002 ISBN 1-4020-0660-8
15. G. Coletti and R. Scozzafava: *Probabilistic Logic in a Coherent Setting.* 2002 ISBN 1-4020-0917-8; Pb: 1-4020-0970-4
16. P. Kawalec: *Structural Reliabilism.* Inductive Logic as a Theory of Justification. 2002 ISBN 1-4020-1013-3
17. B. Löwe, W. Malzkorn and T. Räsch (eds.): *Foundations of the Formal Sciences II.* Applications of Mathematical Logic in Philosophy and Linguistics, Papers of a conference held in Bonn, November 10-13, 2000. 2003 ISBN 1-4020-1154-7
18. R.J.G.B. de Queiroz (ed.): *Logic for Concurrency and Synchronisation.* 2003 ISBN 1-4020-1270-5

KLUWER ACADEMIC PUBLISHERS – DORDRECHT / BOSTON / LONDON

MIX
Papier aus verantwortungsvollen Quellen
Paper from responsible sources
FSC® C105338

If you have any concerns about our products,
you can contact us on
ProductSafety@springernature.com

In case Publisher is established outside the EU,
the EU authorized representative is:
Springer Nature Customer Service Center GmbH
Europaplatz 3, 69115 Heidelberg, Germany

Printed by Libri Plureos GmbH
in Hamburg, Germany